Mouse Behavioral Testing

How to use Mice in Behavioral Neuroscience

Mouse Behavioral Testing

How to use Mice in Behavioral Neuroscience

Douglas Wahlsten

University of Alberta
Canada

AMSTERDAM • BOSTON • HEIDELBERG • LONDON • NEW YORK • OXFORD
PARIS • SAN DIEGO • SAN FRANCISCO • SINGAPORE • SYDNEY • TOKYO

Academic Press is an imprint of Elsevier

Academic Press is an imprint of Elsevier
32 Jamestown Road, London NW1 7BY, UK
30 Corporate Drive, Suite 400, Burlington, MA 01803, USA
525 B Street, Suite 1800, San Diego, CA 92101-4495, USA

First edition 2011

British Library Cataloguing-in-Publication Data
A catalogue record for this book is available from the British Library

Library of Congress Cataloging-in-Publication Data
A catalog record for this book is available from the Library of Congress

ISBN: 978-0-12-375674-9

For information on all Academic Press publications visit our
website at elsevierdirect.com

Typeset by TNQ Books and Journals Pvt Ltd.
www.tnq.co.in

Printed and bound in United States of America
10 11 12 13 14 15 10 9 8 7 6 5 4 3 2 1

CONTENTS

Contents

vi

Contents

ix

Contents

x

This book explains the fine details of how to conduct an experiment involving mouse behavior from the initial planning of the research design through every step of the process until data are ready for analysis. Most of these details are not mentioned in texts on behavioral genetics, neuroscience, or statistical data analysis. Instead, the successful student who has progressed through a series of courses organized around academic textbooks eventually is allowed to begin work in a laboratory that studies mice. Only then does the student experience the complex realities of assessing behaviors of live animals and encounter challenges that were never discussed in the classroom. This book anticipates those challenges.

Part of this book is devoted to practical matters that need to be considered carefully when working with any species of animal, such as how many animals need to be tested and methods to balance the order of tests. These more general features of work with behavior are presented in detail using mice as examples. The other part is devoted to tests and techniques that have been devised specifically for work with mice.

Many texts delve deeply into what is known about the anatomy, chemistry, and electrophysiology of brain function and the neural bases for behavior in a wide range of species. To progress further, it is important to know not just why but how to study behavior. The principles of behavioral psychology can guide this enterprise in a general sense, but the best way to apply those principles depends on the species studied. Two animal species may have remarkably similar brains and even genomes, yet their behavioral tendencies and sensitivities can differ greatly. To be successful in research with mice, one must know the peculiar features of mice.

THE INTENDED AUDIENCE

Prior to the invention of the gene knockout method, mouse behavior genetics was a relatively small field of study concentrated mainly in psychology departments. The situation changed dramatically in the 1990s when molecular biologists became interested in assessing the phenotypic consequences of targeted mutations or transgenic insertions into the mouse genome. Those scientists were very well funded in comparison with others who specialized in the study of behavior, and they often bought equipment from commercial suppliers, unpacked the shiny new devices, and added mice. Then they learned from experience that behavioral testing is just as complex as a biochemical assay and can be influenced by a wide range of factors. The need for guidance on methodology became apparent, yet the information was scattered widely in the literature.

This book brings together many of the specialized methods employed in collecting data on mouse behavior. It should be useful for those who already are familiar with the general principles of research but are new to the realm of behavioral testing of live mice. It is hoped that even people with considerable experience with lab mice will find value in discussions of some of the more daunting challenges in this field. Much of this book explains the basics, but it also delves into some of the thorniest issues and perennial disputes about standardization of tests, ethical practices, and measurement of unseen constructs. Almost every chapter begins with simple examples, then progresses to discussion of situations where no formula or recipe currently exists. In this way, it strives to serve as both an introduction and a guide to advanced topics in behavioral assessment.

SPECIAL FEATURES

Every step of the research process is illustrated with real situations encountered in previous studies. Outstanding examples are highlighted, but many valuable lessons are also drawn from embarrassing disasters that are often hidden from colleagues. We all make mistakes and those who are most active in research probably make more mistakes than others. It is better to learn those lessons by reading about someone else's debacle than repeating it in one's own lab. Because this book is about techniques, it is essential to explain what *not* to do as well as what *should* be done.

All examples are based on real experiments, and extensive details of several published experiments are provided. Essential features of a behavioral test protocol are outlined, then several complete protocols are made available. Methods to balance the order of tests and determine throughput are described, then a completely balanced order of tests in a complex experiment is presented. Only by viewing the full complexity of a behavioral study can the importance of controlling a large number of potentially influential variables be grasped.

DATA

A large amount of original data are presented from the author's mouse labs, some of it not previously published. Data on fine methodological points belong in a book on methods but perhaps not in a journal article. Nobody wants to read long recitations about nit-picking details in article after article in the literature. The fine points should be illuminating for those who are new to the field. Some of the data have already been submitted to the Mouse Phenome Database and several data sets on inbred strain surveys described here will also be included. They will also be made available on the Web site created for use with this book. For more information and the appendix of Excel utilities visit our Web site http://www.elsevierdirect.com/companions/9780123756749.

UTILITIES AND COMPUTATIONAL AIDS

Many of the methods needed to plan and execute a study are implemented with utilities devised for an Excel spreadsheet. They are stand-alone applications designed for a familiar computer environment. They strive to make most calculations as easy as possible by high-lighting where to insert numbers, while preventing the user from altering anything else in the spreadsheet. The utilities are described in each chapter where they are first introduced, and the reader is given access to them via the Web site for this book. Experience has shown that simple utilities are handy for day-to-day work in the lab. A few of the utilities are rather complex and perform feats that simply cannot be done with standard statistical packages; they were constructed out of necessity.

TEACHING A COURSE

This book is designed to aid the investigator who is getting ready to plan an experiment, but it could also serve as a text for a formal course, most likely at the graduate level. In fact, it originated in a graduate course on Research Methods in Animal Behavior offered at the University of North Carolina at Greensboro. The various utilities can be used by students to do exercises assigned by the instructor. The only way to master most of these methods is to do a wide variety of exercises while progressing through the chapters. An instructor might ask the student to prepare a detailed research proposal taking into account almost every issue likely to be encountered in the mouse behavior lab. Even better, a lab course could provide experience with mice in several kinds of test apparatus.

THE AUTHOR

Douglas Wahlsten began scientific research on mouse behavior while completing his Ph.D. dissertation in the psychology of learning at the University of California at Irvine. His thesis research was done with genetically variable dogs of unknown background obtained from a local pound. The allure of genetically uniform inbred mice was great, and, anticipating postdoctoral study in behavioral genetics, he procured several dozen inbred mice for preliminary study. The first mouse ever held by the author, a C57BL/6J male, did not care for the way it was handled and inflicted a nasty puncture wound that bled profusely. Imbued with greater respect for the humble lab mouse, the author did postdoctoral study at the Institute for Behavioral Genetics at the University of Colorado in Boulder, then set up mouse behavior labs at the University of Waterloo, the University of Alberta, the University of Windsor, and finally the University of North Carolina at Greensboro. His research on mice has been done in nine different physical locations, ranging from one crudely improvised room in a cinder block warehouse to a suite of rooms designed explicitly for work with mouse behavior. This experience has taught a number of lessons that should be helpful and in some instances amusing to others.

He has published numerous articles and book chapters on mouse behavior and hereditary brain defects. He previously co-edited with Dan Goldowitz and Richard Wimer a book on *Techniques for the Genetic Analysis of Brain and Behavior: Focus on the Mouse* (Elsevier, 1992) and co-authored with John Crabbe the chapter on behavioral testing for *The Mouse in Biomedical Research* (Elsevier, 2006). In 2006 he received the Distinguished Scientist award from the International Behavioral and Neural Genetics Society. He is Professor Emenitus at the University of Albenta.

Several outstanding Ph.D. students and technicians have not only assisted the author in the collection of behavioral data but also contributed significant innovations to research with mice. At the University of Waterloo, former students Hymie Anisman, Patricia Wainwright, and Barbara Bulman-Fleming did exceptional work. Important contributions to the study of mouse brain and behavior were made by Melike Schalomon, Dan Livy, and Kathie Bishop at the University of Alberta; Mark Bunning at the University of Windsor; Martin Bohlen and Jeremy Bailoo at the University of North Carolina at Greensboro. Behavioral studies were greatly improved by the masterful machining skills of Gerry Blom at the University of Waterloo, Isaac Lank and Sean Cooper at the University of Alberta, Eric Clausen and Louis Beaudry at the University of Windsor, and Stuart Thompson at UNCG. Technical expertise in testing mice has been provided by many skilled individuals over the years, especially Brandie Moisan, Sean Cooper, Elizabeth Munn, Maria O'Neill, Sofia Prada, Erika Hayes, and Richard Jordan.

The content of the book is the sole responsibility of the author, but it has benefited greatly from many discussions with esteemed colleagues who have extensive experience working with mice: Fred Biddle, Valerie Bolivar, Richard Brown, John C. Crabbe, Jacqueline Crawley, Wim E. Crusio, Robert Gerlai, Dan Goldowitz, Hans-Peter Lipp, Peter Nguyen, Robert Williams, and many others.

None of the research done over a long career would have been possible without the financial aid of grants from the Natural Science and Engineering Research Council of Canada, the Alberta Heritage Foundation for Medical Research, and the National Institutes of Health.

The encouragement and good example provided by my mentor and friend Henry Klugh were greatly appreciated throughout the preparation of this book.

Introduction to the Research Process

1

Testing mouse behavior is done as part of a scientific investigation of some larger question, such as the genetic bases for neuropsychiatric disorders or mechanisms of memory formation. Those who begin their career in science with behavioral testing naturally focus on the fine details of the task set for them by a more experienced person. Only gradually does a larger picture form of how their work relates to other parts of the research process. This chapter sketches some of those parts and puts the tests into a context.

THE RESEARCH PROCESS: SCIENTIFIC ASPECTS

Whereas the research environment varies greatly in different countries and institutions, the actual process of doing research with mice is or should be the same everywhere. The main steps are illustrated in Figure 1.1, emphasizing things that are addressed in depth in this book. Conceptualization and literature review are governed by the existing knowledge and theories in a specific field of study. These are usually introduced to the student in an entire course such as behavioral neuroscience, behavioral genetics, psychopharmacology, or animal behavior. Through that kind of advanced study and extensive reading of the literature, the investigator becomes familiar with specialized methods of experimentation and learns what kinds of factors need to be controlled and measured in a good study. Little is said in this book about the first two steps in the process because they are thoroughly dealt with in many graduate training programs and postdoctoral settings. Several key reference works are listed at the end of this chapter for the benefit of those new to the field.

Most researchers working with mice have taken at least an introductory course in statistical data analysis. Because not everyone will have a strong background in these procedures, there is some coverage of statistical methods in Chapter 5, but no attempt is made to condense a one-term

Mouse Behavioral Testing. DOI: 10.1016/B978-0-12-375674-9.10001-1

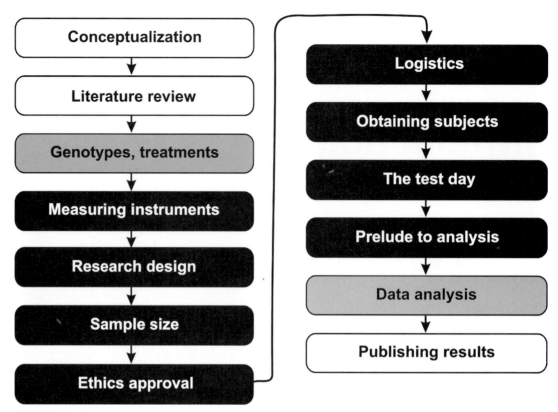

FIGURE 1.1
Broad outline of the research process followed by studies of both human and animal subjects. Shading of the box indicates the depth of discussion each step receives in this book. The darker the shading, the deeper the discussion.

statistics course into one chapter. Statistical principles will not be grasped firmly until the investigator needs to apply them to real data, at which time a review of the basics can be helpful. This is especially true for the planning phase of a project. Many texts on statistics devote chapter after chapter to what should be done with data that have already been collected, while little is said about critical matters such as choosing the proper sample size before the experiment begins. When analysis of data is discussed here, attention is focused on issues that are common in work with mouse behavior.

Adapting the approach to the stage of an investigation

Most long-term research programs involve two stages: an exploratory stage followed by a formal test of well-specified hypotheses. The latter methodology is elaborated in many courses, whereas the former receives little attention. The exploratory stage is perhaps the most important because it can be the time when a discovery is first made or a decisive clue is found. More often than not, an important discovery occurs by accident during the course of some other experiment. Students and technicians are supposed to follow a protocol rigidly and everything is supposed to happen according to well-laid plans, but nobody tells the mice about this, and some of the mice behave unexpectedly. Good researchers take note of the exceptions and surprises. An experienced researcher may know when a surprising result is potentially very important, but this same experience can also make the investigator jaded and too willing to attribute every exception to sampling error. There is considerable benefit in having people new to the field take a fresh look at the procedures used in a behavioral test and also at the mice as they are run through their paces.

It is good practice to conduct the exploratory stage of a project without rigid protocols and controls. The *pilot study* allows different test parameters and procedures to be evaluated informally using just a few animals in any one condition or perhaps using repeated testing of the same animals in practice runs. This kind of study scans a wide range of possibilities in a way that will detect those with very large effects. Later, when the most important variables have been brought under control and optimized for that lab, factors with more subtle effects can be studied in a formal experiment with the stricter controls and larger samples that are needed to detect them. The principal investigator (PI) faces a special challenge in convincing an animal ethics committee to approve a pilot study. Suggestions about how to do this are offered in Chapter 6. The purpose of a pilot study is not to circumvent the ethical principles of research with animals. Those principles should be respected in any kind of study with mice, which imposes limits on what range of test parameters can be tried in a pilot study.

An open-ended exploratory study can be done in some situations. Crowcroft (1973), in his charming book *Mice All Over,* described a study of mice free to run loose in an entire building. At the Institute for Behavioral Genetics in Colorado, Gerald McClearn had a device constructed that he dubbed the "hypothesis generating apparatus." It was a large and complex environment populated with mice of both sexes from several inbred strains. The students and post-doctoral fellows sat in chairs and watched the antics of 5 to 10 mice for hours at a time at different phases of the light–dark cycle (dim light for the dark phase). This was an excellent way to become familiar with the range of behaviors that mice commonly express, and hypotheses flowed freely from this device (e.g., Yanai & McClearn, 1972). David Sternthal, a graduate student at the University of Waterloo, went one step further by filling a small lab room with topsoil covered by grass turf with places to build deep burrows and a hut where an outcast could find shelter away from the main group of mice. Those inbred mice that sometimes seem so dull and listless in a "shoebox" colony cage or boring behavioral tests came to life and utilized every portion of an environment that was vastly larger than the typical colony cage. A complex mouse society emerged quickly and soon the entire environment was transformed with tunnels, nests, and home bases. Hans-Peter Lipp and co-workers (Lipp, Amrein, Slomankia, Wolfer, 2007) at the University of Zürich and Moscow State University took inbred mice to the outer limits of practical research by releasing them into large outdoor pens in a rural research station north of Moscow, where they were exposed to the rigors of natural selection. To the amazement of almost everyone, inbred mice survived and some of them even thrived in the Russian winter . This was high-risk, exploratory research at its finest, because nobody knew if the study would provide any useful data.

Another source of ideas about testing mouse behavior is the Internet site YouTube. There are fascinating videos of "the smartest mouse in the world" that rapidly navigates an astoundingly complex sequence (probably aided by odor trails) over barriers and through a wide variety of devices that put anything we do in the lab to shame. There are mice on surf boards that exhibit the notorious aversion of mice to water, even at the beach. At the beginning of scientific work with mice early in the twentieth century, laboratory researchers went to mouse fanciers to obtain their subjects and learn about how to breed and maintain mice (Chapter 2). Today there is still a thriving community of mouse fanciers with fascinating Web sites who keep these animals as pets and have a greater interest in behavior than fanciers 100 years ago, who were interested mainly in coat color variants.

One perfectly legal approach to exploratory work is to buy mice at a local pet shop and keep them in a basement or garage. This approach needs to respect the needs of the mice and adhere to regulations that proscribe animal cruelty. Never should a pet be made to suffer just in case it might give a scientist some prize winning idea. At the same time, all the things that are done to train cats and dogs for the show ring are reasonable for training mice and testing the limits of what they can achieve. Animal trainers have been able to induce bears to play ice hockey and porpoises to serve cocktails on silver trays at poolside. There is no good reason why a professional scientist who studies mouse behavior in the lab cannot also be an animal trainer or observer after hours, provided friends and family can live with the mouse odors.

3

Mice, mice, and more mice

Pets are fun to watch, but research is done with special mice developed for use in laboratories around the world. Their surprising origins are now fairly well understood. Modern lab mice are a mixture of genes from four wild *Mus* subspecies. Mouse fanciers in Europe, the United States, China, and Japan kept mice as pets, and those animals were convenient sources of research subjects in the twentieth century when the scientific study of mouse behavior commenced. Before long, standard inbred strains were created and mouse genetics progressed rapidly. With the advent of molecular genetic technology to sequence the genome and introduce mutations into known genes, mice became more popular than ever. Today, mouse models of human genetic disorders have become prime subjects for neuroscience research. Chapter 2 reviews mouse history and the many kinds of mice that inhabit labs today.

Measuring instruments: Tests

Behavior is a measurable phenotype (Wahlsten & Crabbe, 2007). The process of measuring it is termed a test. The test involves a physical apparatus and a protocol instructing how to conduct the trials. Virtually every kind of behavior, with the possible exception of some simple reflexes, is complex and can be measured in several ways. The test chosen for a study must pertain to the kinds of behavioral or neural processes that are the subjects of the study. The researcher needs to decide on the specific behavioral tests to employ in a study quite early in the process before there can be sample size estimation or ethics approval. Chapter 3 introduces more than 50 kinds of tests. The entire second half of the book (Chapters 10 to 14) is devoted to an in-depth discussion of many kinds of tests. Chapters on general features of behavioral tests cover a wide range of topics from ways to construct a test battery to methods for assessing reliability. Specialized techniques to provide adequate motivation and obtain good results from computer-based video tracking are explored in considerable depth. Finally, the challenge of obtaining comparable results in different labs is described, and possibilities for standardizing the tests as well as the conditions in a lab are discussed.

The formal research design

Many brilliant "discoveries" in a pilot study, accidental findings in a project aimed at studying something else, or flashes of insight while watching pets at play arise from events that cannot be repeated. Many hunches turn out to be wrong. Good ideas need to survive a rigorous, formal test before the rest of the field needs to take notice. For this purpose, the investigator must adopt a research design that is capable of proving the truth of an idea to a skeptical audience of peers. Many elaborate designs with sophisticated control groups are used for genetic studies of mouse behavior. A few of the main ones are described in Chapter 4. Design of a study is critically important for sample size determination, ethics approval, and logistics of running the experiment. Once those hurdles have been crossed, the proper execution of an elegant design is essential for obtaining good data. It is fairly easy to design a study on a piece of paper so that logic appears to govern all else. Taking that design into a real laboratory, however, soon teaches the researcher to anticipate and respect the many pitfalls and complications that beset the practical implementation of well-laid plans. It is for this reason that most of the remaining chapters address the practical side of research on mouse behavior.

Sample size

Only after the design of the study and the specific behavioral tests are chosen can the researcher rationally decide the appropriate number of mice to be tested. This step is taken before the application for ethics approval and then grant funding, because the number of animals in the study is an important aspect of external review and consent. After the data are collected and the report is submitted for publication, the author may be asked by a skeptical reviewer to justify the sample size. If the number of mice is found to be too small at that stage of the process, the ship of science will break apart on the shoals of bad methodology. It is better by far to plan ahead and take great care to get this number right. Chapter 5 is devoted to this topic, and a series of statistical utilities is provided to make the process as quick and painless as possible.

Ethics approval

Having decided upon the design, the tests, and the number of mice, permission to conduct the study must be given. This is usually done through some kind of peer review by colleagues who also work with animals or have expertise pertinent to the study. The specific rules and procedures differ greatly in different countries, and within a single country an institution may have considerable latitude in implementing ethical principles. The journal to which an article is submitted for publication may also, through its peer review process, judge the ethical aspects of a study. Official bodies sanctioned by a government agency may be given powers to pass judgment on a research proposal, and those bodies provide instructions on how to pass over the hurdles and through the hoops of regulations. The discussion in Chapter 6 does not attempt to replace or reduce the need for an investigator to engage the official mechanism. Instead, it offers a researcher's perspective on the process and advice on how to get the job done well and done quickly. Specific examples of research procedures that are considered good practice or in some cases unethical in work with mice are presented, and a scheme for rating the degree of discomfort or invasiveness for a wide range of behavioral tests is proposed.

Logistics

After funding and permissions are obtained, it is time to plan the details of execution of the experiment. This step could be taken earlier in the process, but everything could change after the sample size calculation or the ethics review. Planning means a list of which mouse is to be tested at what time on which day under what conditions, and it requires other aspects of the lab to be arranged so that the testing can progress smoothly according to protocol. This stage is complete when the researchers have produced a stack of data sheets, one per mouse, in the correct order for the entire experiment. The study should not begin until every detail of every trial for every mouse in the entire experiment has been specified.

There are two major parts to the planning process. First, the throughput for the test, how much time on how many days is required to obtain measures on just one animal, must be determined. Second, the order in which animals from the various genotypic and treatment groups are to be tested must be decided. The order should be carefully balanced and randomized so there is no preponderance of mice from any one group at any particular period of testing. The effects of test order must always be considered. Some mouse must go first or last and early or late each day. A skillful execution of test order causes variance arising from time of day and sequential effects such as odor trails to end up in the within-groups variance, rather than creating spurious between-group effects. Chapter 7 explains how to achieve this and provides several elaborate examples from real experiments.

Obtaining subjects

After everything is ready for the experiment, it is time to bring the mice to the laboratory, if they are to be purchased from a commercial supplier. In labs where the mice are to be bred locally, as when unique genotypes are involved, the planning for obtaining research subjects is far more difficult and requires months of advance notice to animal care personnel. Mice are born in litters ranging in size from 2 to 13 pups, and not all mated females become pregnant, especially in some of the less fertile inbred strains. Commercial suppliers such as the Jackson Laboratory in Bar Harbor, Maine, breed so many mice of the more common strains that average litter size and pregnancy rates can be tuned to the anticipated demand for each genotype, with the happy consequence that the experimenter can simply ask for 10 mice of each strain and sex to be shipped at a narrow age range on a specific day. Obtaining the same numbers in the same age range at the same time locally can challenge the resources of most laboratories when more than two or three genotypes are involved.

Obtaining mice includes things done the day they arrive in the lab, such as unpacking, marking and assigning them to cages placed in specific locations on a rack in a colony room, and everything that happens to them until the first day of testing. The researcher must make some deliberate choices: how much cage enrichment to use, what kind of food and water to provide, the light–dark cycle, and so forth. Life in the colony room can have a significant impact on behavioral test results. Even when there is little the experimenter can do to control these phenomena, it is very important to be cognizant of their presence when planning the study and interpreting results. Chapter 8 discusses many of these issues, and Chapter 15 explores them in greater depth.

The test day

After all the planning is completed and everyone is ready for the first day of testing, the real action begins. Everything should go smoothly if the conditions have been properly prepared. In reality, however, something almost always goes wrong and adjustments need to be made on the fly. A major study simply cannot be put on hold while the researcher ponders what went wrong and what to do next. It can be terminated early, and this is a disaster. To avoid such disasters, it is wise to have contingency plans. This is best done by looking closely at exactly what is done during the test day and identifying vulnerable points in the protocol where a small mouse can run a large experiment right off the rails. This happened in the author's lab when the roster of strains in a large study for the Mouse Phenome Project included four wild-derived inbred mouse species. Inbreeding had not bred the wildness out of them, and more than once the testing schedule was imperiled when the technician was led on a merry chase after an escapee around a cluttered testing room.

There are events too large to accommodate in any plan, such as major weather calamities or terrorist attacks. Sadly, the mouse lab can be a target for deliberate disruption. Security remains a concern at many institutions. Options are briefly considered here, but rare events are difficult to anticipate. In a long career, the author has never experienced a truly major disruption of an experiment, but others have. Gordon Harrington (personal communication) at the University of Northern Iowa maintained eight unique inbred strains of rats, but the entire building housing his collection burned to the ground. His students responded quickly to the emergency and carried every one of the rats in their cages onto the lawn, saving a valuable genetic resource. Harrington had not organized them in advance into fire brigades with little rat resuscitation apparatuses; instead, the students exercised good judgment and acted spontaneously. What Harrington had done was teach his students the importance of unique strains of research animals. They knew there was something housed in that building that was really worth saving.

Prelude to data analysis

After reading current texts on research design and data analysis, it appears that analysis with sophisticated multivariate methods begins the day after data collection is complete. What a false impression this is. In reality, the study may appear to progress smoothly to a brilliant finale, but the data can later prove to be gravely flawed or even worthless. All behavioral test data need to be subjected to a rigorous process of quality control. This is usually done after data collection is completed because the tight testing schedule leaves little spare time. In recent studies in the author's lab involving video tracking of mice in complex apparatus, almost as much time was spent checking for and correcting tracking errors as was spent doing the study in the first place. Those errors were obvious from the bizarre patterns on the computer screen

(Bailoo, Bohlen, & Wahlsten, 2010); a mouse simply could not have jumped from one arm of a maze to another in 0.04 sec. More insidious are those occasional errors or oddities in the data that generate outlier data points and have a major impact on the statistical analysis. Hunting down exceptional events requires good tactics and a keen eye, and deciding when an outlier is a real event or an outright error requires wise judgment. This stage in the process is critically important. Chapter 9 suggests ways to detect and expunge bad data. The finest design and perfect balancing of test order can be defeated by poor quality control after data collection is finished. This stage includes the backing up of data in whatever form it was collected. Hard disks fail; it has happened to me more than once.

Data analysis

Almost everyone who conducts a thorough statistical analysis of behavioral test data will do so after taking at least one formal course on statistics. Nevertheless, it is seen time and again that lessons from a formal course are not permanently implanted into the mind until the researcher is faced with the task of making sense of his own real data and then trying to convince other scientists that the conclusions are justified by those data. It is suggested in Chapter 9 that the researcher should first compile descriptive statistics and graphs so that the major results of the study are evident, then inquire whether the apparent effects of treatments are indeed statistically significant. The first part of this process can be immediately rewarding, even fun, while the second generates severe pains in the frontal lobes.

Publishing results

An experiment does not become part of the body of scientific knowledge until it is published or somehow made available to other scholars. On occasion a student may conduct an experiment with mice to obtain a grade in a course or even an advanced degree but then fail to follow through with this last step in the process. The study then is not a part of science, and is rendered utterly worthless. This is not a happy outcome. The data must reach others interested in the same topic. Publishing is a difficult art usually learned from a master through close collaboration on a project.

Those are the major steps in research on mouse behavior from conceptualization through to publication. This book emphasizes aspects of the process that generally are not explained in-depth in texts on behavioral neuroscience or research design and statistical data analysis. This information will be most helpful to those entering the field and hopefully will provide some added insights for experienced researchers as well. The field is continually changing and there are always new things to learn.

THE INSTITUTIONAL CONTEXT OF RESEARCH

Scientific research is done in an institutional context consisting of people, facilities, and policies. The personnel involved in research can be divided between those in the academic or scientific world and those in the administrative or service realm (Figure 1.2). In a very small institution, the researcher may be in charge of almost every aspect of the enterprise, whereas at a large university he or she may feel remotely connected to officials in the upper administration and service sector. Whatever the case, there are essential roles in research and steps in the process that are the same almost everywhere. Here we focus on those parts that are important for testing mouse behavior.

The research usually follows a series of steps that is almost the same for work with both human and animal subjects. These steps are introduced in this Chapter and become the basis for organizing most of this book. The principles of science are universal and apply to all domains of investigation, while the realization of the scientific study is achieved in a local context using a kind of organism that may require slight modifications in the approach.

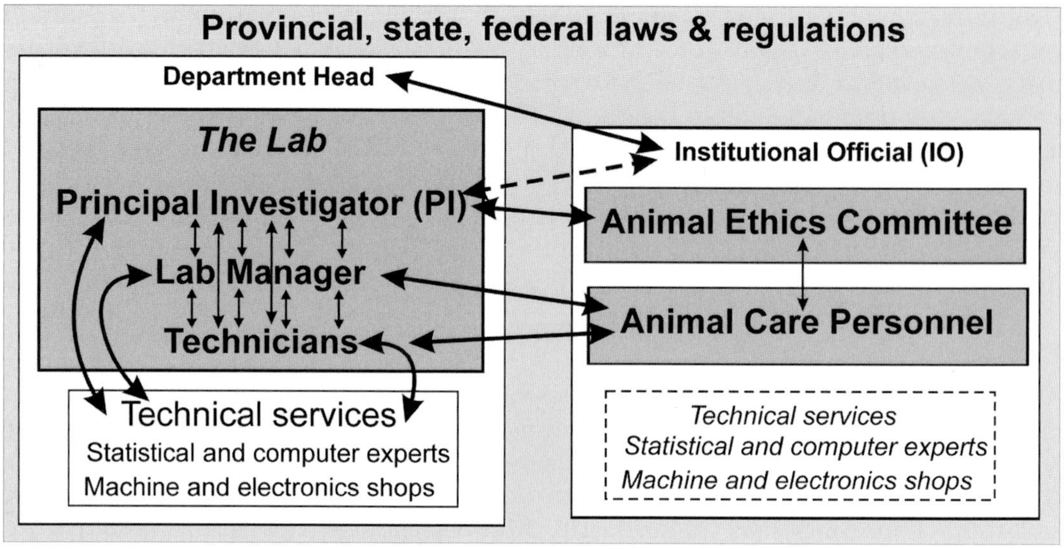

FIGURE 1.2

Organization of a research laboratory operating in an institutional and regulatory context. The configuration can differ greatly among institutions. Arrows indicate the approximate frequency of interactions based mainly on the author's experiences at four universities in Canada and the United States. In some institutions the technical services are based in a department and responsible to the PI, whereas in others they are centralized.

People

Every laboratory exists in a larger setting governed by a director or head of a department or institute, but these individuals, who may be eminent scientists, are usually involved very little in a research project when a PI is in the lab. The PI is responsible for all aspects of the research, from conceptualization of the problem, gaining ethics approval, applying for a grant, hiring and training staff, analyzing data, and writing articles for publication. With sufficient grant funds, the PI can have a laboratory manager, possibly someone with a Ph.D. in neuroscience or psychology, to organize the daily work of the lab. The actual testing of the mice may be done by career technicians with a bachelor's degree or a technical certificate in lab animal science. A postdoctoral fellow may do the work of both a PI and lab manager, whereas a graduate student may function as both lab manager and technician. Students sometimes believe they are doing the work of the PI too because the boss is traveling to conferences so much of the time. Whatever the titles and academic status of the people doing the research, the roles of PI, lab manager, and technician are essential. All of these individuals need to have a good understanding of the methodology of behavioral testing.

Every institution has some kind of administrative apparatus that governs all research operations and work with animals. This apparatus will be under the sway of some high institutional official (IO), perhaps a vice president of research or an associate provost, who is rarely involved directly with researchers. The domain of the IO includes three offices of great importance for work in the behavioral test lab: the animal ethics committee, the office of the IO, and animal care personnel. No research can be done until the project is approved by the animal ethics committee (Chapter 6). After it is approved, funds must be obtained to support the work, which usually requires applications to external granting agencies. The office of the IO includes people who know where and how to apply. Once funds are available, care of the animals will often be done by animal care staff that report to the IO, not the PI. This staff will include a professional veterinarian as well as workers who maintain, change, and wash the animal cages. These people are a critical part of the laboratory environment (Chapter 15), a component that is especially difficult for the PI to alter. Although it would be wonderful if the IO, animal ethics committee, and animal care personnel knew about behavioral testing principles, this is unlikely to happen. These people need to deal with a very wide range of research projects and cannot be expected to read an entire book on behavioral testing. They have power over crucial aspects of the research but lack the expertise to plan and conduct a program of behavioral research. The PI and everyone in the lab, therefore, need to develop educational and diplomatic skills when working with colleagues who report to the IO. If a good working relationship cannot be established among all the participants, the research process will not progress smoothly, and in extreme cases will not generate good data.

Technical services such as statistical consulting, computer and Internet technology, and machine and electronics shops are of great importance for the mouse lab. If the lab is very large or the department has many well-funded researchers, these people may be employed to serve one or a few PIs and their staff. Otherwise, the institution may have a central service used by people in all fields of research as well as classroom instruction or medical practice. In some settings, these services are obtained externally from private companies. These people are unlikely to know much about mice; therefore, the researchers need to be able to clearly explain the problems and available options related to the special features of the test animals.

Facilities

A behavioral testing facility usually consists of rooms devoted to the tests, colony rooms for housing the animals over several weeks, and rooms for cleaning the cages and storing supplies. Offices for personnel often are not part of the test facility per se and are not allowed in the animal facility in many institutions. The researcher should have considerable control over who enters the testing rooms and how things are arranged there, but the colony rooms and cleaning areas often are part of a central animal facility managed by an independent professional staff. This can create challenges if routine operations like cage changing, washing the floors and cage racks, or inspections are done without regard to the behavioral testing schedule.

Both federal government and granting agencies provide extensive recommendations on the design of animal facilities. For example, the Canadian Council on Animal Care (CCAC) likes to see all animal housing and procedure rooms contained in one large, centralized facility with a secure means to prevent entry by unauthorized people. These facilities are also supposed to prevent intrusions by wild mice, but mice can be ever so devious and find ways to breach all but the most solid barrier. Hollow walls and ceilings where there are unfilled gaps along heating ducts and water pipes sometimes provide ready access. For example, on one occasion in the author's lab at the University of Alberta, a technician was surprised to find a wild deer mouse (*Peromyscus maniculatus*) sitting atop a mouse cage dining on the lab chow.

From the standpoint of the researcher, there are major advantages to having mice in a small satellite facility under the control of the PI. The author designed one such facility (Figure 1.3) at the University of Windsor that was housed in the Great Lakes Institute for Environmental Research where there were no other mammals.. The entire lab consisted of only four rooms in one suite having a single unmarked door to the hallway. All air was vented to the outdoors, so that nobody walking in the hallway even knew it was a mouse lab. Two side-by-side colony rooms could be maintained on opposite light–dark cycles. It was a very short distance from colony to test room, with transit through the cleaning area. Cage washing was scheduled when no behavioral testing was being done. We hired undergraduate students at a modest wage to do the cage changing and cleaning. We got the job done with a small commercial dishwasher (Hobart 7i) instead of the humongous, expensive, and temperamental cage washer often found in central facilities. The Hobart 7i could accommodate only three cages at a time but had a 2 minute cycle time with a bleach rinse at the end, yielding cages that were clean enough for people to eat from. The net result of these features was high quality data and healthy mice.

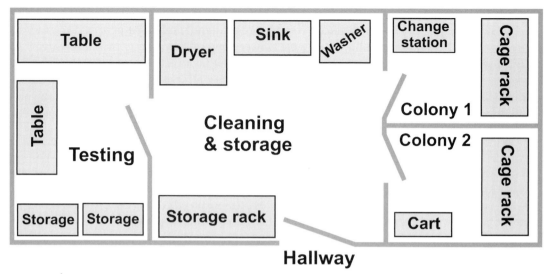

FIGURE 1.3

Layout of a compact, self-contained laboratory for behavioral testing that was a satellite facility remote from a central animal housing facility. Operations were highly efficient, and the lab director had almost complete control over every aspect of the daily routine. The layout was not optimal, and was dictated in part by location of pipes and hallway doors in the building. It would be better to have the cleaning room on the far side of the colony so there could be direct access from colony room to testing room.

In comparison with the central facility, our lab offered a number of real advantages.

1. There were no other mice nearby to transmit disease. All holes were sealed.
2. Noise was totally under our control.
3. Distance from colony room to test room was very short and internal to the lab.
4. Costs were much lower using undergraduate helpers.
5. Security was very high. Everybody knew everybody else in a small crew.

Even for our work, the setup was not ideal. It would have been better to locate two testing rooms at one end of the colony rooms and the washing area at the other, but space, existing doors to the hallway, and utility connections did not permit this without added expense, and we were required to minimize costs. Most labs are created through a process of negotiation and compromise. Without knowing the history of the place, current reality may not make much sense.

Agents of the CCAC did not like the location of this lab and directed that it be moved into a central facility far away from our offices when renovations on the main campus were completed, but their objections were not based on any identifiable shortcomings in care of the mice.

Lucky indeed is the scientist who can specify the layout of a behavioral testing lab and have it built expressly for that purpose. The author has conducted his research with mice in nine laboratories at five institutions over the course of a career, ranging from one small room (10 × 12 ft) next to his office in a warehouse at the University of Waterloo that encompassed the mouse colony on two racks and all the behavioral test apparatus on two tables, to a spacious, modern facility with controlled access at the University of North Carolina at Greensboro. Not one of these labs was ideal for the purpose, but almost all were sufficient to obtain good data from inbred mice. To what extent these different laboratory environments can influence the scores on behavioral tests is discussed at length in Chapter 15. Mouse behavior is sometimes exquisitely sensitive to the animal's surroundings. We now have sufficient knowledge to recommend best practices for some aspects of facility design, yet a wide range of options is likely to be adopted at different institutions and to remain an important part of the behavioral testing landscape for many more years. The enduring gap between what is optimal and what is achievable is just one of those stubborn facts of life in science.

Policies

All of the work of the lab and the administrative apparatus is governed by a maze of institutional and government laws and regulations. It is the job of the IO to know these in great detail and make sure that those involved with animal research have an adequate knowledge of the most relevant rules. The PI and the lab staff usually learn about these policies as their work progresses, often when somebody with greater knowledge of official policies notes deficiencies or a grant proposal needs to be revised. At some sites there may be regular inspections of the lab and colony rooms by an expert veterinarian who can call attention to places where improvements can be made. Brief tutorials and manuals are available to obtain an overview, usually via a Web site. Because work with mice is done in so many countries and laws vary considerably across countries, the legal and regulatory aspects of animal research are discussed only briefly in this book, mainly in Chapter 6.

The process of obtaining funds to support the research is labyrinthine and requires a great deal of experience and/or luck to engage it successfully. Larger institutions usually have an office of research services that keeps up to date on where and how to obtain grants. More experienced colleagues can provide invaluable advice in this area. This book only indirectly grapples with this difficult and sometimes mysterious part of science. The proposal usually has a strictly defined length limit and scarce space is typically devoted to the scientific substance of the project while omitting details. Mastery of behavioral testing methods presented here should enhance the chances of success in competition for funds. Especially when applying to agencies that ask for a detailed justification of the budget, sometimes with no page limit, the applicant can demonstrate this mastery with detailed discussions of logistics, sample size estimations, and discussion of special challenges and advantages for a wide variety of tests.

One feature of the grant process has special importance for methodology. There are three broad categories of research that differ in scope and duration: the experiment, the project, and the program. A single *experiment* or *study* is a self-contained entity with a clear beginning and end that, once begun, must be continued to completion with no substantial changes in methodology. When complete, its results can be analyzed statistically and described with a few graphs and tables, but interpretation depends on comparisons with other experiments. A *project* is typically a series of carefully planned experiments that aim to answer a limited set of specific questions. These aims can be enunciated clearly and the design of the experiments should be adequate to answer the questions, if results concur with the hypotheses. If all goes as planned, the questions will be answered at the end of the experiments outlined in the project. A research *program*, on the other hand, is built around a larger question. For example, what is the role of neurogenesis in the loss of memory capacity with aging? Are there several fundamentally different kinds of fears and anxiety, or are fears specific to the stimulus while anxiety transcends the situation? Can we discover or construct a mouse model for childhood autism? No single research project is going to answer these questions. Inevitably, things learned in the course of one project will have a major influence on how the next phase of the program is conducted, and the investigator will not be able to describe with confidence what experiments will be done several years into the future.

The strategy and tactics of research must be adapted to the funding environment. Local institutions often have some limited resources to support a single experiment that may help to win external support. Many large granting agencies prefer to fund multiyear projects but not research programs. A few countries and foundations do support longer term programs, and applications to them are likely to focus on the larger scientific issues as well as the track record of the PI. In genuine program funding, the request for continued support is structured around a progress report describing what was actually accomplished. Program support typically is awarded to experienced investigators who have demonstrated mastery of the research process through a series of projects. At some stage of this journey, those who grant the money may finally recognize that the scientist requesting support actually knows more about the area of science than those making decisions about funding. This is the ideal situation, as described by a working scientist. Unless one is independently wealthy, it will always be necessary to approach on bended knee those who control the flow of funds. How to do this successfully is critical for career success, but it entails many issues far from the realm of behavioral testing.

General References on Mice

Eisen, E. J. (2005). *The mouse in animal genetics and breeding research.* London, UK: Imperial College Press.

Fox, J., Davisson, M., Quimby, F., Barthold, S., Newcomer, C., & Smith, A. (2007). *The mouse in biomedical research.* Amsterdam: Elsevier/Academic Press.

Hedrich, H. J., & Bullock, G. (2004). *The laboratory mouse.* London, UK: Elsevier.

Papaioannou, V. E., & Behringer, R. R. (2005). *Mouse phenotypes. A handbook of mutation analysis.* Cold Spring Harbor, NY: Cold Spring Harbor Laboratory Press.

Suckow, M. A., Danneman, P., & Brayton, C. (2001). *The laboratory mouse.* Boca Raton, FL: CRC Press, Taylor & Francis Group.

Books on Behavioral Testing

Buccafusco, J. J. (2001). *Methods of behavior analysis in neuroscience.* Boca Raton, FL: CRC Press, Taylor & Francis Group.

Crawley, J. N. (2000). *What's wrong with my mouse? Behavioral phenotyping of transgenic and knockout mice* (1st or 2nd ed.). New York: Wiley.

Crawley, J. N., Gerfen, C. R., Rogawski, M. A., Sibley, D. R., Skolnick, P., & Wray, S. (2007). *Short protocols in neuroscience.* Hoboken, NJ: Wiley.

Crusio, W. E., & Gerlai, R. (1999). *Handbook of molecular-genetic techniques for brain and behavior research (techniques in the behavioral and neural sciences).* Amsterdam: Elsevier.

Jones, B. C., & Mormede, P. (2007). *Neurobehavioral genetics: methods and applications* (2nd ed.). Boca Raton, FL: CRC Press—Taylor & Francis Group.

Martin, P., & Bateson, P. (2007). *Measuring behavior. An introductory guide* (3rd ed.). Cambridge, UK: Cambridge University Press.

Whishaw, I. Q., & Kolb, B. (2005). *The behavior of the laboratory rat.* Oxford, UK: Oxford University Press.

Chapters and Review Articles on Mouse Behavior

Crabbe, J. C., & Morris, R. G. M. (2004). Festina lente: Late-night thoughts on high-throughput screening of mouse behavior. *Nature Neuroscience, 7*(11), 1175—1179.

Crawley, J. N. (2008). Behavioral phenotyping strategies for mutant mice. *Neuron, 57*(6), 809—818.

Lipp, H.-P., Amrein, I., Slomankia, L., & Wolfer, D. P. (2007). Natural genetic variation of hippocampal structures and behavior — an update. In B. C. Jones, & P. Mormède (Eds.), *Neurobehavioral genetics* (2nd ed.). (pp. 389—410) Boca Raton, FL: CRC Press, Taylor & Francis Group.

McClearn, G. E. (1982). Selected uses of the mouse in behavioral research. In H. L. Foster, J. D. Small, & J. G. Fox (Eds.), *Experimental biology and oncology* (pp. 37—49). New York, NY: Academic Press.

Peters, L. L., Robledo, R. F., Bult, C. J., Churchill, G. A., Paigen, B. J., & Svenson, K. L. (2007). The mouse as a model for human biology: A resource guide for complex trait analysis. *Nature Reviews Genetics, 8*(1), 58—69.

Schellinck, H., Cyr, R., & Brown, R. E. (2010). How many ways can mouse behavioral experiments go wrong? Confounding variables in mouse models of neurodegenerative diseases and how to control them. *Advances in the Study of Behavior, 41,* 225—271.

Tarantino, L. M., & Bucan, M. (2000). Dissection of behavior and psychiatric disorders using the mouse as a model. *Human Molecular Genetics, 9*(6), 953—965.

Tecott, L. H., & Nestler, E. J. (2004). Neurobehavioral assessment in the information age. *Nature Neuroscience, 7*(5), 462—466.

12

van der Staay, F. J., Arndt, S. S., & Nordquist, R. E. (2009). Evaluation of animal models of neurobehavioral disorders. *Behavioral and Brain Functions, 5*, 11.

Vitaterna, M. H., Pinto, L. H., & Takahashi, J. S. (2006). Large-scale mutagenesis and phenotypic screens for the nervous system and behavior in mice. *Trends in Neurosciences, 29*(4), 233–240.

Wahlsten, D., & Crabbe, J. C. (2007). Behavioral testing. In J. G. Fox, S. Barthold, M. T. Davisson, C. Newcomer, F. Quimby, & A. Smith (Eds.), *The mouse in biomedical research: Vol. 3. Normative biology, husbandry, and models* (pp. 513–534). Amsterdam: Elsevier.

Mice

15

A few words about mice are needed at the outset, because they are the focus of every other chapter. For those who plan to work with the kinds of tests addressed in this chapter, the choice of whether to use mice or rats is a major decision.

MICE VERSUS RATS

Almost everyone working in science rejects the urban myth that mice are just small rats that grow up to become bigger rats. Nevertheless, vestiges of this opinion can be found when some apparatus manufacturers take a test designed and perfected for use with rats and scale it down proportionally to make it suitable for mice. This geometric approach may not be the best one. When testing animal behavior, the test should be adapted to a wide range of animal features including its behavioral penchants and proclivities, not just its size.

Rats and mice are both rodents, but their ancestral lines diverged about 8 million years ago and they became tuned to different ecological niches (Morse III, 2007). Only recently did the Norway rat (*Rattus norvegicus*) and house mouse (*Mus musculus domesticus*) move into dwellings provided by newly evolved humans and then spread rapidly around the globe. Both species, having become commensal with humans and accustomed to our presence, were eventually tamed and adopted as pets. Mouse and rat fanciers preserved and bred certain variants because of their unique coat colors and peculiar motor behaviors, and these variants were the source of subjects for scientific research on rodent behavior in the early twentieth century (Yerkes, 1907).

Rats evolved along the edges of water bodies and are adept at foraging for food by diving and swimming. They are predators that kill and eat mice as a food source. Mice evolved in drier climates. They tend to avoid water and, because of their higher ratio of surface area to body volume, are very buoyant and generally cannot dive below the surface. When a mouse is swimming in water it is a target for scaly predators, and it readily becomes panic stricken. If a mouse cannot escape from the water quickly, it may float motionless (Wolfer, Stagljar-Bozicevic, Errington, & Lipp, 1998), something rarely seen in rats; mice often are not proficient in the Morris water maze that has been so successful with rats (Whishaw & Tomie, 1996). Mice do not like the odor of rats and show anxiety-like symptoms when exposed to rat odor (Blanchard, Griebel, Rodgers, & Blanchard, 1998; Merali, Levac, & Anisman, 2003). These and other features need to be taken into account when devising behavioral tests.

Many people perceive mice as being cute and mischievous (Figure 2.1), and they may enjoy watching the antics of a cage full of mice, while they find rats to be gross, ugly, and threatening. Furthermore, in the contest for popularity, carrying the black plague did not help the case for rats (Epling, 1989). At the same time, wild mice sometimes harbor the deadly hantavirus. Perceived cuteness may arise from neotony, in which the head and eyes are relatively large with respect to the body. Whatever its source, it provides a modicum of pleasure to scientists who work with mice.

In the laboratory, there are some clear advantages to working with mice. Size is a major factor. An adult rat is more than 10 times larger than a mouse. Some of the most important features of mice as research subjects are listed here.

1. Several can be housed in a *social group* in a relatively small cage. This avoids isolation-induced behavioral abnormalities such as fighting and hyperreactivity.
2. Many kinds of mouse cages provide abundant space to add *enrichment* devices and allow for considerable *exercise.*
3. Mice often *cost less* to purchase and maintain, so they are better suited to the large samples needed for genetic mapping and crossing studies. The sample size requirements for research (Chapter 5) are essentially the same for all species. There is no lowering of statistical standards just because a species is large.
4. The smaller size of mice also allows for *smaller animal facilities*, including both colony space and laboratory testing rooms.
5. A *greater genetic diversity* of standard inbred strains is readily available from commercial breeders, including strains derived from wild species, and *extensive genetic* and *phenotypic data* have been compiled for many of these strains.

FIGURE 2.1
Mice taking a test. (*Drawing by Sara Tripp. With permission.*)

6. A much greater number of *recombinant inbred* (RI) *strains* is available for analysis of complex traits, and a large volume of phenotypic data has already been compiled for certain strains, especially the 99 BXD RI.

7. Transgenic technology for creating targeted mutations is more advanced, and many more *gene knockouts* are available. A project to knock out every mouse gene has made great progress.

To be fair, rats have their advocates among scientists, and for good reasons. Several kinds of rats have been selectively bred for tameness and ease of handling by humans. They make good pets that like to be petted, and they are well suited to undergraduate teaching labs. Much of the psychology of animal learning is based on the laboratory rat. For many years, the study of rodent behavior was done with rats in psychology departments. The Sprague-Dawley albino rat in particular was bred selectively for tameness over many generations and served well as a subject for explorations of the laws of learning by undergraduate students in the "rat lab." Rats were the subjects of classical breeding studies of learning by MacDougall, Tolman, Tryon, Heron, Hebb and Thompson in psychology, and until very recently the array of behavioral tests used in work with rats (Whishaw & Kolb, 2005) was far more extensive and sophisticated than with mice. The field of psychopharmacology focused on studies of rats for many years.

Although it may be true that a wider range of behavioral tests is currently available for use with rats, especially those for evaluating learning and memory of complex tasks, this situation is changing as more researchers work with mouse models of human biomedical disorders. Many excellent tests for mice are already available (Crawley et al., 1997; Crawley, 2000), and it is hoped that this volume will help to develop more and better tests for mice.

While mouse researchers are improving the behavioral testing facet of their work, rat researchers are devoting more attention to improving the genetic knowledge of rats. Inbreeding and transgenic techniques can be applied with rats, and the rat genome has been sequenced. Box 2.1 lists several sources of information about mice and rats. The reader is invited to peruse them to see what is available.

2.1 SOURCES OF INFORMATION ON MICE AND RATS

Mice

> Mouse Genome Informatics (Jackson Lab)
> Mouse Genome Database (MGD)
> Mouse Phenome Database (MPD)
> National Center for Biological Information (NCBI)
> Federation of International Mouse Resources (FIMR)
> International Knockout Mouse Consortium (IKMC)
> Complex Trait Consortium & The GeneNetwork
> Mouse Brain Library

Rats

> NCBI & Rat Genome Database (RGD)
> Rat Resource & Research Center (RRRC)
> Knockout Rat Consortium (KORC)
> Programs for Genomic Applications (PhysGen)
> National BioResource Project for the Rat (Japan)
> Rat & Mouse Club of America

Note: Web sites are not given because they often change. A search engine will almost always find these sources via the names listed above.

TABLE 2.1 Salient Features of Genetic Knowledge of Mice, Rats, and Humans

Feature	Mice	Rats	Humans	Source
Chromosome pairs	19 + XY	20 + XY	22 + XY	NCBI[a]
Genome size (nucleotides)	2,716,965,481	2,826,224,306	3,101,788,170	NCBI
Genes	>26,000 ??	27,782 ??	21,470	MGI, RGD
Known orthologs with mouse	—	16,768	17,802	MGI
Polypeptides sequenced	36,629	?	?	MGI
Inbred strains	450	234	—	NCBI, MGI
Strains with SNPs	86	48	—	MGI, RGD
SNPs	10,089,692	?	26,293,003	MGI; IHMP
Genes with targeted mutations	7,385	176	—	MGI, RGD
Mutated genes with ES lines	13,411	?	—	IKMC

[a]Abbreviations for sources are given in Box 2.1. These can readily be used to access databases via the Internet. Other acronyms include: IHMP, International HapMap Project (see http://snp.cshl.org); SNP, single nucleotide polymorphism, a change in one base at a specific location in the DNA; and ES line, embryonic stem cell line perpetuated in vitro that can be used to generate knockouts or conditional mutants. An ortholog is a gene that is clearly the same kind as a gene in another species, most likely via descent from a common ancestor. Gene counts for humans are based on Clamp et al., 2007.

The current state of genetic knowledge of mice, rats, and humans is summarized in Table 2.1. Some of the numbers need to be viewed with caution. Only a few years ago, it was common practice to announce confidently the number of genes possessed by each species in a running tally on the Internet. Now recent discoveries about the complexity of the genome have raised serious questions about what is a gene and the role of regulatory DNA sequences beyond the boundaries of the DNA transcribed into mRNA and then translated into protein or polypeptides. Gene counts are changing daily, and sometimes contract. A careful examination of the human DNA sequence recently identified 2787 alleged genes in the human list that are not real genes (Clamp et al., 2007). This kind of computationally intensive editing will be required for many other species, including the mouse.

MODELS OF HUMAN FUNCTIONS

Whether or not the lab rat or mouse can serve as a better model for human disorders is a very interesting and important question. The similarity of the mouse and human genomes is quite remarkable, with most genes of one species occurring in the other, arranged in large blocks (haplotypes) of DNA that have many of the same genes in the same order in both species (Waterston et al., 2002). Among the 20,470 human genes that meet a stringent criterion for gene identity, only 168 are not found in mice or dogs (Clamp et al., 2007). Mice and humans are so similar at the genetic level that the entire human chromosome 21 can be incorporated into the mouse germ line and its genes can be expressed using mouse molecular machinery (Wilson et al., 2008). Transgenic technology has allowed specific human disease alleles to be inserted into the mouse genome to create remarkable models for research, such as a segment of exon 1 of the human Huntington disease allele (Mangiarini et al., 1996), which is used to study potential drug therapies for the human condition (Giampa et al., 2009). Considerable success has been reported for treatment of amyotrophic lateral sclerosis (ALS) with microRNA in a mouse model (Williams et al., 2009), and hopes are high that this can be beneficial for humans suffering from ALS. The mouse models have limitations; one is the substantial difference between regulatory portions of the mouse and human genomes (Odom et al., 2007).

As the genetic knowledge of mice burgeons far beyond that of rats, mouse models are becoming more prevalent in the scientific literature. Popularity is no guarantee, however, that the current species of choice in neurobehavioral genetics will forever be the mouse. Mouse models are not always so successful in illuminating the human condition (Benatar, 2007; Gawrylewski, 2007; Kalueff, Wheaton, & Murphy, 2007). For example, when the human and mouse genome sequences were completed (Lander et al., 2001; Waterston et al., 2002), it became evident that the base sequence of the disease allele in the human cystic fibrosis gene is actually the same as the normal allele in mice. Many factors must be taken into consideration when evaluating an animal model of a human function or disorder (Bourin, Petit-Demouliere, Dhonnchadha, & Hascoet, 2007; Finn, Rutledge-Gorman, & Crabbe, 2003; Kas, Kaye, Foulds, & Bulik, 2009). Of particular importance is the careful definition of a condition to be modeled, such as depression in humans (Anisman & Matheson, 2005). It is also interesting to note that knowledge of human genetics may be able to aid in the search for genes influencing mouse behavior (Willis-Owen & Flint, 2007); the human can sometimes serve as a model for mice.

It seems likely that the depth of our understanding will be greater if more than one species is used to address similar research questions. There is a remarkable diversity of social behaviors among various species of rodents (Fink, Excoffier, & Heckel, 2006), and it is now possible to adapt transgenic technology to the study of wild rodent species (Donaldson, Yang, Chan, & Young, 2009). Excessive concentration of work on laboratory mice is probably not wise, no matter how cute they may be. Of one thing we can be certain: no animal model is going to tell us much about the human condition if the behavioral test used for the study is not well suited to that animal species.

ORIGINS OF STANDARD MOUSE STRAINS

Early studies of mouse behavior utilized whatever mice could be readily acquired from fanciers and pet breeders. The stocks were not maintained in any standard way, and behavioral tests were devised de novo in a way that was idiosyncratic to each investigator. The masterful study of behavior of the "dancing" mouse by Yerkes involved the first extensive test battery applied with any animal (Yerkes, 1907), but it is not entirely clear which neurological mutation was involved. The large maze used by Yerkes for testing learning was not replicated by Halsey Bagg when he studied mouse learning using albino mice from a local pet store (Bagg, 1916). Bagg's lasting contribution was to create a standard inbred strain (BALB/c for Bagg Albino) that is commonly used today. Little and others initiated standard breeding of the DBA and C57 strains during the same period (Morse III, 2007).

When the Jackson Laboratory was established in Bar Harbor, Maine, systematic inbreeding was practiced to create and maintain many standard strains that were then made available to other researchers. The first edition of *The Biology of the Laboratory Mouse* by Little et al., (1941), emphasized anatomy, immunology, physiology, and cancer biology while saying virtually nothing about the origins of mouse strains or their behaviors. One chapter described results of a cross between the "Japanese waltzer" and the "common mouse." The field had not even adopted standard nomenclature for inbred strains and mouse genes. Instead, the authors tried to make a case for the practice of inbreeding for purposes of research.

Many of what we now consider standard strains were preserved for posterity because of some special characteristic such as a tumor or diabetes that made them ideal for biomedical studies. As a consequence, behavioral studies of mice were more often than not done with sick animals or those that would soon develop significant health problems, or they involved subjects that had serious sensory deficits that posed no disadvantage for cancer research but imperiled psychological analysis. Mutations causing interesting behavioral effects were usually severely abnormal — what are now regarded as neurological defects.

Much of the early research on mice was done with non-standardized mouse stocks such as the ubiquitous Swiss-Webster mouse. Several commercial breeders employed that name but the so-called Swiss mice studied in different labs were simply not the same animals genetically (Lynch, 1969). Finally some breeding stock from a Swiss-Webster colony was used to generate the standard inbred SWR/J strain at the Jackson Laboratory, while a non-inbred Swiss-Webster stock (CFW) is available from Charles River Laboratories. The amazingly fertile CF-1 stock

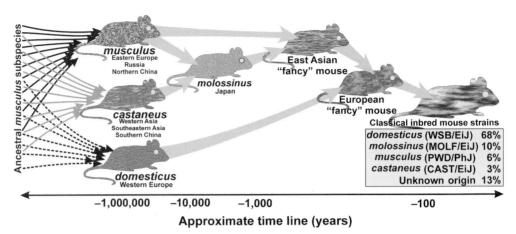

FIGURE 2.2

Genetic origins of contemporary laboratory mice from four wild *Mus* subspecies originating in different geographic regions, indicating the domestication of house mice and "fancy" mice from 1000 to 300 years ago. This pattern has inspired some geneticists to regard the lab mouse as a unique subspecies, *Mus musculus laboratorius*. *(Redrawn with permission from Frazer et al., 2007).*

from Carworth Farms, also of Swiss origin, was acquired by Charles River in 1974, inbred for over 20 generations, and is now outbred to form a new stock.

As the field of mouse genetics grew and became well established, more attention was devoted to the history and origins of the better known strains and mutations. Nevertheless, ideas about the sources of ancestors that gave rise to specific laboratory mice prior to 1900 were murky at best. Only recently have advances in DNA sequencing made it possible to assemble a crude and surprising picture of the origins of lab mice. Several wild mouse species exist today, and lab mice appear to have genes derived from four of them (Figure 2.2). The three main ancestors of today's lab mice are *Mus musculus domesticus*, *Mus musculus musculus*, and *Mus musculus castaneus*, along with *Mus musculus molossinus*, which appears to be a cross between the latter two (Frazer et al., 2007). So intermixed are the genes from these diverse sources, and so closely joined with modern human migrations and industry, that some scholars now regard the lab mouse as an entity unto itself: *Mus musculus laboratorius*. No violence is done to evolutionary biology by adopting this colorful term. The animals are indeed adapted to the contemporary environment of mouse laboratories.

The work on genetic origins of inbred strains focused primarily on autosomal DNA, but mice also have DNA in the mitochondria that entails 16,299 base pairs, and this mtDNA is always transmitted by the female parent. Researchers were startled to learn that mtDNA is not an amalgam of species but instead appears to have a single origin in all modern lab strains of mice (Goios, Pereira, Bogue, Macaulay, & Amorim, 2007). By sequencing the entire mtDNA genome in 12 common inbred strains and four wild-derived strains, they found that almost all lab mice were very similar to each other and must have been descendants from a *musculus domesticus* female, whereas the wild-derived strains were very different. From estimated mutation rates, it appeared that some of today's lab mice are from lines that diverged as much as 500 to 3,000 years ago. That is a time period known to include the domestication of "fancy" mice in China and Japan. Over the ensuing centuries, there were introductions of chromosomal genes from other species via males that perhaps crept into the tents of early mouse fanciers on their long journeys across the steppes.

KINDS OF BREEDING SCHEMES

There are five kinds of breeding schemes available with mice and rats, and certain ones are usually applied in a grant-supported lab but not by commercial breeders. For those working

TABLE 2.2 Kinds of Mouse Strains

Kind of Strains	Generations to Create New Strain	Genetic Variation Within Strain	Genotype at One Locus	Difference Among Strains	Fertility	Availability From Suppliers	Replicability of Genotype Across Labs
Inbred	20	None	Homozygous	Many genes	Often low	Very good	Excellent
F_1 hybrid	1	None	Heterozygous	Many genes	High	A few crosses	Excellent
Coisogenic	1	One gene	Heterozygous	Many genes	Low	Limited supply	Excellent
Congenic	10 to 20	A few genes	Heterozygous	A few genes	Low	Limited supply	Excellent
Consomic	10	None	Homozygous	One entire chromosome	Low	Limited supply	Excellent
Recombinant inbred	20	None	Homozygous	Many genes	Low	Good for BXD and AXB	Excellent
Selectively bred	5 to 10	Substantial	Variable	Many genes	High	Sometimes present in one lab	Good
Outbred	?	Substantial	Diverse	Many genes	High	Very good	Questionable

with mice, by far the most common is full-sib mating that characterizes all standard inbred strains. Some salient features of the different kinds of breeding schemes are summarized in Table 2.2. A concise discussion of mouse breeding is provided in a classic text (Silver, 1995) that is now available free at http://www.informatics.jax.org/silver/.

Outbred animals

Outbred stocks reside in a breeding colony at one institution where mating pairs may be chosen haphazardly or by selecting some of the more fertile animals. Inevitably, even if the breeder avoids pairing close relatives, there will be inbreeding and fertility will eventually decline unless a very large number of breeding pairs is employed. One solution is to import fresh breeding stock from other colonies of the same general stock from time to time, as is usually done for Sprague-Dawley rats. For example, Taconic Farms refreshed its breeding stock of Sprague-Dawley rats with animals from the National Institutes of Health Genetic Resource facility in 1998, long after 1945 when NIH originally obtained its breeding stock (http://www. taconic.com). Charles River maintains several colonies of CD-1 mice and uses "backward and forward migrations between international production colonies" to forestall inbreeding (http://www.criver.com). With outbreeding, there are likely to be more than two alleles at many loci and it will be almost impossible to trace a new mutation to its originator.

Closed colony

The breeder of a closed colony never imports breeding stock from outside the colony. The manager of the colony has complete control over the choice of matings and, hopefully, maintains an extensive pedigree dating back many generations. Provided there are many breeding pairs each generation, the progress of inbreeding will be slow. This progress can be further retarded by a small amount if the breeding scheme entails a random choice of breeders with the constraint that no second cousins or closer relatives are ever paired.

Inbreeding makes it likely that any two mice to be mated share a common ancestor somewhere in the pedigree. The consequence is that they have a higher likelihood of being homozygous for an allele than is expected purely on the basis of random sampling from a large population. Suppose there are two alleles (A, a) in the original population and their frequencies are p_0 and q_0, respectively, such that $p_0 + q_0 = 1$. The probability of any particular individual having genotype AA is p_0^2 when there is random mating and the population is very large. When the population has a finite size, the coefficient of inbreeding (F) is defined as the probability that an individual inherits two identical alleles by descent from a common ancestor. If the population size is infinite, then F = 0. If it is finite, F will increase by a small amount ΔF every generation because there are more and more common ancestors in the individual's past. Presuming that F begins at zero, then after t generations of random mating, it is expected that $F_t = 1 - (1 - \Delta F)^t$, a quantity that gradually approaches the upper limit of 1.0 when one allele or the other is finally lost forever. The probability of homozygotes for the dominant A allele in any generation will be $p_0^2 (1 - F) + p_0 F$, the excess over p_0^2 being the increase from inbreeding. The rate of inbreeding ΔF depends on the effective population size N_e, which is the number of individuals that actually breed and generate offspring. In general, $\Delta F = 1/2N_e$, but if one male and one female are chosen from each family to breed, the effective population size is doubled and $\Delta F = 1/4N$, where N is the number of breeding parents. Suppose a company or lab perpetuates its stock with 50 breeding pairs each generation. Then $\Delta F = 1/400 = 0.0025$. After 10 generations the inbreeding coefficient will be only $F_{10} = 1 - (1 - 0.0025)^{10} = 0.025$, but after 30 generations it will be 0.07 and after 100 generations it will reach 0.22, causing an appreciable increase in homozygotes. For mice, 100 generations might be attained in 30 years. If a single laboratory attempts to maintain an outbred stock with only 10 breeding pairs, $\Delta F = 1/80 = 0.0125$, and after 100 generations F = 0.72, a substantial degree of inbreeding.

The number of alleles at a locus will depend on the nature of the *foundation population*. If it all started with a hybrid cross of two inbred strains, then the closed colony will generally have only two alleles at a locus. If it began with crosses among eight inbred strains, there could be considerably more than two alleles at a locus, depending on the specific strains. Unless a huge number of mating pairs is used, it is highly likely that one or more of the alleles will disappear from the colony purely by chance. On the other hand, it is highly unlikely that all but one allele will vanish. One example of a closed colony is the HS control strain that was created at the Institute for Behavioral Genetics in Boulder, Colorado, by intercrossing eight inbred strains. It has been in existence for more than 70 generations. These mice are proving to be valuable for genetic mapping studies because the large number of meioses and crossover events since the inception of the colony has broken the genome into many very small segments, each containing only a few genes (Talbot et al., 2003). Another useful stock for fine mapping is MF1 (Darvasi, 2005), derived from mice of unknown parentage and now maintained by Charles River Labs. These kinds of stocks make it much easier to identify which specific genetic polymorphism is affecting the phenotype when the gene is responsible for only a small, quantitative change in the behavior (McKay, Stone, & Ayroles, 2009).

Inbred strains

An inbred strain is systematically propagated by mating brother by sister each generation. The effective population size each generation is therefore only two animals, and this soon leads to the loss of all but one allele at every locus, thus the strain consists entirely of homozygotes. The process is explained in greater depth in the next section.

F$_1$ hybrids

Hybrids are made by crossing two highly inbred strains. The offspring will be heterozygous at every gene where the parent strains have different alleles, but all mice in the F_1 hybrid will have the same genotype. These animals can be valuable for behavior work because of their excellent ability to reproduce and relative freedom from bizarre abnormalities of the brain. Most commercial suppliers offer a few different hybrid crosses to the researcher, and it is fairly easy to create an F_1 hybrid in one's own lab.

Selectively bred lines

Artificial selection is a valuable technique for studying specific behavioral domains. One drawback of inbred strains is that two strains may differ markedly in a phenotype of interest, perhaps preference for a 10% ethanol solution versus tap water, but they will also differ in many other neural and behavioral phenotypes that have nothing at all to do with appetite for ethanol. If one first generates a population that has a large amount of genetic variation, perhaps by intercrossing several inbred strains for a few generations, selective breeding can then accumulate alleles in one selected line (the HIGH line) that favor the specific phenotype and the alternative alleles in the LOW line. Provided a large number of animals is used each generation, the difficulty of spurious associations of gene differences with phenotypic differences should be considerably reduced. The problem cannot be eliminated entirely, and it is therefore wise to create two or more replicate HIGH lines selected for high phenotypic values and two or more LOW lines. In some experiments the investigator also maintains replicate CONTROL lines that are not selectively bred at all; matings are done through the random choice of breeders.

A very ambitious selection experiment for high and low levels of motor activity in a brightly lit open field was done by DeFries, Gervais, and Thomas (1978) at the Institute for Behavioral Genetics at the University of Colorado (Figure 2.3). The change in activity levels was so large that there soon was virtually no overlap between the HIGH and LOW lines, and the LOW lines moved very little at all in the test chamber. The ancestral inbred strains included several that were albino, and albinism results in a moderate reduction in open field activity (OFA). As expected, the HIGH lines eventually lost the albino allele, while the LOW lines became 100% albino. In the unselected CONTROLS, one line gradually accumulated albino alleles and after 30 generations consisted of about 85% albinos, whereas the other CONTROL line declined to about 10% albinos. These changes are attributable to random genetic drift in a moderate sized population. Once the extreme phenotypic line differences had stabilized, the lines were propagated with random matings a few more generations and then deliberately inbred to preserve the allelic differences and reduce the colony size (Toye & Cox, 2001). The resulting inbred strains are now proving valuable for mapping genes that influence mouse activity and anxiety (Turri, Datta, DeFries, Henderson, & Flint, 2001; Turri, Henderson, DeFries, & Flint, 2001).

23

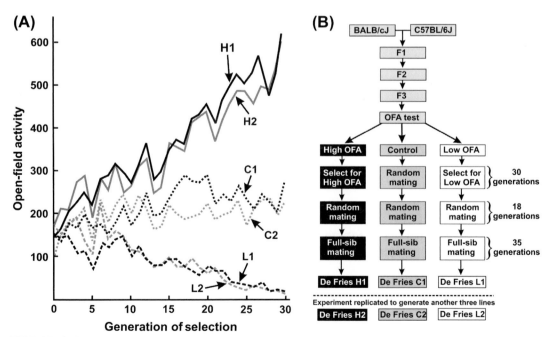

FIGURE 2.3

(A) Selective breeding of replicate high, low, and control lines of mice for open field activity (OFA), showing the gradual but dramatic divergence of the selected lines over 30 generations of selection (Redrawn with permission from DeFries, Gervais, & Thomas, 1978). (B) Strategy for selection followed by inbreeding to fix extreme genetic differences between lines. *(Reprinted with permission from Toye & Cox, 2001).*

The rate of progress of selection depends on the extent of genetic variation relevant to the phenotype in the foundation population and the degree of selection pressure (Falconer & MacKay, 1996). The *selection differential* in any generation is the difference between the mean phenotypic score of all animals in the line and the mean of those chosen to serve as parents for the next generation. The greater the selection differential within each line, the more rapid the progress in creating divergent HIGH and LOW lines. A problem arises, however, if the researcher tries to use a very large selection differential, because this will occur when only the most extreme mice are chosen as breeders. If only a few actually breed, the inbreeding co-efficient will rise more quickly. The problem can be overcome by producing a markedly large number of mice each generation so there are dozens in the extreme tail of the distribution that can serve as breeders. This becomes expensive indeed.

Many generations are required to create widely divergent selected lines, and considerable resources are required to maintain them over many years. Commercial breeders generally do not maintain genetically variable selected lines, so the interested researcher usually must obtain them from the one laboratory where they are propagated. Fluctuations in grant and institutional support can place the lines in peril, as can hazards such as fire or disease.

Several investigators have opted for creating selectively bred lines that will yield important information about genetic influences on behavior after only a few generations of selection (Belknap, Richards, O'Toole, Helms, & Phillips, 1997). When genetic variation influencing a trait is substantial, rapid divergence of the HIGH and LOW lines can occur even when the number of animals chosen to perpetuate the lines is modest. The lines will inevitably succumb to inbreeding if they are continued for many generations and the population is small. It may be more efficient to terminate them after a period of time and re-create new, selected lines using the same phenotype as a criterion.

THE PROCESS OF CLOSE INBREEDING

The theory of inbreeding as applied to lab mice is thoroughly explained in standard works on mice (Green, 1981) as well as quantitative genetics (Falconer & MacKay, 1996). When there are only two animals, one male and one female, they can possess at most four different alleles. Two of the four will be lost very quickly, while the third will probably persist in the strain for several generations. Once there are only two alleles (call them A, a), there can be only three genotypes: AA, Aa, aa. Furthermore, there can be only five possible values of the frequency of the a allele in the entire population: 0, 0.25, 0.50, 0.75, 1.0, as shown in Table 2.3.

Suppose that in generation n the gene frequency is 0.25 and the two parents have genotypes AA and Aa. The probabilities of genotypes among their offspring will then be 50% AA and 50% Aa because each pup must get A from one parent and has an equal chance of getting A or a from the other. From the offspring, two will be chosen to produce the next generation. If both happen to have genotype AA, then the a allele will be lost forever from that strain, purely by chance. Because choices of the two parents are independent events, the probability that both are AA is

TABLE 2.3 Possible Mating Combinations When There Are Two Alleles of Autosomal Gene

Parent 1	Parent 2	% *A* Allele	% *a* Allele	Offspring Genotypes
AA	*AA*	100	0	All *AA*
AA	*Aa*	75	25	50% *AA*, 50% *Aa*
AA	*aa*	50	50	All *Aa*
Aa	*Aa*	50	50	25% *AA*, 50% *Aa*, 25% *aa*
Aa	*aa*	25	75	50% *Aa*, 50% *aa*
aa	*aa*	0	100	All *aa*

the product of each being *AA*: $0.5*0.5 = 0.25$. In general, only three mating combinations are possible with just two parents (*AA* × *AA*, *AA* × *Aa*, and *Aa* × *Aa*), and their probabilities are 0.25, 0.5, and 0.25, respectively. The gene frequencies (of the recessive allele) in the population in generation $n + 1$ that correspond to these matings are 0, 0.25, and 0.5, respectively.

In general terms, there are six possible mating combinations according to genotype, shown in Table 2.3, and for each of them a specific frequency of the *a* allele obtains. The key feature of full-sib inbreeding is the mating of siblings that have the same two parents. The effective population size is therefore just two mice, and there can be only four alleles in the population at any one locus. Consequently, once an allele is lost from the breeding population, it is gone forever. Eventually the strain will become genetically *fixed* for one surviving allele, so that every animal will be homozygous at that locus. The probability of this happening depends solely on the gene frequency in the preceding generation. Using the basic laws of probability, one can compute the probability of observing any of the five possible gene frequencies in one generation, given the frequency in the preceding generation. This result is shown in Table 2.4. In mathematical terms, this is a finite-state Markov process with absorbing barriers, and Table 2.4 shows the transition matrix for jumps from one state to another. For this situation, after the first two generations, the probability of losing one of the alleles in any one generation settles down at a constant value, and the probability of population fixation for either allele inexorably rises, as shown in Figure 2.4. After 20 generations of full-sib mating, the probability of one allele having been lost is 0.98, and this has been adopted as the convention for designating a new inbred strain.

Any researcher can generate a new inbred strain by crossing two or more strains and then mating full sibs for 20 generations. Provided that the two alleles at the outset do not influence reproductive fitness, the process is purely probabilistic, and it proceeds independently at all loci unless they are closely linked on the same chromosome. The chance of losing one specific allele is 0.5, the same as for the other allele. Applied across the entire genome at the same time, this process will generate a novel combination of homozygous genotypes at all loci that were initially segregating. To achieve genome purity, wherein the strain is homozygous at every locus, requires considerably longer than 20 generations. A new strain should be pure or nearly so after 60 generations of full-sib mating. Many of the common strains in use today surpassed 60 generations long ago. The most venerable of inbred strains, DBA/2J, has undergone more than 223 generations of full-sib mating since it was initiated in 1909 (www.jax.org).

Whereas the theoretical basis of allele distribution during inbreeding is probabilistic, its realization in practice is not. It has been estimated that more than half of all attempts to generate a new inbred strain fail when the strain is lost because of infertility. When there is epistatic interaction among loci, certain combinations of alleles at different loci will function

TABLE 2.4 Transition Matrix Probabilities for Generations n to n + 1 with Full-Sib Inbreeding

		% *A* Allele in Generation n + 1				
		100	75	50	25	0
% *A* Allele in Generation n	100	1.0	0	0	0	0
	75	0.25	0.5	0.25	0	0
	50	0.031	0.125	0.687	0.125	0.031
	25	0	0	0.25	0.5	0.25
	0	0	0	0	0	1.0

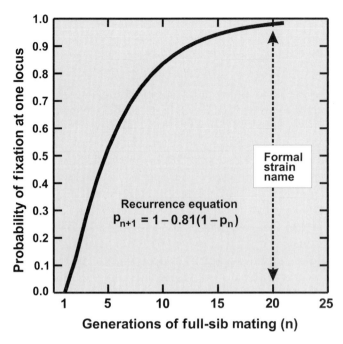

FIGURE 2.4
Probability of fixation for a single allele after n generations of brother by sister inbreeding. A new inbred strain can be recognized formally after 20 generations of full-sib mating, at which time the probability of fixation for one allele surpasses 98% at one locus. Purity of the entire genome requires 60 or more generations.

26

better than others, and certain combinations may not work well at all, leading to an animal that will not breed or cannot rear its young successfully. Thus, the reality of close inbreeding is that it usually involves a certain degree of selective breeding as well. Natural selection happens when fertility is impaired, whereas deliberate selection by the experimenter may occur if a new strain does not have interesting characteristics that make it worth continuing.

Reputable suppliers of inbred mice, such as the Jackson Laboratory, maintain separate colonies for the nucleus of a strain and the expansion stocks. The nucleus colony perpetuates the strain across generations and is very carefully documented and protected from disease to ensure the genetic integrity and health of the strain. Strictly stated, an inbred strain is propagated by only one breeding pair each generation. It is wise practice, however, to mate more than one pair in any one generation, just in case the designated parents do not leave offspring or something unfavorable happens, such as all male offspring. Thus, the ancestry of an inbred strain can be traced back through just one breeding pair each generation, even though more than one pair is actually mated each generation. Some of the spare offspring not needed to perpetuate the strain are then moved to the expansion colony where much larger numbers are generated for distribution to labs across the globe.

Once pure, a strain tends to remain so because it is reproductively isolated from carriers of novel alleles, but spontaneous mutations are inevitable. With only two mice in the breeding population, any new mutation will start at a gene frequency of 0.25 and most likely will be lost, but from time to time it will survive and become fixed in the strain. This process makes *substrain differentiation* likely if two labs obtain breeding stock of a strain from the same supplier and then perpetuate them separately (Taft, Davisson, & Wiles, 2006). The rate of genetic evolution by random mutation can be estimated by comparing substrains whose time of separation from a highly inbred parent strain is well documented. This has been done for the C57BL/6NTac strain that is maintained by Taconic Farms (Tac) after obtaining stock from the NIH (N) substrain that was in turn derived from the Jackson Lab strain C57BL/6J in 1951. After being separated for 150 generations, the two substrains differed at only 12 of 342 microsatellite loci examined (Bothe, Bolivar, Vedder, & Geistfeld, 2004). Closely related substrains sometimes show interesting differences in behavior, such as C57BL/6NCrl and C57BL/6JOlaHsd that differ in ultrasonic vocalization (Wohr et al., 2008).

By convention, a lab can designate its own substrain after 20 generations of separation from a bona fide inbred strain. The principal investigator can then apply for a unique substrain symbol, usually based on his or her name or institution. For example, the fascinating strain CDS/LayBid maintained by Fred Biddle at the University of Calgary has little or no preference for paw usage (Biddle, Jones, & Eales, 2001), unlike the C57BL/6J strain where most mice are either strongly right or left pawed in a test of reaching for flakes of food in a narrow tube (Collins, 1975). Biddle (Bid) obtained the strain from W. M. Layton (Lay) who originally selectively bred some CD-1 mice for teratogenic response to acetazolamide. Biddle also maintains the euphonious C3H/HeHaBid originally developed by Heston (He), then split from the main strain by Hauschka (Ha) and eventually established in Biddle's lab.

The general rules for strain names are summarized in Box 2.2 where a Web site for the official guidelines is indicated. This Web site also provides forms to fill out when applying for official recognition of a new strain or substrain. To qualify as a new inbred strain, there must have been at least 20 generations of full-sib inbreeding and some indication that the strain is genetically unique. To be recognized as a new substrain, the substrain must have been derived from a recognized inbred strain and then reproductively isolated from that strain with continued inbreeding for another 20 generations, or it must show clear evidence of genetic divergence.

2.2 NAMING INBRED STRAINS

General form: STRAIN/SubstrainBreeder(s)
Examples: C57BL/6J, DBA/2NCrl

Unique **strain name** *before* the / is given in capitals and integers. It must be approved by a committee.

Substrain *after* the / can be an integer followed by registered code(s) of the official breeder(s), showing line of descent, with the most recent breeder given last.

Rules are set by the International Committee on Standardized Genetic Nomenclature for Mice.

To see the rules or register a new strain, go to www.informatics.jax.org/strains_SNPs.shtml

INBRED STRAINS FOR RESEARCH

Among the more than 450 inbred strains maintained in various laboratories and commercial breeding facilities, a core of 36 in 3 priority tiers has been designated as top priority for inclusion in the Mouse Phenome Database (MPD) (phenome.jax.org; Grubb, Churchill, & Bogue, 2004), now part of the Mouse Genome Database (MGD; Bult, Eppig, Kadin, Richardson, & Blake, 2008). Those and other strains have been genotyped at numerous marker loci to establish genetic similarities and lineages among the more common strains (Petkov et al., 2004). The strains designated for Tier 1 were chosen to represent a wide diversity of mouse genotypes with many interesting characteristics that make them valuable for research (Figure 2.5). Thus, an investigator wanting to conduct a survey of inbred strains for research on a new phenotype is advised to employ the top priority strains that, by common agreement among mouse researchers, are most likely to have a large amount of data already available. The 19 strains in Table 2.5 are listed by order of the number of data sets already in the MPD that involve each strain. Table 2.5 presents some of the salient

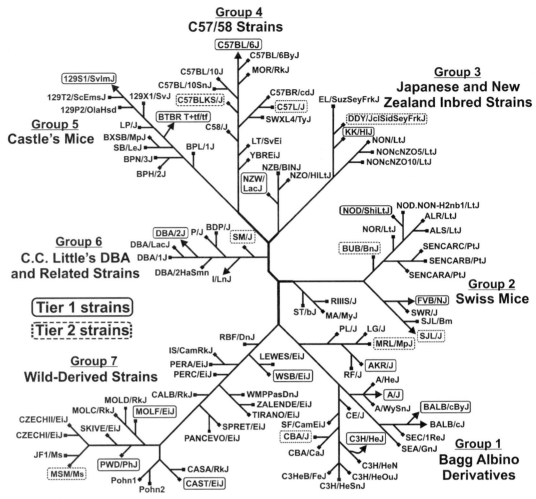

FIGURE 2.5

Origins and genetic similarity across a large number of marker loci for a wide range of inbred strains available for research on behavior, most of which are maintained at the Jackson Laboratory. Circled strains have been given the highest priority (Tier 1) for collecting phenotypic data for the Mouse Genome Database. Genetic similarity is roughly proportional to the length of the path joining two strains, such that bona fide substrains are relatively close on the same large branch of the tree. *(Redrawn with permission from Petkov et al., 2004).*

features that make the strains important for research on behavior and warns about sensory defects such as albinism, retinal degeneration, and age-related hearing loss that afflict so many of these creatures. This vast trove of information should be consulted by anyone seeking to collect more data on popular strains. It is always wise to know one's animal before starting a project, and today it is very easy to begin this process with a review of what mouse researchers have already learned.

Great care is warranted when choosing certain strains for research models. While some have well-documented, pure, and distinguished pedigrees extending to the early days of mouse genetics, others have been out-crossed to other strains over the years. Most notorious is the situation with the 129 strains that are so useful for transgenic knockout studies because they are well adapted to in vitro stem cell research. Research crossed descendants of the classical 129 strains, 129/J and 129/ReJ, with other strains, and today there are 15 strains designated 129 (Simpson et al., 1997; Threadgill, Yee, Matin, Nadeau, & Magnuson, 1997), as summarized in www.informatics.jax.org/mgihome/nomen/strain_129.shtml. The strains are sufficiently different genetically to warrant formal strain status and are not merely substrains. Accordingly, part of the strain name occurs to the left of the "/" in the official name: 129P1, 129P2, 129P3,

TABLE 2.5 Mouse Phenome Database Priority Strains[a]

Strain	MPD Sets	MPD Priority	Coat Color	RD[a]	HL[a]	Special Characteristics
C57BL/6J	113	1	Black	No	Yes	Prefers 10% ethanol; Type II diabetes; obesity; high motor activity
C3H/HeJ	86	1	Agouti	Yes	No	Prior to 1999 carried Mouse Mammary Tumor Virus
A/J	85	1	Albino	No	Yes	Low motor activity; lung tumors; induced asthma
DBA/2J	83	1	Light brown	No	Yes	Audiogenic seizures; aversion to ethanol; glaucoma
BALB/cByJ	75	1	Albino	No	Yes	Low frequency of absent corpus callosum
FVB/NJ	75	1	Albino	Yes	No	Convenient for transgenic injection
129S1/SvImJ	69	1	Agouti	No	No	Embryonic stem cells for knockouts; absent corpus callosum
SJL/J	68	3	Albino	Yes	No	Model of Hodgkin's disease
AKR/J	66	1	Albino	No	No	Leukemia; adrenal lipid depletion
CBA/J	61	2	Agouti	Yes	No	Thyroiditis
SWR/J	61	3	Albino	Yes	No	Lung, mammary tumors; avoid sucrose oxyacetate
PL/J	56	3	Albino	Yes	No	Leukemia; handling-induced seizures
CAST/EiJ	53	1	Agouti	No	No	Inbred from wild *Mus musculus castaneus*; hard to handle
C57L/J	52	2	Gray	No	Yes	Prone to gallstones
SM/J	52	2	Agouti	No	No	Diet-induced obesity
BTBR T+ tf/J	51	1	Black/ tan	No	No	100% absent corpus callosum; aberrant social behavior
BALB/cJ	51	None	Albino	No	No	Absent corpus callosum; fighting in males
LP/J	48	3	Piebald agouti	No	Yes	Audiogenic seizures; many tumors
NOD/ShiLtJ	47	1	Albino	No	Yes	Non-obese diabetic

[a]Based on Mouse Phenome Database (http://phenome.jax.org) on October 30, 2009. RD denotes retinal degeneration caused by the Pde6b[rd1] phosphodiesterase 6B rod receptor mutation. HL indicates hearing loss caused by the Cdh23[ahl] otocadherin mutation.

29

129X1, 129S1 to 129S9, 129T1, and 129T2. For each of the distinct strains, there may be several substrains in labs around the world. This diversity poses a challenge for the behavior geneticist because some of the 129 strains show notable behavioral differences (Cook, Bolivar, McFadyen, & Flaherty, 2002).

SPECIAL GENOTYPES DERIVED FROM INBRED STRAINS

Inbred strains are very valuable for assessing phenotypic differences among a wide range of genotypes, and they are also valuable as the source of several more specialized kinds of strains. Beginning with a strain that is homozygous at every locus, specific genetic differences can be introduced to examine effects of one or a few genes on behavior.

Coisogenic strains

Coisogenic mice are created when a spontaneous mutation, perhaps a point mutation that substitutes one nucleotide base for another at one location in a gene, occurs in a strain that is already homozygous at all loci (Figure 2.6). For example, a mutation in the *reelin* (*reln*) gene occurred in 2006 in the C57BL/6J strain at the Jackson Laboratories, the seventh such spontaneous mutation discovered in that gene at that lab. The strain is now designated C57BL/6J-Reln[rl-7J]/J. Mice homozygous for the 7J allele (Reln[rl-7J]/Reln[rl-7J]) have a severely reduced cerebellum and radically abnormal motor behavior. Once the mutation (-allele) is detected

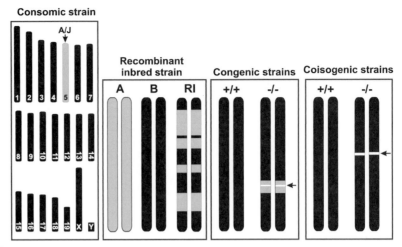

FIGURE 2.6

Four kinds of strains derived from an inbred strain. A *consomic strain* has one chromosome from another strain backcrossed onto a uniform inbred background, usually C57BL/6J; only the haploid set is shown. A *recombinant inbred strain* (RI) is formed from an F_2 hybrid cross of two or more inbred strains and then inbred for 20 generations to create homozygosity at most loci. The RI strain will have each chromosome comprised of blocks of genes from the parent strains, with the length and location of the blocks determined by random crossover events during meiosis. A *congenic strain* is formed by backcrossing a specific mutant allele of a gene onto a uniform inbred background; after 10 to 20 generations of backcrossing, there will still be a number of flanking alleles at other nearby genes that have the same strain origin as the mutation. A *coisogenic strain* is formed when a spontaneous or induced mutation occurs in an inbred strain already homozygous at all loci. The new strain differs from the parent strain at only one gene and may differ by a single nucleotide base in the case of a point mutation.

and the gene involved is identified, a very elegant experiment can be done to compare $+/+$, $+/-$, and $-/-$ genotypes at that gene. If the mice are littermates, then every other genetic and almost every environmental factor will be the same for the three genotypes. This allows a remarkably clear view of the role of the gene without other confounding factors. Its effects will probably depend on the kind of local environment and the strain genetic background, but these effects will not be evident in the data. In the case of recessive mutations, distinguishing between carriers $(+/-)$ and normal homozygotes $(+/+)$ can be done through breeding tests or by direct analysis of the DNA sequence using marker sequences adjacent to the nucleotide polymorphism.

Another interesting case is the *mdx* allele of the X-linked *Dmd* gene that mutated spontaneously shortly before 1977 in the C57BL/10ScSn inbred strain maintained at the Poultry Research Centre in the UK. *Dmd* is the *dystrophin* gene, and mice homozygous for the *mdx* allele have a murine form of Duchenne muscular dystrophy that is less severe than the human form. Both hemizygous males and homozygous females can breed, and the C57BL/10ScSn-Dmd^{mdx}/ J strain now maintained at the Jackson Labs consists entirely of mutants. Accordingly, the ancestral strain C57BL/10ScSnJ is used as a control.

Congenic strains

Congenic strains are created by the investigator to place a mutation onto different genetic backgrounds. Suppose an interesting neurological mutation occurs in a rare strain maintained by just one lab. To compare its developmental consequences with other kinds of mutations affecting the cerebellum, for example, it would be very helpful to have each mutation placed on the same genetic background such as C57BL/6. This is done by first breeding an F_1 hybrid between a mouse from the RARE strain that carries the mutation and C57BL/6. Suppose the mutation is recessive and the $-/-$ homozygote has markedly reduced fertility. Most likely the mating would be RARE($+/-$) by C57BL/6, and half of the offspring would be carriers of the mutation. A test cross with a known carrier can detect new carriers when any pup in the litter shows the aberrant motor behavior. That animal from the F_1 generation would then be crossed to C57BL/6 and the test for carriers done again. Each generation a mouse (or several of them) heterozygous for the mutation would be backcrossed to C57BL/6. Very rapidly the congenic strain would consist entirely of chromosomes from C57BL/6 at all but the one chromosome on which the mutation resides. Backcrossing would place not just the mutant gene but a fairly large piece of that chromosome onto the C57BL/6 background. As recombination events during germ cell formation break that piece of chromosome into smaller segments, only the piece with the mutant allele would be crossed onto C57BL/6. Gradually, the number of *flanking alleles* or *passenger genes* that are carried along with the mutant allele through backcrossing will become smaller and smaller with repeated generations of backcrossing. By genotyping the congenic mice for several SNPs in the vicinity of the mutation, one can determine precisely the size of the segment from the original RARE strain that remains after many generations of backcrossing. When the number of passenger genes is very small, one can also sequence their DNA to see if there are any polymorphisms that might have phenotypic consequences.

Official names of congenic strains have two parts: an abbreviation for the strain that serves as the genetic background onto which the mutation is backcrossed, separated by a period (.) from the strain that was the source of the mutation, and the mutation name. To receive official recognition as a new congenic strain, there must have been at least 10 generations of backcrossing. For example, the congenic strain SWR.129X1(B6)-$Apob^{tm1.1Zc}$/J represents a targeted mutation of the mouse apolipoprotein B gene that was created by Zhouji Chen (Zc) in 129X1/SvJ embryonic stem cells, where modified cells were injected into a C57BL/6 embryo, viable offspring were crossed with a C57BL/6 (B6) female, and then descendants were backcrossed onto the SWR/J background. That strain is now preserved as frozen embryos. The same mutation was also backcrossed onto the C57BL/6J background and live B6.129X1(B6)-$Apob^{tm1.1Zc}$/J are now available from Jackson Labs.

Congenic strains are very important in the study of induced and targeted mutations because the embryonic stem cells that work well for in vitro methods usually come from one of the 129 inbred strains, and that strain has a number of peculiarities of its own. Researchers then backcross the knockout to the C57BL/6 background or some other strain to make comparisons of the consequences of different gene knockouts on the same genetic background. This also allows a test of interaction between the mutant gene and the *genetic background*, either C57BL/6 or 129.

It is unfortunate that many brilliant molecular geneticists who created novel mutations with chemical mutagenesis or transgenic technology have paid insufficient attention to the problems of interpretation posed by flanking alleles and the genetic background. In many studies, it is unclear just what phenotypic change is caused by the mutation as opposed to the genetic background and flanking alleles, all of which can interact with each other. The editors of *Genes, Brain and Behavior* (G2B) recently observed that "a significant proportion of mutant studies suffer from serious methodological problems, which are the most frequent reason for rejection without review for manuscripts submitted for publication to G2B" (Crusio, Goldowitz, Holmes, & Wolfer, 2009). They further point out that the problems have been recognized by mouse geneticists for many years (Crusio, 2004; Gerlai, 1996; Wolfer, Crusio, & Lipp, 2002), and they set down several guidelines for good research design when working with a new mutation.

The minimal experiment needed to document the effects of a new mutation adequately entails four genetic groups: the original strain in which the mutation was created, that strain with the mutation, a second background strain, and that background with the mutation backcrossed onto it for at least 10 generations. In Box 2.3 these groups are outlined for a hypothetical situation where a gene is altered by a targeted mutation (tm1) in the embryonic stem cells of the 129X1/SvJ strain, then backcrossed onto C57BL/6J. This design does not conclusively rule

2.3 DESIGN TO STUDY TARGETED MUTATION (ᴛᴍ1) IN A KNOWN GENE

Group 1: 129X1/SvJ
Group 2: 129X1/SvJ-*Gene*^tm1
Group 3: C57BL/6J
Group 4: B6.129X1-*Gene*^tm1

out possible effects of flanking alleles near the mutation on the B6 background, but those problems decline if backcrossing is continued for more generations. It is often complicated by the low viability of 129X1/SvJ, as well as the lack of live 129X1/SvJ mice with the mutation. In such circumstances, genetic background effects can only be evaluated by backcrossing the mutation onto an additional inbred strain such as DBA/2J.

Some of the difficulties that arise from use of strain 129 embryonic stem cells may be remedied by the recent creation of the highly viable cell line JM8 from the C57BL/6N strain that has been altered to express the agouti coat color rather than the solid black of the parent strain (Pettitt et al., 2009). Cell lines from that strain are now used in an ambitious project by the International Mouse Knockout Consortium that creates targeted mutations in all mouse genes (Collins, Rossant, & Wurst, 2007).

Consomic strains

Consomic inbred strains are also created through successive backcrossing to a different inbred strain, but the choice of breeders is based on an entire chromosome, not just one mutant gene. This approach has been employed to create a valuable set of 20 consomic strains from the C57BL/6 strain, where each strain has just one chromosome from the A/J strain. For example, the C57BL/6J-Chr $1^{A/J}$/NaJ consomic strain has the A/J chromosome 1 crossed onto C57BL/6J by Nadeau (Na). If a characteristic differs greatly between C57BL/6 and A/J inbred parents, and the difference arises from relatively few major gene effects, then a survey of the 20 consomic strains should quickly point to which chromosome harbors one of the culprit genes. These strains are especially valuable for the analysis of complex traits (Kas et al., 2009; Singer et al., 2004). An effective strategy is to first assess the phenotype in the panel of 20 consomic strains plus the two inbred progenitors, locate one or two chromosomes that seem to show the largest effects (Figure 2.7), and then map markers associated with the phenotype in an F_2 hybrid cross between C57BL/6J and the consomic strain having that specific chromosome (Singer, Hill, Nadeau, & Lander, 2005).

Benson Ginsburg (Bg) and Steven Maxson used successive backcrossing to place the non-recombining part of the Y chromosome from males of the C57BL/10Bg strain onto the DBA/1Bg background. Comparisons of fighting behavior of D1 and D1.B10-Y males revealed significant influences of genes on the B10 Y chromosome, including those mediated by olfactory signals in the urine (Monahan & Maxson, 1998). Backcrossing where the parent with the donor chromosome is always male does not transfer the entire chromosome, and the resulting strain is best regarded as congenic rather than consomic. The C57BL/6J-Chr $Y^{A/J}$/NaJ strain has the entire A/J Y chromosome on the other strain background.

Conplastic strains

A variation on the theme of consomic strains was used by Roubertoux and co-workers (2003) to place the entire mitochondrial genome (mtDNA) from the NZB/BINJ strain onto the CBA/H background and vice versa. The mtDNA is found only in the cytoplasm of the ovum, never in the sperm, and it does not recombine with chromosomes in the nucleus. The mtDNA influenced a wide range of behavioral phenotypes in that study. More recently,

(A)

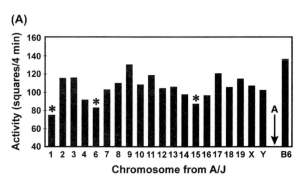

(B)

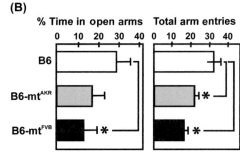

FIGURE 2.7

(A) Open field activity of a set of consomic strains, each with a single strain A/J chromosome placed onto the C57BL/6J background. The A/J strain has exceedingly low motor activity, and the three consomic strains indicated by the asterisk (∗) are significantly lower than the C57BL/6J parent, indicating they harbor one or more genes from A/J that reduce activity. (Redrawn with permission from Singer, Hill, Nadeau, & Lander, 2005). (B) Performance on the elevated plus maze by C57BL/6J and two conplastic strains that have had the mitochondria from AKR/J or FVB/NJ backcrossed onto the C57BL/6J background. Genetic variants in the mitochondrial DNA of AKR and FVB evidently increase the levels of anxiety-related behaviors. *(Reprinted with permission from Yu et al., 2009).*

a comprehensive study of mtDNA in 52 common inbred strains further documented strain ancestries in which NZB/BINJ is unique, and identified several new mtDNA mutations (Yu et al., 2009). They backcrossed the mtDNA of 16 common inbred strains onto the C57BL/6J background and demonstrated influences of the AKR/J and FVB/NJ mtDNA on anxiety-related behaviors in the elevated plus maze and activity in the open field (Figure 2.7B) using the C57BL/6J-mt$^{AKR/J}$ and C57BL/6J-mt$^{FVB/NJ}$ conplastic strains. Other studies have highlighted the importance of mitochondrial genes for a wide range of physiological functions and the potential importance of mouse models for understanding human mitochondrial disorders (Tyynismaa & Suomalainen, 2009).

Recombinant inbred strains

Recombinant inbred (RI) strains are formed by crossing two inbred strains and obtaining a genetically segregating population, perhaps an F$_2$ or more advanced hybrid cross, then establishing new inbred lines, each from a single breeding pair. After 20 generations of inbreeding, the new strains can be registered, and used especially for genetic mapping research. Because different groups of investigators can work with the same genetic material, research with RI strains tends to be cumulative and collaborative. Alleles at many loci in the two inbred parents were originally fixed through random processes during inbreeding, and many phenotypes in the two strains may appear to be correlated when there is no strong causal or physiological relation between them. This kind of fortuitous correlation is greatly reduced in RI strains. Crossover events during meiosis and random recombination of chromosomal material generate new strains in which each chromosome is a mosaic of haplotypes derived from one parent strain or the other. This makes the strains valuable for establishing linkage with a marker locus, a key technique in identifying quantitative trait loci (QTLs) that alter a complex behavioral trait by a relatively modest amount.

Several sets of RI strains are readily available for work with mice, including the BXD strains derived from a C57BL6J mother and DBA/2J father. Forty-two BXD strains were established by Taylor at the Jackson Labs and then more than 50 new strains were added by Williams at the University of Tennessee. Extensive data on these animals, including 250 measures on more than 70 RI strains (Philip et al., 2009), are available via The GeneNetwork and WebQTL. Other RI sets for mice include AKXD, AXB/BXA, BXH, CXB, and LXS (ILSXISS at Jackson Labs).

33

This abundant resource contrasts with the situation in rats where only the HXB/BXH RI strain set has limited data available on WebQTL and can only be obtained from its creator in the Czech Republic. *The Laboratory Rat* cites two other recently developed sets (FXLE/LEXF, PXO) that are represented by only one or two citations in PubMed.

DESIGNER MICE: TRANSGENIC METHODS AND TARGETED MUTATIONS

Until the 1990s, virtually all work with genetics and mouse behavior employed the *forward genetic approach* in which existing genetic differences among inbred strains or selected lines were shown to influence behavior (Takahashi, Pinto, & Vitaterna, 1994). Then a long and sometimes futile hunt ensued for the specific gene or genes responsible for the phenotypic differences. This approach was accelerated in experiments with random mutagenesis, where mice are treated with the mutagenic chemical *N*-ethyl-*N*-nitrosourea (ENU) that creates many new mutations throughout the genome (Nadeau & Frankel, 2000). The methodology is well described at www.mouse-genome.bcm.feu.edu/ENU/ENUHome.asp. The major challenge in the forward approach is to discover which specific gene(s) are involved.

The science of mouse neurobehavioral genetics was transformed in a major way in the 1980s through discoveries of methods to create mutations in specific genes using the *reverse genetic approach*. Two investigators, Mario Cappechi and Oliver Smithies, devised ways to engineer homologous recombination between the native mouse DNA and a small DNA targeting vector containing a fragment of foreign DNA. This work was most effectively done in tissue culture using mouse embryonic stem cells, a technique perfected by Sir Martin Evans. In 2007, Cappechi, Evans, and Smithies shared the Nobel Prize for this important advance in molecular and cell biology (Box 2.4). Excellent summaries of their contributions are available from www.nobelprize.org.

2.4 NOBEL PRIZE IN PHYSIOLOGY AND MEDICINE

Alfred Nobel, who made a fortune from his invention of dynamite, stipulated in his will that up to five prizes should be given to "those who, during the preceding year, shall have conferred the greatest benefit on mankind." Several scientists whose work is important for genes, brain, and behavior have won the prize in physiology and medicine since it was first conferred in 1901, including Ivan Pavlov, T. H. Morgan, Rita Levi-Montalcini, Eric Kandel, and Susumu Tonegawa. The Web site www.nobelprize. org provides a wealth of information on each of them, including the presentation speech by a member of the Karolinska Institute, a biography, the laureate's speech at the ceremony, and, in recent years, figures and references on their discoveries.

These methods are high tech and entail advanced knowledge of molecular genetics as well as tissue culture. Several convenient descriptions of the methods are available (Alberts, Johnson, Lewis, Roberts, & Walter, 2008; Jackson & Abbott, 2000).The first nine steps in the process are described in the Nobel Prize slides.

1. Perpetual lines of mouse embryonic stem (ES) cells are established in a glass dish. Usually these are derived from one of the 129 inbred strains.
2. A targeting vector is created with a bacterial gene for neomycin resistance (neo^r) inserted between two mouse-specific DNA sequences homologous with sites in a specific mouse gene (Figure 2.8).
3. The DNA vector is injected into the nucleus of the ES cells using a fine glass needle.
4. In a few ES cells, the vector recombines with the mouse gene, such that the neo^r sequence is inserted into the mouse genome.

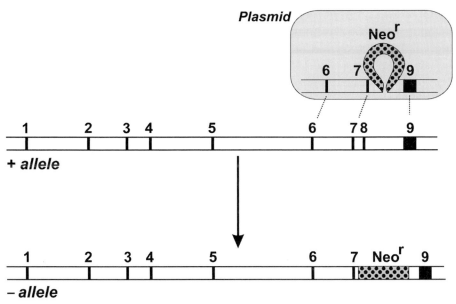

Plasmid

Neo^r

6   7   9

1   2   3   4   5   6   7 8   9

+ allele

1   2   3   4   5   6   7   Neo^r   9

– allele

FIGURE 2.8

Scheme for creating a targeted mutation by homologous recombination. First a targeting vector is created in which a bacterial neomycin resistance gene is inserted into a known location in a mouse gene. Then the vector is injected into a cell nucleus of mouse embryonic stem cells and in a few cases recombines with the resident allele so that the bacterial transgene is incorporated into the mouse germ line. This effectively prevents transcription of that gene in future generations of mice that are homozygous for the mutation, and the gene is thereby "knocked out." The hypothetical gene in the figure consists of nine exons (black bars) separated by long segments of non-coding introns (*Redrawn with permission from apecchi (2007)*).

5. ES cells are treated with neomycin, which kills most of those lacking the neo^r gene.
6. ES cells with the altered gene are injected into a blastocyst of strain C57BL/6 origin.
7. The blastocyst, with a few cells from the ES cells, is placed into a surrogate C57BL/6 mother and allowed to implant in the uterus and develop to term, when chimeric mice are born that contain a mix of mainly C57BL/6 cells plus a few altered 129 strain cells. Because C57BL/6 mice are solid black and 129 mice have lighter coats, the successful experiments result in chimeras showing patches or bands of pigmentation.
8. After reaching the age of reproductive maturity, chimeric mice are mated with a normal C57BL/6 mouse.
9. Those chimeras that have the altered 129 ES cells incorporated into their germ cells give rise to a few offspring that have a lighter coat color that is not black because they have one coat color gene from strain 129 that is dominant over the C57BL/6 allele at the agouti locus. Those animals are carriers of the new mutation in the targeted gene. They are F_1 hybrids between the 129 and C57BL/6 strains at other loci. If the new mutation is recessive, the mice will not show major abnormalities.
10. Further steps in the process entail conventional transmission genetics and crossbreeding. Mating two carriers of the new mutation gives rise to 25% of offspring that are homozygous for the altered gene. Coat color is no longer a reliable indicator of genotype at the mutated gene. The bacterial neo^r sequence inserted into the mouse gene usually prevents transcription of that gene altogether, so that the animal cannot synthesize any of the protein normally encoded by the mouse gene. In such a case, the targeted mutation is said to be a complete "knockout" of the normal mouse gene and may have substantially altered brain chemistry and behavior. Interpretation of effects is difficult at this stage of the investigation if each targeted gene resides in a fairly long stretch of 129 strain DNA that did *not* recombine during the meioses that generated sperm and ovum. Thus, the knockout mice will be homozygous for quite a few other 129 strain genes, including the flanking alleles near the targeted gene. The first generation knockout mice will also be

homozygous for numerous other 129 alleles across all of the other chromosomes. Although the targeting technology is extraordinarily precise in tissue culture, recombination of DNA during live mouse reproduction is not precise or well controlled.

11. Then begins the long process of backcrossing the new mutation onto the C57BL/6 strain background and hopefully some other inbred background as well. Within 10 generations of breeding, which may require three years, all chromosomes except the one where the targeted mutation resides should be pure C57BL/6. Most of the chromosome with the targeted mutation will also be from C57BL/6, but there will still be a portion of the 129 strain genome on either side of the mutation that may contain several non-mutated 129 genes. The flanking alleles create difficulties for the interpretation of many experiments, as discussed in the section Congenic Strains.

A most important issue for the behavioral neuroscientist working with transgenic and knockout mice is how to obtain these precious animals for their research. Relatively few labs have the capacity to generate new knockouts by going through every step previously listed. It would be very helpful if the researcher could simply order them and expect shipping within a reasonable time, as can be done for many inbred strains of mice. This is indeed possible for a rather large number of genetically engineered mice through the Jackson Laboratory (www. jax.org). Over 500 transgenic and knockout mice are currently available as live animals and another 1200 are available as frozen embryos that need to be thawed and transferred to a pseudopregnant female to obtain live offspring. Some of the more popular knockouts such as B6.129P2-$Apoe^{tm1Unc}$/J are readily available in moderate quantities for shipping at a moderate cost of $80 to $100 per mouse, a small price to pay in comparison with what it would cost to manufacture them locally. Others that are less viable or less in demand will cost more and take longer to ship. For the cryopreserved embryos, the investigator will need to wait about 4 months and pay about $1,900 to obtain two pairs of mice that include at least two mutants. If the mutation is viable or recessive, the investigator can then establish a local breeding colony to obtain the required numbers for a research project.

More than 500 kinds of live knockout mice or carriers of a targeted mutation are also available through the institutions participating in the International Knockout Mouse Consortium (IKMC). The IKMC collaborators have created more than 13,000 mouse gene knockouts in the form of embryonic stem cell lines that have the targeting vector in place, ready for injection into a mouse blastocyst at the previously listed step 6. Production of live mice from existing ES lines can also be provided on a fee-for-service basis from the Jackson Laboratory and Charles River Laboratory.

The new technology made it possible for neuroscientists to disable or knock out any gene whose DNA sequence is already known, and the methods were widely adopted to unravel intricate mechanisms of neural activity. Because the methods worked especially well with mice, there was a spectacular upsurge in the number of neuroscientists working with mice. Once a new knockout was created, its creators sought to document which phenotypes were altered. Many mutations had effects on behavior, and this gave rise to an urgent demand to run the mutants through batteries of behavioral tests. Several commercial manufacturers of behavioral test equipment quickly produced diverse apparatus sized for mice. There were too few specialists in mouse behavior to service the needs for mass screening, so the experts sought to teach the molecular biologists about behavioral testing. A foremost example is the book *What's Wrong with My Mouse?* by Jacqueline Crawley (2000) that is now in use in many mouse labs.

More than any other development in the field of behavioral neuroscience, the advent of reverse genetics in mice has given rise to a need for improved methods of mouse behavioral testing. Although the enthusiasm of neophytes in this field is laudable, many instances of inadequate behavioral test methods are finding their way into the scientific literature. Thus there is a need for greater depth in the training of scientists and technical staff who want to work with mouse behavior. The stunning diversity of targeted mutations and transgenic insertions of human

DNA into ·the mouse genome also poses a challenge for existing behavioral tests. Many tests were designed to study behavioral processes of interest to psychologists and pharmacologists, whereas the field has increasingly become interested in mutations in mice that provide good models for human neuropsychiatric disorders. Some of the current tests are sensitive to processes that seem to be involved in human anxiety, depression, or autism, but there are no valid tests directly aimed at assessing the murine equivalents of any human disorder. There is a need for devising new and better behavioral tests. Mouse behavioral testing requires its own advances in technology.

Tests of Mouse Behavior

Along with the genotypes of the subjects and treatments applied to various groups, the choice of how many and which tests to use determines the structure of the experiment. The specific tests have a pervasive influence on the ordering, balancing, and randomization of test administration. The tests even play a major role in determining sample size, because every study is constrained by funds and time. This chapter provides a brief overview of the tests that are commonly used in work with mice. Later chapters (Chapters 10 to 14) examine several of these in greater depth and discuss the shortcomings of several tests, as well as potential remedies.

POPULARITY OF TESTS

Tests such as the open field and Morris water maze are ubiquitous in neuroscience. Many other tests are used with mice, as described in several reviews (Brooks & Dunnett, 2009; Brooks, Pask, Jones, & Dunnett, 2004, 2005; Crawley, 2000; Crawley, 2008; Crawley & Paylor, 1997; Rogers et al., 1997). From those and other sources, a list of tests was compiled and then searches of PubMed were done to tabulate citations. The search always began with the term (mouse OR mice), then joined a phrase or keyword using AND to the Boolean string. For some tests that are described in various ways in the literature, several phrases and keywords were evaluated. Results of this survey are shown in Table 3.1. Five tests were cited in the title or abstract of an article more than 1,000 times: tail flick hot plate, open field, elevated plus maze, Morris maze, and rotarod. Those tests are often included in test batteries to screen new mutations. The large number of articles indicates the considerable growth of the field of neurobehavioral genetics of mice since the discovery of transgenic methods (Wahlsten & Crabbe, 2007).

Mouse Behavioral Testing. DOI: 10.1016/B978-0-12-375674-9.10003-5

TABLE 3.1 Tests and Number of Articles (#A) in PubMed (Searches Done in May 2010, for Phrase or Key Words in Title or Abstract)

Domain	#A	Test	Domain	#A	Test	Domain	#A	Test
Anxiety	1452	Elevated plus maze	Learning	1037	Morris water maze	Motor	1046	Rotarod
	374	Light–dark box		725	Conditioned place preference		237	Hole board
	77	Elevated zero maze		334	T maze with food reward		205	Grip strength
	28	Vogel conflict test		263	Active shock avoidance		104	Staircase test
	2	Elevated square maze		261	Fear conditioning to context		50	Screen test
Arousal	480	Prepulse inhibition		238	Conditioned taste aversion		39	SHIRPA reflex tests
	36	Fear-potentiated startle		234	Inhibitory shock avoidance		32	Hanging wire
Learned helplessness	482	Tail suspension test		202	Radial maze		29	Balance beam
	78	Learned helplessness		78	Eye blink conditioning		24	Grid test
	57	Porsolt forced swim test		74	Operant conditioning	Pain	2770	Tail flick hot plate test
Exploration	1944	Open field test		44	Barnes maze		467	Formalin test
	247	Y maze spontaneous altern.		32	Shuttle shock avoidance		54	Tail flick, hot or cold water
	39	Object exploration		29	Lashley 3 maze	Social	178	Agonistic; resident—intruder
	9	Barrier test		15	Hebb-Williams mazes		102	Social recognition/preference
Ingestion	70	Food preference		7	IntelliCage		52	Social dominance
	70	Paw preference food reaching					36	Pup retrieval
	54	Two-bottle choice test					22	Agonistic; standard opponent

Several of the more popular tests are described briefly in this chapter. There is no attempt to review the use of each in current research, because the literature is vast and varied. Anyone wanting further information on any test in Table 3.1 can enter all or part of its name into PubMed or Google Scholar and quickly download a long list of articles. Any list will rapidly become out of date as new papers are published. For the most popular tests, more than 100 new articles appear in the literature each year. In almost all cases, the articles are not about the actual test but instead use it to study effects of drugs, mutations, and alterations of the brain.

COMMERCIAL DEVICES

Another indicator of popularity of behavioral tests is the number of companies that manufacture each kind of apparatus (Table 3.2). The company names provided in Box 3.1 may be used in a search engine to obtain details of the current menagerie of tests offered for sale. Table 3.2 lists companies that exhibit their wares at the annual meeting of the Society for Neuroscience and offer apparatus for assessing more than two domains listed in Table 3.1. Only tests offered by more than two manufacturers are shown. How many labs have purchased which items from each supplier is a closely held trade secret that would tell much more about the present state of the field than compilations of tests from published articles that describe only successes. Companies undertake production on a commercial scale only after a test has piqued the interest of many investigators and a market is thought to exist for the product. Inevitably, in the rapidly evolving field of neuroscience the array of tests for sale includes several that are no longer favored by investigators. The abbreviations in Table 3.2 are utilized to make the presentation more concise, but they are not universally accepted in the field.

3.1 MANUFACTURERS

AccuScan Instruments
Columbus Instruments
Coulbourn Instruments
Med Associates Inc.
Panlab—Harvard Apparatus
San Diego Instruments
Stoelting
TSE Systems
Ugo Basile

A further caveat concerning the list arises from the rat—mouse distinction. In some cases, it is not clear from descriptions on Web sites if a device originally tuned for use with rats has been appropriately scaled and refined for work with mice. Some manufacturers are very explicit about this matter, whereas others are ominously silent. Whether a device yields valid data with mice is generally not apparent from the Web site. The manufacturer is not responsible for the validity of test results. No warranties about the suitability of any test for mouse research are offered, nor should they be.

Commercial suppliers of apparatus want to sell physical objects that are difficult for many labs to make themselves. There are several tests of behavior, however, that do not require elaborate equipment, such as the tail suspension (Ripoll, David, Dailly, Hascoet, & Bourin, 2003), formalin, and resident—intruder tests. Many tests are described in published protocols (Crawley et al., 2005; 2007) and can be done in almost any lab without sophisticated apparatus.

TABLE 3.2 Tests and Manufacturers (See Box 3.1)

Domain	Abbrev	Test name	AccuScan	Columbus	Coulbourn	Med Assoc.	Panlab	San Diego.	Stoelting	TSE	Ugo Basile
Anxiety	LD	Light–dark box	✓				✓		✓	✓	
Anxiety	EPM	Elevated plus maze		✓		✓	✓	✓	✓	✓	
Anxiety	ZM	Elevated zero maze		✓						✓	
Anxiety	VCT	Vogel conflict test	✓	✓		✓	✓	✓	✓	✓	
Arousal	FPS	Fear potentiated startle		✓	✓			✓		✓	
Arousal	PPI	Prepulse inhibition			✓					✓	
Exploration	HCA	Home cage activity	✓	✓	✓		✓			✓	✓
Exploration	NP	Nose poke/hole board	✓	✓	✓					✓	✓
Exploration	OFA	Open field activity	✓	✓	✓		✓	✓		✓	✓
Exploration	YMSA	Y maze spontaneous alt.	✓	✓	✓					✓	✓
Learning	CPP	Conditioned place pref.	✓		✓	✓	✓	✓		✓	
Learning	FCA	Fear conditioning to cue	✓		✓	✓	✓		✓	✓	✓
Learning	OPC	Operant conditioning		✓	✓	✓	✓	✓		✓	✓
Learning	IA	Inhibitory avoidance	✓	✓	✓		✓	✓		✓	✓
Learning	SHA	Shuttle avoidance	✓	✓		✓	✓	✓		✓	✓
Learning	BM	Barnes maze	✓	✓						✓	✓
Learning	MWM	Morris water maze	✓	✓		✓	✓	✓	✓	✓	
Learning	RAD	Radial maze — 8 arms	✓	✓			✓	✓	✓	✓	✓
Motor	ARR	Accelerating rotarod	✓	✓	✓		✓	✓		✓	
Motor	WHR	Wheel running	✓	✓			✓			✓	
Motor	GRIP	Grip strength		✓			✓			✓	
Motor	TRDM	Treadmill	✓	✓			✓	✓		✓	
Pain	TFHP	Tail flick — hot plate		✓	✓		✓	✓	✓	✓	✓

The more popular tests presented in this chapter are grouped into broad domains mainly for convenience, and many tests are clearly pertinent to more than one domain. The seemingly simple open field test, for example, assesses exploratory activity as well as anxiety-related behavior (wall hugging), motor capabilities (rearing, grooming), and even memory (habituation of activity) to some extent. Although the researcher may design a test to tap capabilities in a single domain, the mice often do not respect the boundaries of our categories.

EXPLORATION

Tests of exploratory activity are generally very simple to build, administer, and score (Figure 3.1A). The level of baseline activity expressed by a strain of mice is very important to consider when interpreting other tests. Animals that move very little in a novel environment (e.g., the A/J strain) may never even discover regions that are of great interest to the experimenter. Tests of anxiety-related behavior, for example, can be clouded or undermined by inactivity. Some genotypes (e.g., the various 129 strains) habituate very rapidly to novelty and will show moderate levels of movement in an apparatus only for the first few minutes unless some significant event occurs.

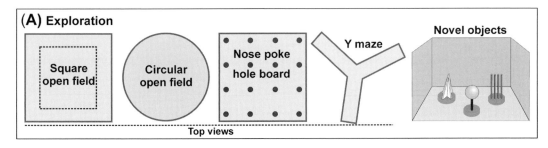

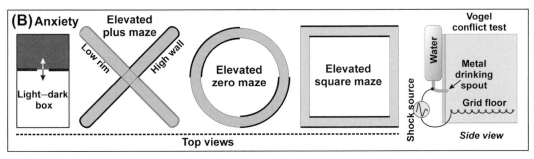

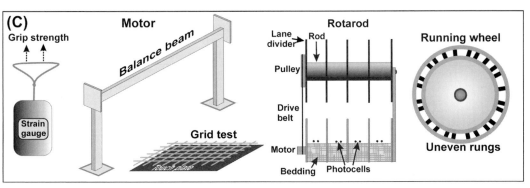

FIGURE 3.1
(A) Tests of motor activity that rely on spontaneous exploration of the apparatus. All but the object exploration task are top views. (B) Tests of anxiety-related behavior. The first four are top views. The three mazes are elevated above the floor and have open regions that are avoided by anxious mice. (C) Tests of motor coordination and strength. The rotarod that accommodates four mice at the same time can accelerate or turn at a fixed speed. The running wheel with uneven rungs is a better test of coordination than a wheel with regular spacing.

Open field

A featureless square box or circular tub barren of all structure yields a convenient index of the tendency to move in an environment (Stanford, 2007). Overall activity level is represented by the length of the path traveled by the mouse, as assessed by photocells, video tracking, or counts by an observer of squares entered when the box has lines drawn on the floor. Some mice tend to hug the walls and avoid the center of the chamber; wall hugging is an indicator of anxiety-related behavior (Lipkind et al., 2004). Behaviors such as leaning against a wall or rearing when away from a wall are correlated with activity levels and are sensitive to many drugs. Cocaine and moderate doses of ethanol, for example, tend to increase motor activity in the open field but reduce rearing.

Nose poke hole board

When the floor of the square open field has small holes, photocells below the floor can detect occasions when the mouse pokes its nose into a hole. The nose poke frequency provides an additional indicator of exploration. When the animal is hungry and small pieces of food are placed in wells below the holes, learning to find food can be observed. The apparatus can also be built with the holes and food cups in the wall of the box.

Symmetrical Y maze

A maze with three identical arms evokes active exploration, and the pattern of arm entries as well as the total number of entries may be informative (Mamiya et al., 2000; Ukai, Watanabe, & Kameyama, 2000; Wall & Messier, 2000). The mouse must return to the hub of the maze after each arm entry, and it can then return to the arm it just visited, the arm it visited on the previous entry, or the one least recently visited. The spontaneous alternation index, expressed as percentage alternations, only considers those instances when it moves to a new arm. A "prior" entry occurs when it goes to the arm that was the most recently other arm visited, and "alternation" occurs when it goes to the third arm, the least recently visited. The short-term working memory interpretation of alternation asserts that preference for the least recently visited arm indicates memory for the history of arm entries. The validity of this interpretation is challenged by the fact that a simple radial strategy (123123123123) will yield 100% alternations, even though it will occur if the mouse recalls nothing but always makes a left turn when leaving an arm.

Home cage activity

Ambulation in the home cage is a primary indicator of circadian rhythms (Benstaali, Mailloux, Bogdan, Auzeby, & Touitou, 2001; Kopp, 2001). Activity can be detected with varying success by photocells, video tracking, ultrasonic reflections, or slight jiggles exerted by moving animals (de Visser, van den Bos, Kuurman, Kas, & Spruijt, 2006; Goulding et al., 2008). This kind of measure from non-stressed animals can serve as a standard for comparison with treatments that either stimulate or suppress activity.

ANXIETY

Anxiety is an unseen construct much like fear but different (Belzung & Griebel, 2001; Bourin et al., 2007; Clément, Calatayud, & Belzung, 2002; File, 2001; Finn et al., 2003; Rodgers, 1997; Treit, 1994). Fear is evoked by a specific object or stimulus. It can be conditioned to a specific stimulus paired with pain or distress. If an object or situation evokes fear, the mouse will move away from it or prefer to spend less time near it than with some other object. Anxiety is less specific. It is a more general state that influences behavior in many kinds of situations and causes the mouse to avoid places that entail risk. No test measures anxiety directly. Instead, the tests quantify behaviors that are altered as the mouse becomes more or less anxious. Anxiety-related behaviors can be quantified, whereas anxiety is ephemeral and intangible. In mice its reality is sometimes demonstrated by administering anxiolytic drugs that reduce anxiety-related behavior in people and mice alike.

For a small mammal that is easy prey, the greatest risk is exposure to attack in an open area with no shelter. Lab mice have never experienced attacks from a predator, yet they sometimes avoid exposure to open spaces. In some labs, they are markedly reluctant to explore all parts of their surroundings. The differences between labs in this respect can be quite large (Chapter 15).

Most tests of anxiety-related behaviors (Figure. 3.1B) compare time spent in exposed areas of the apparatus with time while in contact with a wall or under a sheltering roof. Table 3.1 identifies several such tests. Every test entails different stimulus conditions; no one test can provide an adequate index of anxiety. Averaging scores across several tests of anxiety-related behaviors reduces the influence of specific sensory and motor factors, emphasizing the role of anxiety. Four types of tests are presented next.

Light–dark box

Mice are placed into the brightly lit compartment and allowed to explore freely for several minutes, entering the dark compartment through a hole in the middle of a wall (Hascoet, Bourin, & Dhonnchadha, 2001). Mouse strains differ greatly in the level of motor activity and distance traveled in this and almost any apparatus, so that activity per se is not a good index of anxiety. Percentage time in the dark compartment is directly related to the perceived aversiveness to the bright light. The task involves a conflict between a tendency to avoid brightly lit areas and a tendency to explore a new environment. Accordingly, the most salient indicator of anxiety is thought to be the number of transitions between light and dark compartments. Mice low in anxiety tend to explore the environment more thoroughly and make many transitions. The principal difficulty with the task is presented by mice that are very low in activity. On occasion an animal will sit quietly in the light box and do just about nothing. If the mouse never enters the dark box at all, it cannot experience the difference between the two compartments, which places its conduct off the continuum of anxiety. Data from such animals are sometimes dropped from the analysis.

Elevated plus maze

The mouse is placed on a small center square at the junction of four arms, two with high walls and two with a low rim (Walf & Frye, 2007). The rim is essential to prevent the mouse from falling from the apparatus. Anxious mice avoid exposure on the open arms, so a good indicator of anxiety is percentage of time spent in the open arms. Certain patterns of behavior, such as the head dip over the rim of an open arm and stretch-attend between arms, also suggest anxiety. Some strains in some labs spend almost half the time in the open arms, whereas many published studies show very low levels, such as 5% open-arm time. A high level of baseline anxiety and therefore a low level of open-arm exploration is helpful when the investigator is studying drugs that may reduce the behavior. If the baseline is too extreme, however, interesting differences among genotypes may be obscured. A rim of 5 mm is commonly employed, but a higher rim can help to reduce the baseline index of anxiety, and this may aid interpretation of data (Wahlsten, Rustay, Metten, & Crabbe, 2003). The higher the rim on the open arms, the more exploration of those arms will occur. The principal difficulty with this test is the ambiguous status of the center square. Many mice spend a substantial portion of a trial at the center square. From time to time a mouse will sit there and make no foray into any arm. Should its percentage time in the open arms therefore be scored as zero or regarded as indeterminate? Hypoactivity can also present a challenge when a mouse enters one open arm and then sits there frozen for the remainder of a trial (see Figure 9.2). Its score of 100% will point to very low anxiety, while its failure to move at all implies strong fear of an open space suspended above a table or floor. If only one animal in a group behaves this way, its data may be excluded from the analysis because of ambiguity.

Elevated zero and square mazes

The circular configuration avoids the problem of time in the center square; however, it cannot overcome the problem of hypoactivity. The square configuration is very similar to the elevated zero maze, but it is easier to construct from plastic.

Vogel conflict test

This test is sometimes used to study drug effects on behavior, and it has been classified as a test of anxiety-related behavior because it is affected by anxiolytic drugs such as benzodiazepines, which increase open-arm exploration in the elevated plus maze and similar tests. A thirsty mouse deprived of water for 24 h or 48 h is placed in a box with a grid floor and allowed to drink from a water spout. An electronic circuit counts the licks and delivers an electric shock to the tongue on every twentieth consecutive lick. The suppression of licking behavior when thirsty is taken as the indicator of sensitivity to conflict. Two studies explored parameters of this test and found that the lowest intensity shock (0.1 to 0.14 mA) suppressed drinking (Mathiasen & Mirza, 2005; Umezu et al., 2006), whereas a third reported a threshold of 0.39 mA (Liao, Hung, & Chen, 2003). Interpretation of data is difficult because any treatment that simply reduces pain caused by shock tends to lessen the suppression of licking. Furthermore, shock to the wet tongue is inevitably accompanied by shock to the dry feet standing on a grid in the box, and shock levels reported in most articles are quite sufficient to cause pain to the feet and a vigorous flinch or jump (Chapter 11). To what extent an electric shock to the tongue contributes to the suppression has not been determined.

Geller conflict test

Mice are deprived of food and then trained to press a bar for food rewards on a fixed-ratio schedule where every twentieth press earns a small pellet of food. During the test, there is a 5 min period with food reward only, then a 5 min period during which every twentieth bar press is followed by a strong electric shock to the feet. Bar pressing rates during safe and shocked periods provide an index of sensitivity to conflict (Umezu et al., 2006). Like the Vogel test, treatments that alter shock sensitivity will also affect bar pressing rates.

MOTOR FUNCTION

Several tasks are sensitive primarily to motor coordination in mice (Brooks & Dunnett, 2009) Brooks, Pask, Jones, & Dunnett, 2004; Crabbe et al., 2003; Crabbe, Metten, Cameron, & Wahlsten, 2005) and are often used to assess drug effects on motor behavior (Figure 3.1C). Results from motor tests can help interpret data from complex tests of functions such as learning, which may be impaired because of motor deficits.

Ataxia observation

Mice are placed in an empty plastic box with low walls and observed by a well-trained researcher (Metten et al., 2004). Many kinds of deficits in motor function can be observed under the influence of ethanol or drugs, such as loss of righting reflex and splayed hind limbs, behaviors that are rarely apparent in normal mice.

Grip strength

When a mouse is suspended while the tail is held by an experimenter, it will grasp almost any object it touches with both forepaws and hold tightly. As the experimenter pulls the animal away from the object, eventually the grip will be broken. When the object, usually a metal rod, is attached to a strain gauge, the maximum force on the gauge provides a measure of grip strength.

Balance beam

The balance beam for a mouse is almost the same as the version employed by humans seeking Olympic fame. Rarely will a mouse fall off the beam. Unimpaired by drugs or ethanol, most mice can cross the beam almost perfectly, whereas under the influence of ethanol, for example, many show slips of the hind paws off the beam. Usually the experimenter watches and counts the number of slips.

Grid test

The mouse is placed on a wire grid and allowed to explore the box freely. From time to time, a paw may slip through a hole in the grid and touch a plate. An electronic circuit then counts the number of foot slips.

Rotarod

Each mouse is placed atop a round rod in a separate lane. In the accelerating rotarod task, the rod is motionless until all mice are in place and then begins to move slowly at first, then faster and faster until the mouse falls into a trough of bedding material (Jones & Roberts, 1968; Rustay, Wahlsten, & Crabbe, 2003a, b). Time until the fall is the index of motor capability. For the fixed speed version, the rod always moves at the same, relatively slow speed, and the mouse is placed onto a moving rod.

Running wheel

Mice can run prodigious distances of more than 10 km in one day on a running wheel (Petree, Haddad, & Berger, 1992; Shyu, Andersson, & Thorén, 1984). The usual wheel with equally spaced bars can assess running speed or enthusiasm for running, but is not a dependable indicator of motor coordination per se. When some of the rungs are removed so that there is an irregular pattern that requires rapid adjustments of paws, the wheel seems to be more sensitive to coordination defects (Liebetanz et al., 2007; Schalomon & Wahlsten, 2002).

Treadmill

The treadmill can be operated at speeds set by the experimenter and is primarily a device to force animals to exercise for long periods of time. As such, it pertains to endurance rather than coordination.

LEARNING

The study of learning and memory has a long history in research with mice, and it experienced a resurgence after the invention of methods to create targeted mutations in mice. The array of tests for this domain is vast and varied, involving diverse motor requirements, motives, and rewards. A few of the more common tests are presented in Figures 3.2 and 3.3. Some tests entail appetitive motivation that inspires a mouse to seek some kind of positive reward or reinforcement, whereas others impose aversive stimuli that provoke alarm, cause pain or distress, and stimulate attempts to escape or avoid them. How best to motivate mice is presented in detail in Chapter 11.

Operant learning

The classical device for studies of rat learning is the "Skinner" box immortalized by B. F. Skinner, a pioneer of behavioral psychology. It can be used with mice as well (Padeh, Wahlsten, & DeFries, 1974). In an automated device that keeps the experimenter out of the equation, a hungry animal explores the novel chamber and eventually learns that depressing a lever causes food to fall into a cup. Many variations in the method are possible. Continuous reinforcement delivers one food pellet for every lever press, while fixed-ratio reinforcement

Appetitive learning

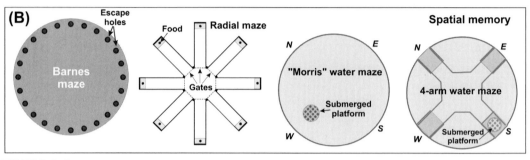

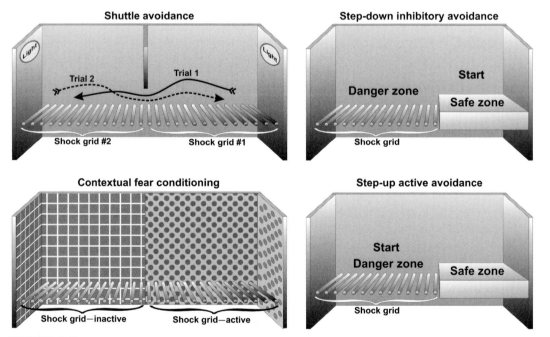

FIGURE 3.2

(A) Tests of learning where the hungry mouse must perform a specific response to obtain a food reward. In the Skinner box it must press a lever. In the tunnel maze it must always go forward and never retrace its path. In the Lashley 3 maze it must avoid blind alleys. The correct arm in the T maze may be indicated by visual stimuli. (B) Tests of spatial memory lack any local cues on the apparatus to indicate the correct choice, and the mouse must utilize cues in the room beyond the device. In the Barnes maze, the mouse must locate an open hole to escape bright light. The radial maze provides food rewards to a hungry mouse, whereas the Morris and 4-arm tasks require swimming to escape from the water.

FIGURE 3.3

Four tests of learning that involve electric shock to the feet. In shuttle avoidance, shock comes on a few seconds after a light or tone is presented, and the mouse must run to the other chamber quickly to avoid shock. On the next trial, the safe and danger zones reverse. In inhibitory avoidance, the mouse must refrain from entering the shock zone, whereas in one-way active avoidance it must move quickly from the danger zone before the shock is turned on. Fear conditioning to context presents the shock in a compartment with unique visual or tactile cues, and later the animal's preference for one box or the other is tested with no shock.

delivers one pellet after some preset number of presses, and a variable-ratio schedule delivers pellets after different numbers of presses so that reward is less predictable. The device to be actuated by the mouse does not have to be a lever; it can be a plastic panel in one wall (Williams, Daigle, & Jacobs, 2005), a chain suspended from the ceiling that must be pulled, or even a hole into which the snout must be inserted. Animals can be required to make a choice between two levers or snout holes, and there may be lights that signal which one is currently correct.

Mazes

Rodents are notoriously competent at finding their way through tunnels or walls and over furniture to obtain food. Many configurations of mazes have been used to study learning of paths by mice, beginning with the thorough studies by Yerkes of the "dancing" mouse (Yerkes, 1907). The Yerkes apparatus had four alleys, three of which contained dead ends. In Figure 3.2A, two variations are shown with and without blind alleys. In the "tunnel" maze, the animal must simply learn to keep going forward, and many mice, even those with retinal degeneration, do this very quickly (Chapter 11). The Lashley 3 configuration involves a blind alley at each choice point so that the mouse must learn which direction to turn at each doorway. When the animal is required to choose between specific stimuli, a simple T maze is often used. The mouse may be rewarded for always making a right turn, for going to the arm that is black instead of white, or for choosing the arm with vertical instead of horizontal stripes. Movement of the animal through all kinds of mazes can be monitored with video tracking. Mazes are vexed by odor trails that persist from one trial to the next and can provide cues about the location of food.

Spatial memory

A large body of research has demonstrated the importance of the hippocampal formation for memory of spatial locations in the environment. Accordingly, several radially symmetric tests have been designed for assessing spatial learning and memory (Patil, Sunyer, Hoger, & Lubec, 2009). Spatial cues are by definition remote from the apparatus and are termed distal cues. If cues in or very close to the apparatus indicate the location of reward, then these proximal cues form the basis for simple visual discrimination learning without a spatial component. To be a truly spatial task, it is important that the distal cues do not align neatly with the correct location, because the animal could then convert the task into a straightforward cue discrimination. A good spatial arrangement of cues has several different objects located at different distances from the apparatus but not directly in line with the correct location. This is done so that the animal must rely on the *relative* locations of objects to locate the correct place in the apparatus. The configuration of spatial cues is critically important for good results on these tasks, but it is rarely described in published methods sections and there is no kind of standardization for use of cues in different labs. Only the physical apparatus in contact with the mice is usually described.

The *Barnes maze* is a dark plastic disk elevated above the floor with a series of escape ports along its perimeter (O'Leary & Brown, 2008). The mouse starts in the center under a very bright light, and it can scurry into a chamber via one or two open holes to escape both the light and exposure to a wide open space. There is no local cue to indicate which hole is a portal to safety. This task is plagued by inactivity by many mouse strains, and the frequent reluctance of mice to climb down into the escape chamber. The *8-arm radial maze* uses hunger and food reward to motivate exploration (Dubreuil, Tixier, Dutrieux, & Edeline, 2003). Two kinds of configurations can be used (Crusio, Schwegler, & Brust, 1993). In some tests, every arm has one food pellet, and the challenge is to visit each arm only once. This puzzle can be "solved" with a simple radial strategy of always proceeding in the same direction when exiting an arm, so the maze is sometimes equipped with gates that confine the mouse to the center hub for a few moments, which disrupts radial strategies. The maze can also be provided with just one or two baited arms, and the mouse must then use spatial room cues to find the correct arm. Odor from the food is a menace to the validity of this version of the test. The familiar *Morris water maze* is not a maze at all; there are no walls or alleys (Klapdor & van der Staay, 1996; Prusky, West, & Douglas, 2000). It is a submerged platform water escape task. In the spatial version of the task, the animal cannot see the platform directly and must locate it by learning its location in relation to spatial room cues. A cued version can be used to verify that the mice have adequate vision (Stavnezer, Hyde, Bimonte, Armstrong, & Denenberg, 2002). A variant of this task has four plastic arms inserted into the circular tank (Wahlsten, Cooper, & Crabbe, 2005). This version and the Morris task are discussed at length in Chapter 13.

Given that the four spatial tasks are similar in size and can be done in a room full of the same cues on contiguous days, it will be interesting to compare results when three of four tasks are run with identical spatial cues using the same genetic strains of mice (Patil et al., 2009). The spatial factor will be common to all tasks, while motivations (escape from bright light and exposure, escape from water, reduction of hunger) and motor actions (ambulation vs. swimming) differ.

Electric shock

When motives are weak, the variability of mouse behavior can be frustratingly large, such as in the Barnes maze or novel object exploration. The effects of electric shock, on the other hand, are consistent and compelling. Accordingly, it is highly convenient for the experimenter. The mice are not quite so fond of these kinds of tasks, and ethics committees may also look askance at them. How to apply electric shock and the many pitfalls in the use of shock with mice are covered in depth in Chapter 11.

Classical or *Pavlovian conditioning* is done by pairing an originally neutral stimulus with a painful shock (Anagnostaras, Josselyn, Frankland, & Silva, 2000). If there is no way it can escape or avoid the shock, the mouse will usually exhibit freezing behavior in the presence of cues that precede shock. This phenomenon is the basis for *contextual and cued fear conditioning* (Figure 3.3). Two versions of the task are commonly employed (Stiedl et al., 1999). In the one-compartment contextual conditioning paradigm, the mouse is shocked in a box with distinctive cues on the wall and then later returned to the same box. If it freezes, this indicates it remembers what happened there previously. The test cannot prove the mouse is utilizing visual cues on the wall; it could simply be reacting to the tactile cues from the grid floor that was the source of its troubles. The cued version of the task entails a distinctive stimulus, either a light or sound, that precedes shock onset. The test trial then compares freezing with and without that cue present. In a second paradigm, a naive mouse is confined to one distinctive half of a two-compartment box and then given an electric shock to the feet via a metal grid floor. Later, it is returned to the box with both compartments open. If it recalls the previous experience, it should spend most of its time in the compartment with cues different from the one where shock was received. The test is somewhat difficult to interpret when both compartments have grid floors and are somewhat aversive on that account, so the grid is sometimes replaced by a smooth floor during non-shock test trials.

Fear conditioning is indicated by simple freezing or lack of action, whereas other tasks require rapid actions. Actions to get away from the shock are termed escape behaviors, whereas things done prior to shock onset constitute avoidance attempts. The early stages of avoidance learning always involve learning to escape a strongly aversive stimulus, and escape learning usually proceeds quickly in most mice. If a genotype has difficulty learning the correct escape response, there is little likelihood that it will ever solve the avoidance conundrum. In *inhibitory avoidance learning*, the mouse learns that going to a specific location with a grid floor earns it a painful shock to the feet (Cimadevilla, Fenton, & Bures, 2001; Geller, Robustelli, Barondes, Cohen, & Jarvik, 1969). Hence a good learner will not return to that place on future trials. The same apparatus can be employed for *active avoidance learning* by starting the trial with the mouse on the grid and turning on the shock a few seconds later. If the mouse scurries quickly to the safe zone, it will not receive a shock at all. This paradigm involves a conflict of motives. As in contextual fear conditioning, there will be a tendency to freeze in the compartment having the shock grid, whereas successful avoidance requires the mouse to move rapidly. An even more difficult version is *shuttle avoidance*, wherein a light or tone signals impending shock and then the mouse flees one box and runs to a seemingly safe compartment (Clark, Vasilevsky, & Myers, 2003; Peeler, 1995). A few seconds later the "safe" box becomes dangerous, and the animal must run back to where it was previously shocked. As discussed in Chapter 11, performance with this paradigm is often very poor. Some mice never learn to avoid the shock at all and wait each trial for the shock to occur before making a wild dash to the other side. Some investigators regard this failure to act before shock onset as a sign of "learned helplessness."

OTHER TESTS

The tests presented briefly in this chapter represent only a small, albeit popular, minority of tests employed with mice in the fields of psychopharmacology and behavioral genetics. There are several tests to assess pain by exposing either the tail or paw of a mouse to a hot plate or hot or cold water (Mogil, 2009; Mogil et al., 1999a; Wilson & Mogil, 2001). Sensitivity to pain is measured by the latency of the animal to react to the noxious stimulus by withdrawing its tail or limb. As discussed in Chapter 15, these tests seem to be especially sensitive to different experimenters, probably because the mouse must be held in the experimenters' hand to get the tail or paw into position. The flinch reaction to a loud sound from a speaker is the basis for the auditory startle test (Willott et al., 2003). In a box where shock was received previously, a mouse may show a larger flinch response to the tone (termed fear-potentiated startle). If a loud sound is preceded by a weaker warning signal, on the other hand, the flinch may be considerably reduced, a phenomenon known as prepulse inhibition (Powell, Zhou, & Geyer, 2009). This test plays an important role in the search for mouse models of neural functions related to schizophrenia. Not everything done to mice in the name of science is unpleasant to the animal. In the conditioned place preference test, a mouse is injected with a drug that evidently has some kind of positive, even pleasurable effect (Cunningham & Phillips, 2003). When the injection is given in a compartment with distinctive cues, including a distinctive floor, the mouse tends to prefer being in that compartment on future test trials when there are two compartments with different floors but no drug is given. This test is used extensively in the study of addictive drugs.

This overview of mouse behavioral tests should help the reader who is new to the field understand many of the methodological issues discussed throughout this book. It is presumed that anyone wanting to work with a specific test will conduct a thorough review of the literature and hopefully visit a laboratory where the test is done (Crawley, 2008). No book can adequately prepare anyone for the surprises that lurk in even the simplest tests. No protocol, no matter how long and detailed, can anticipate everything. It is hoped that a good understanding of the ways tests are designed and refined will help the reader do them the right way and perhaps even improve them for the benefit of others in the field.

Designs

The design of an experiment consists of the groups of animals to be tested, the treatments they will receive, and the kinds of measures that will be collected for each animal. Designs can range from a simple comparison of two groups, one of which receives the treatment while the other serves as a control, to elaborate schemes with multiple factors or a variety of reciprocal hybrid crosses. Some common designs are shown in Figure 4.1. The nature of the study determines the specific genotypes or treatments, and the design follows from the logic of the research. It is best practice to pose a question or hypothesis and then devise a suitable experimental design to find an answer. The same general kinds of designs are used in many fields of study, and nothing about designs in the abstract is unique to research with mice. The specific features of groups and kinds of treatments will depend on the domain of science in which one works.

A complete study usually consists of several experiments, each with its own design, and the control groups deemed necessary in later experiments may be determined by the findings of prior experiments. Consequently, the investigator usually dictates the design of the first experiment with total confidence, whereas later experiments in the series may need to have their designs altered after the first experiment is analyzed. Ethics and grant review panels require that all future experiments in the study, sometimes three to five years into the future, are described in detail, but there are almost always changes in design as the work progresses. During the course of a single experiment, however, the design must never be changed. If a change is clearly required because of unexpected results, it is wise to halt the experiment and begin a new one with an improved design.

Design includes the *study factors* that are of central interest in the experiment as well as a variety of other potential influences that the researcher seeks to control or limit to improve the quality

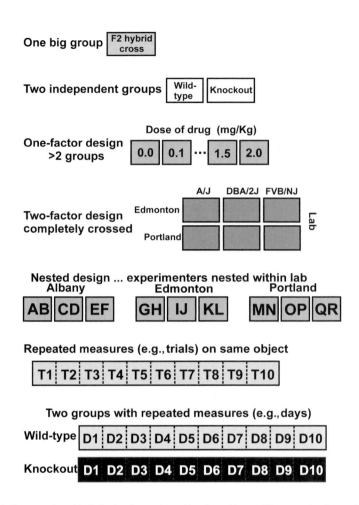

FIGURE 4.1

Common research designs employed in behavioral genetics with mice with specific examples for each one. In the first five designs, separate mice are used in each group, whereas with a repeated measures design one mouse experiences two or more treatment conditions.

of the data. In every study with mice there are dozens of environmental variables that affect the score of an individual, and it is not feasible to examine them all in one huge experiment. Instead, it is good practice to avoid confounds between the study factors and other potential influences, termed *control factors* (see Box 4.1). For example, behavior might depend on the month when the mice are shipped, the time of day when the test is given (Benstaali et al., 2001), the location of an animal's cage on a rack in the colony (Greenman, Bryant, Kodell, & Sheldon, 1982), when the cages are changed (Gray & Hurst, 1995), which specific experimenter administers the tests (Chesler, Wilson, Lariviere, Rodriguez-Zas, & Mogil, 2002),

4.1 TWO KINDS OF TREATMENT FACTORS

Study factors	Control factors
Genotype	Shipment
Drug	Shelf or cage rack
Brain lesion	Session
Training regimen	Replication
Early environment	Experimenter
Age	Sex
Sex	

temperature in the room (Gaskill, Rohr, Pajor, Lucas, & Garner, 2009), and kind of diet and water, as well as many other factors (Deacon, 2006).

Certain conditions can serve as either a study or a control factor. If the researcher is explicitly interested in differences between males and females (Jazin & Cahill, 2010), sex will clearly be a study factor. Other investigators may not expect a major sex difference, but may want to control for any sex effect by including both sexes in the study, making sex a control factor. Having data on both sexes increases the generality of the findings. This distinction becomes very important when deciding on sample size for the experiment (Chapter 5), because sample size usually pertains to major groups that differ on the study factors but not the control factors. For example, if sex is a control factor and calculations indicate that the sample size should be 20 mice per group, it would be appropriate to use 10 of each sex. On the other hand, if sex is a study factor, it may be important to examine 20 mice of each sex to be sure a meaningful sex difference can be detected.

There are two ways to make sure control factors do not affect the results attributable to study factors.

1. Many things can be equated or *standardized* within the scope of the experiment so that every animal experiences the same diet and water, the same kind of caging, the same animal caretaker, etc.
2. There inevitably are things that cannot be equated for all animals, and these potential influences need to be carefully *balanced and randomized* so that they cannot exert a significant influence on one or a few specific groups in the study. How this can be done is addressed in depth in Chapter 7. Here the focus is on the major study factors.

The lexicon of research design distinguishes two kinds of variables. The study factors are termed the *independent variables* by statisticians. They are deliberately chosen by the researcher and can be listed in a large spreadsheet before the start of data collection. They are independent because they are not affected by treatment conditions; instead, they denote the actual treatment conditions. Study factors such as genotype, strain, age, and drug treatment group are independent variables. All animals in the same group have the same value for each independent variable that represents a study factor. Control factors equated for all animals in the experiment are not independent variables in a design sense because they do not vary. Balanced or randomized control factors do qualify as independent variables, even if the investigator does not intend to analyze them as such; they can differ among mice of the same major treatment group therefore contributing to within-group variance in test scores. The *dependent variables* are the measures taken of an animal. The specific values of a variable are different for different individuals. The kinds of dependent variables to be observed are part of the research plan, but their values cannot be specified in advance of the experiment because they depend on the experiences of the animal prior to and during the behavioral test. Both independent and dependent variables need to be considered when deciding on a research design.

ONE GROUP

In some situations, a study may not involve distinct treatment groups. Instead, there may be just one large collection of individuals, and the study is termed *correlational*. Correlations among measures can be observed, but which thing caused another will be difficult to judge because the investigator has no control over the variables. It may be possible to group the individuals into categories after the fact to aid analysis, but just because the groupings show phenotypic differences does not mean that the cause of the differences is known. Ideas conceived on the basis of a correlational study will eventually need to be tested using experimental manipulations of key variables.

A *correlational design* might be used to assess the relations among body size, brain size, and sizes of various parts of the brain in a genetically homogeneous group of mice. Figure 4.2

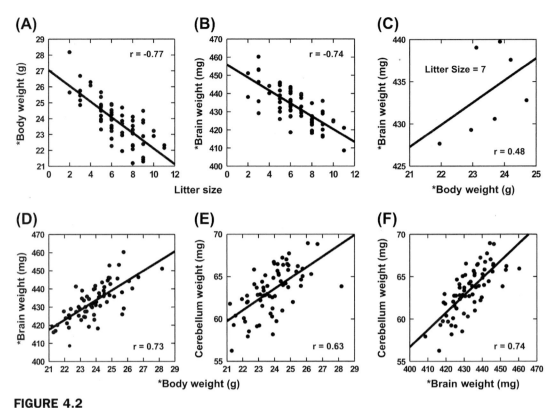

FIGURE 4.2

Results from 67 litters in one group of mice, all having the BALB/cWah genotype. Each point is a mean of one litter. Straight lines of best fit and Pearson correlation coefficients (r) indicate the direction and strength of the linear relationships. Where body weight has an asterisk (A, C, D, E), it represents body weight–brain weight. Brain with asterisk (B, C, D, F) is brain weight–cerebellum weight. A large portion of the within-strain variation in body and brain sizes is produced by differences in litter size (A, B). *(Based on data from Wahlsten & Bulman-Fleming, 1987.)*

shows relations among 67 litters of adult inbred BALBcWah mice (Wahlsten & Bulman-Fleming, 1987). Each point is the mean of one entire litter, giving equal weight to male and female offspring. Interpretation could be clouded because the brain is a component of the body and the cerebellum is part of the brain. This tends to inflate the correlations. Consequently, for the data shown in Figure 4.2, brain weight was first deducted from body weight and cerebellum weight was deducted from brain weight. Litters all of one sex are not shown. Data like these show that mice from larger litters tend to have smaller bodies and brains, and that larger mice tend to have larger brains, while those with a larger brain tend to have a larger cerebellum, as expected from the principle of allometry (Gould, 1966; Schmidt-Nielsen, 1984). Because the correlations among parts and the whole generally are not very high, it is evident that many other factors important for nervous system growth are involved but were not measured in this study. Litter size appears to be a major influence on growth, and a cursory glance at the figure might suggest it is the only influence. Within litters of the same size, however, there is still a correlation between body and brain size (Figure 4.2C). There might also be other maternal factors that covary with litter size, and the relation might not reflect only nutritional influences. No matter how sophisticated the statistical analysis or how many parts of the system are measured, a simple correlational study cannot demonstrate causal contributions of one thing to the development of another. Instead, the correlational study may provide some promising hints about what factors are most important, and then the investigator can design an experiment to manipulate certain factors experimentally and test appropriate control groups to prove what causes what. Altering litter size at birth to establish many small or large litters could help to determine causality, for example. Ablating some of the ovarian follicle cells prior to mating can generate small litters. Fostering mice from a small litter at birth into a much larger litter could help to separate pre- and postnatal influences on brain growth.

When data are collected from both males and females, a correlational study might be regarded as a form of two group study. Sex is determined at the time of fertilization, long before there is any rudiment of a nervous system. The set of chromosomes of the embryo must exert a causal influence on brain development, but the pathways through which chromosomal sex shapes the body and brain will not be found merely from the correlations. Multiple regression analysis can be used to assess whether the brain versus body relationship is the same for males and females. If statistical analysis reveals that females tend to have a large specific region of the hypothalamus relative to brain size, the difficult task of deciphering causal pathways will remain a challenge. One always needs to keep in mind that correlations provide clues, not final answers.

Many genetic linkage studies seek correlations between a phenotype and numerous marker loci scattered widely across all chromosomes, and in mice this is often done in an F_2 hybrid cross, a more advanced intercross, or a large set of RI strains that is just one large population. These designs are especially important in behavioral genetic studies of quantitative trait loci (QTLs). Many QTLs influencing mouse behavior have been identified (Belknap & Atkins, 2001; Belknap, Dubay, Crabbe, & Buck, 1997; Flint, 2003), but progress in identifying the specific genes responsible for the QTL effects has been frustratingly slow (Willis-Owen & Flint, 2007).

TWO GROUPS

When an experiment has two groups of subjects where each individual contributes data to just one group, then there will be independent groups. Independence in this situation means that something done to change the scores in just one group will have no effect on scores in the other group. A classic example is coisogenic mice that are created when a spontaneous mutation occurs in an isogenic inbred strain (Chapter 2). The animals have the same genotype at every other locus and differ only in their alleles at the one locus. If it is well established that the mutation is completely recessive, then a study of brain or behavior might involve just two groups, the $-/-$ mutants and their $+/?$ littermates (both $+/+$ and $+/-$). If the genotype of the non-mutants is known, then the study might have three groups ($+/+$, $+/-$, $-/-$). A fine example is provided by the *diabetes* mutation ($Lepr^{db}$) that occurred in the C57BL/KsJ strain (Coleman & Hummel, 1967). The phenotypic differences between the genotypic groups must arise from a causal effect of the mutation, because nothing else differs between the groups. Even then, tracing the pathways of physiological processes involved in the phenotypic difference is a major task, and along the way it will be necessary to do some experiments to manipulate key factors.

A simple comparison of two inbred strains also qualifies as a two group design. All variation within a strain must be non-genetic (apart from the sex chromosomes), whereas the difference between the strains could arise directly from offspring genotype or might reflect a difference in the maternal factors provided by the two strains or, most likely, could be some combination of both influences.

Research in human behavior genetics sometimes seeks to emulate a design with two independent groups. For example, researchers in France identified two groups of children who differed greatly in their early environments but had very similar heredities (Schiff, Duyme, Dumaret, & Tomkiewicz, 1982). Several mothers with little education and low income had given up one or more children for adoption but also kept one child. In the adopted group, each child was placed into a home with well-educated, high socioeconomic status parents. The fathers of the adopted-away and mother-reared children were not the same in all cases, but they did not appear to differ greatly. Thus, the genotypes of the two groups of children were similar, although not identical,

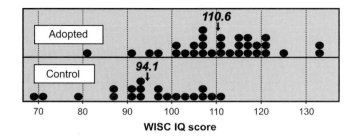

FIGURE 4.3

Frequency distributions of IQ scores of groups of children having the same mother but different fathers. Control children were raised by biological mothers having low education and income, whereas adopted children were reared by families with higher education and income. Because some mothers gave up more than one child for adoption, the study was analyzed as two independent groups. It could also have been analyzed as a matched-pairs design. *(Based on data presented in Schiff, Duyme, Dumaret, & Tomkiewicz, 1982.)*

whereas rearing environments were clearly quite different. Psychological testing after several years of separation revealed a substantial group difference in IQ score (Figure 4.3). The study could be conceptualized and analyzed as two independent groups or as a matched-pairs design, where each pair had the same biological mother. Because the published article did not report the data in pairs, the study is presented here as a two group design.

MATCHED PAIRS

If one is interested in comparing two kinds of individuals, it may be possible to control for a wide array of other influences by studying carefully matched pairs. In this case there will be two groups, but they will not be independent. On the contrary, by design they will have many things in common, things that will differ greatly between, but not within, the pairs. For example, a study in the author's lab sought to measure the longer term consequences for brain growth during a 24 h period of food restriction shortly after birth (Wahlsten, Blom, Stefanescu, Conover, & Cake, 1987). The effect was not expected to be very large, so a within-litter design was employed. The study also employed a between-litters design for purposes of comparison. Mice typically have litters ranging from 2 to 13 pups, and litter size has a major and lasting effect on brain growth. If a between-groups design is used, where some entire litters are deprived of nursing while others are not, the variance within treatment groups tends to be quite large when the litters differ in numbers of pups. Furthermore, removing all pups from a lactating female for a day can lead to reduced or even a shutdown of milk production, so that the deprivation period will actually be longer than specified in the plan of the experiment. A within-litter design addresses both problems. Removing just one pup from the nest and isolating it for 24 h has little effect on its littermates, and comparing it with a matched littermate of the same sex and initial body weight removes the between-litter variation in growth from the data analysis, making the test of deprivation effects more sensitive.

Depriving the entire litter clearly was a more severe treatment. Only 46 of 85 pups in nine deprived litters survived until eight weeks of age, whereas depriving just one or two pups in a litter did not elevate their mortality rate. At eight weeks of age brain weight was lower in the formerly deprived mice in both designs, the effect was only marginally significant in the between-groups design, but it was clearly significant in the matched littermate pairs design (Table 4.1).

REPEATED MEASURES

When the same phenotype is measured for the same individual on several occasions, the experiment involves a repeated measures design. This often is done in a study of learning,

TABLE 4.1 Brain Weight (mg) of BALB/c Mice at Eight Weeks After Birth[a]

Design	Controls (n)	Deprived (n)	Difference	d	t	P
Between litters	486.6 (47)	475.5 (46)	11.1	0.41	1.97	0.026
Matched pairs	501.5 (24)	485.8 (24)	15.7	1.76	4.41	0.0001

Source: Wahlsten, Blom, Stefanescu, Conover, & Cake, 1987. With permission.
[a]*Effect size d is the number of standard deviations by which group means differ (see Chapter 5), and t is the test of statistical significance of the group difference. P is probability of a Type I error or false positive conclusion from the test.*

where each mouse is given several training trials on the same day or across days. Similarly, a study of endocrine development might measure hormone levels at several ages in blood samples taken from the same individuals. Data can then be analyzed with a repeated measures analysis of variance (ANOVA).

Because researchers are almost always interested in a genetic or treatment variable, only rarely would one want to use a repeated measures design with just one group. Instead, there would be two or more independent groups, each with the same number of repeated measures; for example, the ANOVA could then assess whether the rate of learning was different in different genotypes.

It is important to appreciate that repeated measures is not the same concept as multiple measures. When the investigator applies a battery of several behavioral tests, there will be multiple measures, but if each test entails only one trial, there are no repeated measures, as discussed previously. With multiple measures, the phenotypes are most likely measured using different scales, such as escape latency in seconds for one test, percent correct for another, and number of head dips for a third. The phenotypes will usually be assessed with separate analyses. A more advanced method, multivariate analysis of variance (MANOVA), could accommodate several different measures or dependent variables. Unfortunately, the kinds of behavioral data collected with mice often violate the assumptions of that method and require sophisticated mathematical procedures beyond the scope of this book.

When an experiment involves repeated measures, it may be legitimate to simplify the analysis by creating a new dependent variable that represents the phenotype by a single score. An obvious application is a pre- versus post-injection test of a drug effect, where the difference score represents the effect of the drug on behavior. In a study of learning where all mice begin training in a naive state, a single composite score may serve as a useful index of learning. For example, in the accelerating rotarod test, most mice of all strains fall quickly on the first trial because they know nothing about the task. Those that quickly learn to walk atop the rod increase their fall latencies over trials, whereas others may show little evidence of learning. Thus, the average fall latency over all trials or perhaps the average fall latency on the last three trials, can serve as a good indicator of learning (Figure 4.4A). If the task involves choice of arms in a 2-arm T maze, where the probability of a correct choice is 0.5 by chance alone, all mice must begin the task by making approximately 50% errors. Suppose training is administered with four trials per day. Total errors then provide a good indicator of learning rate (Figure 4.4B). Shortcuts in data analysis are most effective when changes across trials are reasonably linear for the various study groups.

SINGLE-FACTOR STUDY WITH MORE THAN TWO GROUPS (ONE-WAY DESIGN)

Few questions can be addressed adequately with only two groups of animals. Suppose a researcher wants to test for effects of a drug given by intraperitoneal injection. There must be

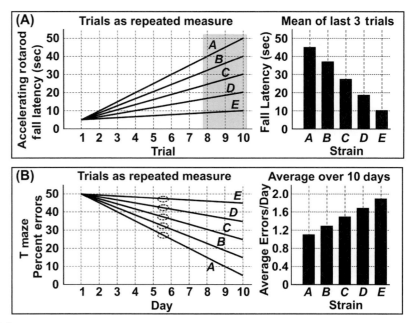

FIGURE 4.4
Two situations involving repeated measures on the same mice of five strains where averaging performance across trials can simplify the analysis. (A) When all mice fall quickly from the rod at the start of training on the accelerating rotarod and then improve gradually, the average of the last three of ten trials is a good indicator of learning. (B) In a T maze for food reward where all strains choose the arms of a maze by chance alone at the start of training and improve gradually, the average number of errors over several trials (dashed ovals) is a good indicator of speed of learning.

a control group that is also injected with saline or whatever solute is used for the drug, but the injection could have a transient effect on a behavioral test given not long afterwards. Hence, a non-injected control might help to interpret the data. A similar logic applies in the case of an invasive procedure such as a brain lesion. It is good practice to use a sham lesion control group that receives every aspect of the surgical procedure except damage to a specific part of the brain, and inclusion of a non-operated control group aids interpretation.

An elegant three-group design was used to study the development of diabetes by *diabetes* ($Lepr^{db}/Lepr^{db}$) mutant mice (Lee & Bressler, 1981). After weaning, normal littermates ($+/?$) and one group of mutants were allowed to eat all the food they wanted, while a third group consisted of mutants that were pair-fed. The freely fed mutants tended to eat considerably more than normal mice, but the pair-fed mutants were given a daily ration limited to the amount consumed the day before by their lean ($+/?$) siblings. The adult mice differed greatly in body weight; mutants weighed far more (50 g) than normal mice (24 g), whereas pair-fed mutants had a normal weight (22 g). Most remarkably, the extremely high blood glucose of the freely fed mutants (490 mg/dL) was not apparent in the pair-fed mutants (79 mg/dL vs. 90 mg/dL for normal mice), and insulin levels as well as kidney damage apparent in freely fed mutants were not seen in pair-fed mutants. Therefore, the mutants that were not allowed to overeat did not develop clinically significant diabetes, even though they were homozygous recessive for a gene named *diabetes*.

More extensive one-way designs are common in neural and behavioral genetics. Inbred strain surveys often involve eight or more strains that are all treated and tested in the same way, so that strain differences reflect differences in their heredities. The Mouse Phenome Database (MPD; www.jax.org/phenome) provides extensive phenotypic data from many such surveys (Grubb et al., 2004). This kind of design is valuable for identifying strains that have extreme values and may be promising subjects for subsequent linkage studies that search for genes having small, quantitative effects on a phenotype (QTLs). A smaller set of strains will be

TABLE 4.2 Study of an Anxiolytic Drug (midazolam) and Its Antagonist (RO-15) Injected into Two Brain Regions[a]

Group # Injection Site	1	2	3	4	5	6	7	8
		Septum				Amygdala		
RO-15	No	**No**	Yes	Yes	No	No	Yes	Yes
Midazolam	No	**Yes**	No	Yes	No	Yes	No	Yes
Contrast coefficient	−1	**+7**	−1	−1	−1	−1	−1	−1
Group mean	17	**51**	21	22	16	23	11	21

Source: Pesold & Treit, 1994. With permission.

[a]Conditions for the one group expected to show a change in exploration are shown in bold. Mean scores are percent time in the open arms of an elevated plus maze. Higher values are indicative of lower rat anxiety. The value of the contrast (ψ) is the sum of each contrast coefficient multiplied by the group mean. For these data, estimated $\psi = 7(51) − (17 + 21 + 22 + 16 + 23 + 11 + 21) = 226$. Its significance can be evaluated with a t test.

helpful for a future experiment that evaluates strain-dependent effects of pharmacological or environmental treatments, because a study of that kind should be more sensitive if it employs strains that are widely spaced across a phenotypic continuum. A one-way design might also be employed to find the dose-response curve when several doses of a drug are given, each to a separate group of animals.

There are situations where a study involves several factors, yet it is most convenient and illuminating to analyze it as a one-way design in which specific groups are compared with *linear contrasts* formulated to answer specific questions (see Chapter 5). Consider a study of exploration of the elevated plus maze (two arms with high walls and two open arms without walls) by rats under the influence of the anxiolytic drug midazolam (Pesold & Treit, 1994) and its antagonist RO-15-1788 injected into either the septum or amygdala. The design entailed three factors: injection site, midazolam (yes, no), and RO-15 (yes, no). The researchers expected only one of the eight groups to show a marked increase in exploration of the open arms of the maze; midazolam in the septum without RO-15. As shown in Table 4.2, one contrast addressed this specific question directly, whereas an analysis of variance with three factors yielded a much more complicated picture but could not answer the specific question asked in the study. Several more specialized research designs presented in the section Specialized Designs (e.g., reciprocal hybrid crosses) are also best analyzed as a one-way design with contrasts.

TWO-FACTOR STUDIES

Many of the experiments done in behavioral and neural genetics vary two or more factors in a systematic way. We want to know not simply whether two genotypes differ in some behavior but whether they differ in how behavior is modified by experience. In a study of memory, for example, one group of animals might be tested for activity level or freezing behavior in a box with distinctive cues that 24 h earlier were paired with shock, while the control group with the same genotype is given the same test but is never shocked in that box. Different levels of freezing would indicate memory for an aversive experience. The main question in such a study is whether the two genotypes differ in their behaviors as a function of prior experience. Thus, there must be four groups in the 2 × 2 factorial design; factor one, genotype and the other, experience.

This simple design helps to compensate for non-associative effects of exposure to the box. On the first day when the control group is exposed to the box, mice explore actively. On the second day, they will probably be somewhat less active in the same box because they have been there before and became habituated to that context. Thus, the difference between activity levels on days 1 and 2 for the controls will reflect habituation. Habituation requires memory. After all, if the mice had no memory, they would be just as active every time they are placed in the same

61

box. What the investigators want to study, however, is associative memory, or recall of the pairing of box cues with shock. They predict that the change in activity level will be greater for the shocked group than the controls. Only the shocked group should have considerable fear-related freezing on day 2.

Although the day 1 and day 2 testing suggests this is a within-group design, it would not be analyzed in that way. Instead, the phenotype is computed as a difference score between day 1 and day 2, and then the experiment is analyzed as a factorial design with four independent groups. The analysis of variance (ANOVA) evaluates three effects. The *main effect* of genotype is the difference between the two genotypes, each averaged over the two treatments. This would reveal whether one genotype generally changed more in activity over the two days, regardless of treatment. The main effect for treatment reflects the consequence of receiving a shock on day 1, averaged over genotypes. In certain cases, knowing these two main effects provides a fairly good impression of the results of the experiment. If there is a large treatment effect, indicating good associative memory but no genotype main effect, then it is probably true that the two genotypes had similar levels of associative memory. Comparing the two main effects, however, does not formally test the hypothesis that the genotypes had the same levels of memory. That test is embodied in the *interaction effect*, which compares the difference between control and shocked groups for the two genotypes. This is the great value of a factorial design: it can determine whether genotypes differ in response to an environmental treatment.

This issue is important when comparing two competing explanatory models. For example, it has been asserted that the profiles of phenotypic change in response to different levels of environment are similar for all or most genotypes, a concept termed the "reaction range" (see Figure 4.5B; Gottlieb, 2007), whereas others cite data indicating that a "norm of reaction" to environments differs substantially among genotypes (Sarkar, 1999), and no genotype is generally high or low across a wide range of environments (Figure 4.5A). The models cannot be evaluated by a single factor study that examines only environment, because both expect environment to be important. A purely genetic experiment also cannot address the dispute, because both expect significant genetic effects. The crucial test requires a genotype × environment factorial design to tell us whether environmental effects depend on genotype (Figure 4.5C).

The two-way factorial design can be applied to much more elaborate studies. Consider a 3 × 8 experiment conducted to determine whether different laboratories will obtain the same results of behavioral tests when studying the same genotypes (Crabbe, Wahlsten, & Dudek, 1999). Three labs joined forces and sought to equate many aspects of their colony and testing rooms. Identical test apparatus and test protocols were used, mice of eight genotypes were shipped from the suppliers on the same day, and testing commenced at

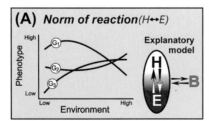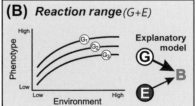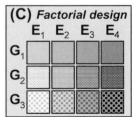

FIGURE 4.5

The concept of the norm of reaction (A) asserts that response to a change in environment depends strongly on genotype and may take almost any form, whereas the reaction range (B) asserts that the form of the response to environment is essentially the same for all genotypes. Predictions from the two competing models must be compared with a genotype × environment factorial design (C). The norm of reaction expects a significant interaction term, whereas the reaction range does not.

the same hour local time on the same day in all three labs. The central question in that study was whether there would be a lab by strain interaction. If not, it would be concluded that the three labs obtained roughly the same results of the genetic experiment. In fact, there was a marked interaction for certain behaviors but not others (Wahlsten et al., 2003).

MULTI-FACTOR STUDIES

When there are more than two factors, every level of one factor will be paired with every level of the other factors. Thus, the number of groups in the design will be the product of the number of levels for each factor. For example, a recent experiment by Wolfer and colleagues (2004) compared behavioral test scores of three genotypes (two inbred strains and a hybrid) that were reared in either standard or enriched cages at three different labs. Thus, the full factorial design involved $3 \times 2 \times 3 = 18$ independent groups of mice. Because they worked only with female mice, there was no sex factor in the design.

Similarly, an experiment done recently in the author's lab compared behaviors of males and females of eight inbred strains housed and tested in a normal or reversed light–dark cycle. It was even more challenging because half of the mice received saline injections and half received ethanol injections prior to testing. Thus, the factorial design had $8 \times 2 \times 2 \times 2 = 64$ independent groups.

It is possible to contemplate experiments with more than four factors, but the logistics of conducting such an elaborate symphony and then doing a thorough analysis are daunting (Chapter 7), and the resources needed to complete the work may exceed the budgets of most labs. An example is shown in Figure 4.6.

Faced with such complexity, it may be wise to simplify the design if at all possible. One approach is to divide the study into a series of smaller experiments. Unfortunately, this may not always be appropriate for the question asked. Consider the $8 \times 2 \times 2 \times 2$ study of light–dark cycle and ethanol effects on behavior. It would be unwise to divide this into two $8 \times 2 \times 2$

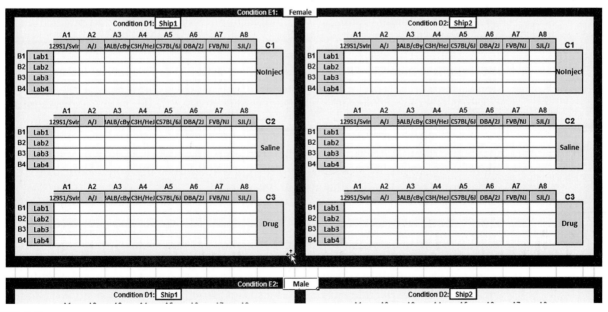

FIGURE 4.6
Top portion of the **Design Example 5way** spreadsheet captured from the computer screen. It shows how names for the various levels of each factor can be typed into the appropriate boxes to generate the complete design list. It is a factorial design because every level of every factor is paired with every level of every other factor. Hence the $8 \times 4 \times 3 \times 2 \times 2$ design has 384 cells.

studies, the first with saline injection and the second with ethanol. There might be uncontrolled and unexpected changes at the supplier, the shipper, the local mouse colony, or the testing lab that intervene between the two studies, and these effects could then masquerade as ethanol effects. From a design perspective, this would be bad practice. The whole point of the study is to determine whether a genetic experiment to assess ethanol effects on behavior will yield the same results in either kind of light–dark cycle. Clearly, all groups in the study need to be tested in the same narrow period of time with proper balancing and randomization of test order.

Another recourse is to conduct the full experiment in all its multifactorial splendor, then try to simplify the analysis and presentation of results so that the reader is not overwhelmed by the complexity. Some unavoidable facts of any large study with mice are that there must be several shipments or test cohorts, the mice must be housed on different racks or shelves, and the test day must be divided into earlier and later sessions. Shipment, shelf, and session can be considered control factors (Box 4.1), and in Chapter 7 methods to balance groups across multiple study and control factors are discussed. Thus, a study designed to compare the responses of both sexes in eight strains to a treatment can be designed as an $8 \times 2 \times 2$ experiment with 32 groups, but using four shipments, five shelves on a rack, and three sessions in a day it turns into an $8 \times 2 \times 2 \times 4 \times 5 \times 3$ monstrosity with 1,920 groups. Faced with such a conundrum where the number of groups exceeds the number of mice to be tested, compromises are necessary. It will not be possible to have exactly the same number of mice in every group. The only solution is to execute the design as well as humanly possible and then omit certain independent variables — perhaps shipment, shelf, and session — from the formal ANOVA and pool any effects of these control variables into the term for within-group variance. Thus, a study designed with six factors might be analyzed as a three-factor experiment. When this is done, it is wise to give at least a cursory glance at possible effects of the control factors in the initial pass through the data. If they do not seem to make much difference, then there is no need to worry about slight imbalances in the design. If they do appear to have substantial impact, then worry.

SPECIALIZED DESIGNS

The two-group, multi-group one-way, and factorial designs employed in research on mouse behavior parallel the designs presented in most texts on research design. Provided there is clear correspondence between the design of the current experiment and a textbook example, data analysis can then follow the recipe described in the text. There are special designs employed in mouse research, however, that will not be found in any text on statistical methods. They require adaptation of general methods to specific needs. The design must always be appropriate for the question asked, whereas the data analysis needs to be done in a way that permits the use of common statistical programs such as SPSS, SAS, or SYSTAT. Several elaborate designs for studies of mice are described in the next sections. Many others may be contemplated to study specific topics. By providing a few examples, it is hoped that some of the basic principles of designing complex experiments will be evident.

Reciprocal hybrid crosses

Two inbred strains reared in adjacent cages on the same rack in the same colony room differ in heredity while having extremely similar surroundings. Heredity is defined operationally as all the things that differ between the strains and are transmitted across generations. These things include more than Mendelian genes. The strains also differ in the mitochondria and other cytoplasmic elements, the Y chromosome, and diverse features of the maternal environment. The effects of these components of heredity can be dissected quantitatively with a set of reciprocal hybrid crosses (Carlier, Nosten-Bertrand, & Michard-Vanhée, 1992; Wahlsten, 1979). If inbred strains A and B are crossed, the result is an F_1 hybrid. Reciprocal F_1 hybrids are crosses of the same two strains that differ only in which strain is the mother and father: A♀ × B♂ versus B♀ × A♂ (genotype of female parent is listed first). The two reciprocal F₁s differ in the source of the Y chromosome in male offspring as well as in the source of the mitochondria. Reciprocal backcrosses (F₁ hybrid mated with an inbred) help to separate the effects, as shown in Figure 4.7 and Table 4.3. Looking closely at this figure and searching through the table, it is possible to identify particular groups that differ in only one component of heredity. Data resulting from this kind of complex breeding design can be analyzed as a one-way factorial

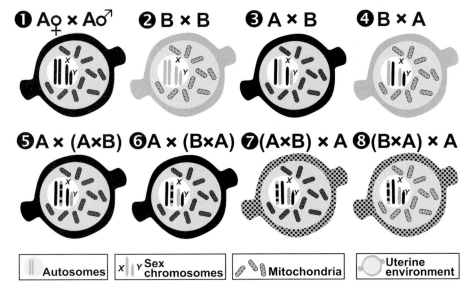

FIGURE 4.7

Diagrammatic representation of reciprocal hybrid crosses to dissect heredity into components. X and Y chromosomes represent the situation in males. Strains A (black) and B (gray) differ in all components, whereas certain crosses differ in only one component and can be compared statistically to assess the importance of that one component, as shown in Table 4.3. *(Redrawn with permission from Wahlsten, 2003.)*

65

TABLE 4.3 Reciprocal Crosses to Dissect Heredity into Components[a]						
Cross	Mother	Father	Mitochondria	Y Chromosome	Autosomes	Maternal Environment
1	A ♀	A ♂	A	A	A	A
2	B ♀	B ♂	B	B	B	B
3	A ♀	B ♂	A	B	F₁	A
4	B ♀	A ♂	B	A	F₁	B
5	A ♀	A ♀ × B ♂	A	B	75% A	A
6	A ♀	B ♀ × A ♂	A	A	75% A	A
7	A ♀ × B ♂	A ♂	A	A	75% A	F₁
8	B ♀ × A ♂	A ♂	B	A	75% A	F₁
9	B ♀	A ♀ × B ♂	B	B	25% A	B
10	B ♀	B ♀ × A ♂	B	A	25% A	B
11	A ♀ × B ♂	B ♂	A	B	25% A	F₁
12	B ♀ × A ♂	B ♂	B	B	25% A	F₁
13	A ♀ × B ♂	A ♀ × B ♂	A	B	F₂	F₁
14	A ♀ × B ♂	B ♀ × A ♂	A	A	F₂	F₁
15	B ♀ × A ♂	A ♀ × B ♂	B	B	F₂	F₁
16	B ♀ × A ♂	B ♀ × A ♂	B	A	F₂	F₁

[a]Specific comparisons to assess effect of one component:

Mitochondria (7 vs 8) (11 vs 12) (13 vs 15) (14 vs 16)
Y chromosome (5 vs 6) (9 vs 10) (13 vs 14) (15 vs 16)
Autosomes (1 vs 6) (2 vs 9) (3 vs 5) (4 vs 10) (7 vs 14) (8 vs 16) (11 vs 13) (12 vs 15)
Maternal environment (6 vs 7) (9 vs 12).

experiment with linear contrasts (Hays, 1988; Wahlsten, 1991; Winer, Brown, & Michels, 1991). Examples include studies of rats (Broadhurst, 1961), mice (Carlier & Nosten, 1987), and fruit flies (Bauer & Sokolowski, 1988).

Manipulations of the maternal environment

A set of reciprocal crosses can reveal the importance of the maternal environment, but it cannot tell us what feature of that environment is most important. The maternal environment includes everything outside of the embryo that impinges on it, from conception to the day when pups are weaned and separated from their mother. An elegant design can be employed to separate the influences of pre- and postnatal maternal environments, as discussed by Carlier et al. (1992). Using ovarian grafting, the ovarian follicle cells of an inbred strain female can be grafted into an F_1 hybrid female formed by crossing that strain with another inbred strain. The F_1 female will accept the graft because she already possesses the antigens from the donor strain. If A strain follicle cells are grafted into an A $\times$ B female and then, after recovery from the operation, the host is mated with an A strain male, offspring will be genetically strain A with an F_1 maternal environment, which can substantially improve brain growth (Bulman-Fleming & Wahlsten, 1988). If offspring are then fostered to a different mother at birth, the influences of A versus A $\times$ B postnatal maternal environments can be compared. A design showing the crucial groups along with controls for effects of the grafting operation and fostering is outlined in Table 4.4. The in-fostering control, for example, compares A strain pups fostered to an A strain mother with A strain pups that are not fostered at all. The autograft of A ovarian follicle cells to the A mother who nurses her live born pups can then be compared with ungrafted A females to assess possible influences of the grafting operation. One of the influences is particularly important: ovarian grafting generally results in smaller litters, and smaller litters result in more rapid brain growth. Thus, for phenotypes such as whole brain size, the grafting

66

TABLE 4.4 Ovarian Grafting and Fostering to Separate Prenatal and Postnatal Maternal Effects[a]

Group	Grafted Host ♀	Donor ♀ Ovaries	Stud ♂	Foster ♀	Offspring Genotype	Maternal Environment	
						Prenatal	*Postnatal*
1	A×B	A	A	A	A	F_1	A
2	A×B	A	A	A×B	A	F_1	F_1
3	A	A	A	A	A	A	A
4	A	A	A	A×B	A	A	F_1
5	A×B	A	B	A	F_1	F_1	A
6	A×B	A	B	A×B	F_1	F_1	F_1
7	A	A	B	A	F_1	A	A
8	A	A	B	A×B	F_1	A	F_1
Ungrafted control mice							
9	A ♀		A	Not fostered	A	A	A
10	A ♀		A	A	A	A	A
11	A ♀		A	F_1	A	A	F_1
12	A ♀		B	Not fostered	F_1	A	A
13	A ♀		B	A	F_1	A	A
14	A ♀		B	F_1	F_1	A	F_1

[a]*Specific comparisons to assess effect of one component:*

Prenatal environment (1 vs 3)(2 vs 4)(5 vs 7)(6 vs 8)
Postnatal environment (1 vs 2)(3 vs 4)(5 vs 6)(7 vs 8)
Interaction of prenatal env. with genotype [(1 vs 3) vs (5 vs 7)] [(2 vs 4) vs (6 vs 8)]
Interaction of postnatal env. with genotype [(1 vs 2) vs (5 vs 6)] [(3 vs 4) vs (7 vs 8)]
Effect of graft operation (3 vs 10)(4 vs 11)(7 vs 13)(8 vs 14)
Effect of fostering (9 vs 10)(12 vs 13)

operation may actually lead to superior development. Because litter size differs among both grafted and ungrafted females, comparisons can be made of neural phenotypes and later behaviors in mice from litters of the same size and overall nutritional status.

Transgenerational influences

Whereas a maternal effect of a treatment can influence the behavior of the offspring more or less directly via the maternal environment the pups experience so intimately, the consequences of some treatments can span more than one generation and affect offspring whose grand-mother but not mother was treated (Holliday, 2006; Jablonka & Raz, 2009; Morgan & Whitelaw, 2008; Youngson & Whitelaw, 2008). Effects of this kind can also occur when maternal behavior toward the pups, a social act, alters the methylation of certain genes and influences expression of genetic information in later generations (Champagne, 2008). An experiment to study this kind of alteration should ask two questions: whether more genera-tions of maternal treatment induce a greater change in descendants and whether effects persist unchanged for several generations or gradually fade away. Two versions of a design to answer these questions are shown in Figure 4.8. Version A starts all of the groups at the same gener-ation and then tests descendants in different generations. It therefore includes an untreated control group to check for extraneous environmental differences among generations that could give rise to differences between treatment groups. There is one major drawback to this design, however. If there are indeed differences between untreated generations of animals, the data for the treated groups may be impossible to interpret or, at best, will not provide a definitive estimate of transgenerational influences. Version B offers a major advantage because all of the crucial comparisons can be made among mice tested in the same time period. Both version A and B entail the same number of groups and generations of testing, but version B is undoubtedly more elegant and interpretable.

Selective breeding with control for environmental influences

Early in the history of behavioral genetics, Lamarckian inheritance of acquired characteristics was widely accepted although hotly disputed. In an experiment by McDougall (1927), for example, rats were trained to find food in a complex maze and then bred to produce the next generation. As the study progressed, successive generations required fewer and fewer trials to find the goal (Figure 4.9A). There were lingering doubts about possible bias in the choice of animals to serve as parents for the next generation. Furthermore, there were no replicated treated and control lines, essential features for any transgenerational experiment (Chapter 2).

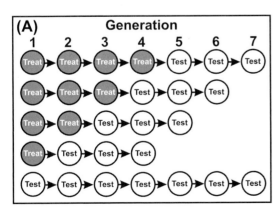
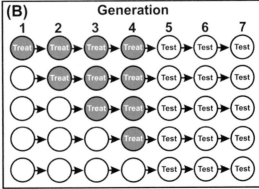

FIGURE 4.8
Two designs to test for transgenerational effects of treatments administered to the mother. (A) The numbers of generations of treatment, and their possible cumulative effects, are confounded with the generation, month, or year in which the untreated offspring are tested. (B) All offspring from families with different treatments of ancestors are tested at the same generation at the same time. Testing subsequent generations can detect longer term consequences.

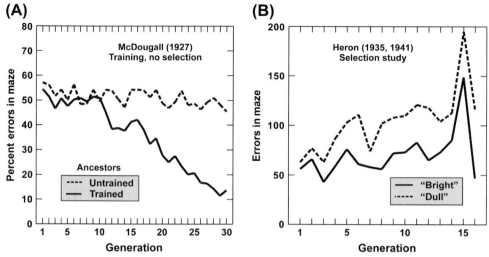

FIGURE 4.9

Two multigenerational experiments that were interpreted in totally different ways by their authors. (A) McDougall (1927) claimed that the effects of training rats in a learning task altered heredity directly such as in a Lamarckian fashion. (B) Heron (1961) bred two lines by mating either the best performing ("bright") or worst performing ("dull") rats. Gradually the two lines diverged, although a remarkable synchrony of environmental effects was also apparent across generations. *(Based on data presented in Heron, 1941 and Fuller & Thompson, 1960.)*

In the same period, selective breeding experiments were reported by Heron (1941; see Figure 4.9B), and Tolman and Tryon in which maze performance improved markedly over generations (Fuller & Thompson, 1960). Those studies also lacked replicated selected and control lines. In view of modern concepts of epigenetic inheritance, however, it is now evident that neither experiment provided the kind of data needed to draw firm conclusions. In both

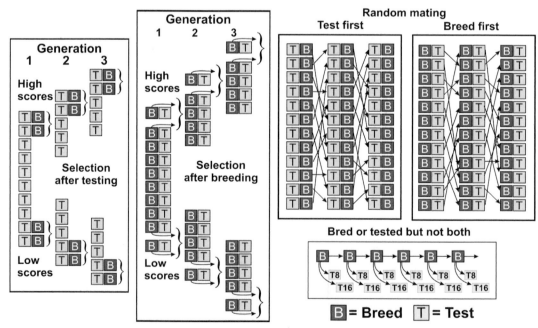

FIGURE 4.10

Designs to control for possible transgenerational effects of a treatment in the course of a selective breeding study. In some lines the testing comes before the breeding, whereas breeding is done before testing in other lines. The two procedures can be compared directly in other lines where parents are chosen randomly. Differences in the time of testing can then be assessed by testing some offspring at 8 weeks of age (T8) and others at 16 weeks (T16).

the Lamarckian and Mendelian versions of the experiment, rats were first trained in the maze and then bred. Thus, there *could* have been a direct effect of training on maternally transmitted influences to the next generation in *both* kinds of experiments, and these could have been cumulative. From the standpoint of experimental design, separate groups of animals are needed in which some are bred first and *then* trained in the maze (Figure 4.10). This design would absolutely and decisively avoid any transgenerational environmental influences. The study should include control lines of randomly mated parents trained either before or after breeding. Because the two conditions require testing at different ages, an additional control group is needed in which parents are never tested but their offspring are tested at different ages. Such a study would entail more groups and more animals than the traditional selective breeding experiment, but recent demonstrations of epigenetic transmission make additional controls necessary to distinguish between the influences of different genomes and changes in the epigenome.

COMPLICATIONS
Fixed and random effects

Statisticians consider factors to be *fixed* or *random*. What determines whether a factor is fixed or random is how the researcher intends to generalize the results. When a factor is fixed, the results of the study should not be generalized beyond the specific levels of the factors that were studied. For example, a study that compares males and females would entail sex as a fixed factor; there are only two sexes in the mouse universe. A study of a normal control versus a gene knockout would entail a fixed factor; knowing the effects of knocking out one gene would not tell us what should happen when another gene is disabled. If a researcher decides to test eight inbred strains from the Tier 1 priority list of the Mouse Phenome Database (MPD), it would not be wise to generalize the results to all mice or even to all the other inbred strains, and strain would then be a fixed factor. Likewise, if three labs collaborate in a study of lab environment effects and they carefully coordinate their efforts and try to make their husbandry conditions very similar, the lab will be a fixed factor, and the results cannot necessarily be generalized to all other labs. A *random* factor is defined as one where the levels are sampled more or less randomly from some larger universe of conditions. Results of a study with random factors can then legitimately be generalized to other conditions. For example, if the 40 strains on the MPD lists are taken as the universe of strains and then a researcher chooses to work with 10 randomly selected strains, results might be generalized to all 40 strains. However, this is virtually never done for a multiple strain study. If several dozen labs were to register for a study and then perhaps six were randomly chosen and funded to participate in a multi-lab replication experiment, results would provide a good test of generalizability to many other labs. Whether a factor is fixed or random is a difficult thing to judge, and deciding how it will affect the design and especially the statistical analysis requires considerable sophistication in ANOVA methods (Winer et al., 1991). Unless otherwise stated, examples considered throughout this book presume each factor to be fixed, not random.

Nested factors

Most factorial designs used with mice are completely crossed designs in which every specific level of one factor is paired with every level of another factor. If one strain of mice is subjected to 0, 12, and 24 h of food deprivation prior to a test of learning, then the same deprivation times will be used with every other strain. There are factors, however, for which the levels of one factor cannot be completely crossed with other factors because of the unique features of those levels. Suppose that three labs decide to compare results for two technicians at each site. If the same two technicians are transported to each lab to do the tests, then the technician is crossed with the lab. If instead each lab uses two local techs, the technician is nested within the

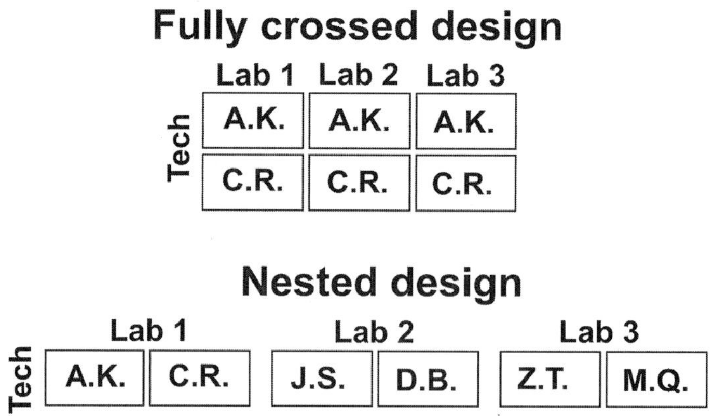

Fully crossed design

Nested design

FIGURE 4.11
Two designs to study experimenter effects on mouse behavior. In the fully crossed design, two individuals (A.K., C.R.) travel and spend time testing mice in each of three labs, whereas in the nested design, data are collected by two or more individuals unique to each lab.

lab (Figure 4.11). The design is still a 3 × 2 factorial, but the variance will be partitioned and interpreted in a different way during the data analysis. In a crossed design, it would be possible to conclude that the difference between technicians varied across labs (a lab by tech interaction) because of the lab environments, whereas in a nested design the same differences in means could easily occur merely because of the unique individuals assigned to test the mice in each lab. One could not know whether the lab environment influenced how the technicians ran the mice through their tests.

Litter as unit

Many treatments applied with mice are administered separately to each individual and then analyzed using the individual mouse as the unit of analysis. This clearly is appropriate when a drug is given as mg/kg body weight, because the volume of solution injected depends on each individual's body weight on the test day. There are situations, however, where the treatment is applied to an entire litter of mice by doing something to the mother either before or after birth but prior to weaning. Suppose a stressor is applied to the mother each day for the last week of gestation, and later the offspring are tested for anxiety-related behaviors at eight weeks after birth. The mother mediates any stress effect on offspring behavior, and the different females in the treatment group may react somewhat differently to the stressor. All the mice in the same litter have the same mother and therefore are not independent observations. In this circumstance, it is perhaps most appropriate to use the mean behavioral test score of the litter at eight weeks of age as the unit of analysis rather than scores of individual mice (Abbey & Howard, 1973). This will reduce the sample size considerably, and a larger number of litters will need to be studied than when individuals are the unit of analysis.

Strictly speaking, litter is a nested factor in a study where treatments are applied to the mother, and it is possible to analyze the data using scores of individual mice and testing treatment effects against the variance among litters. This may require special software, because litter size will not be the same for all mothers. Designs where treatments are applied to entire litters can be addressed with mixed effects models considerably more sophisticated than the usual ANOVA models and can reveal interesting interactions with litter size (Wainwright, Leatherdale, & Dubin, 2007).

All mice are part of a litter that shares the same mother, but it is not necessary to analyze the litter mean in every study. It is quite impossible to do when purchasing mice from a commercial breeder, because a shipment will combine several litters without identifying who is a sibling of whom. This is acceptable for any study where the treatments are applied to individual mice after they arrive in the lab. If mice are bred in the investigator's lab and littermates are known, the individual can serve as the unit of analysis, provided the treatment was not applied to the entire litter as one unit. It is good practice, when litter membership is known, to be sure littermates are distributed across different treatment groups rather than having them all in the same group. In this case, litter could be used as a control factor. Because litter size inevitably varies over a wide range from 2 to 13 pups, a study cannot be perfectly balanced for litter membership unless smaller litters are discarded, which is wasteful. Avoiding confounds of treatment group and litter membership is paramount.

Consulting experts

The design of experiments is not overly complicated in most studies of mouse behavior, but there are situations that call for special design features and advanced analytical methods. Consulting local experts or a more advanced text on design is well advised.

UTILITIES

This book is accompanied by several utilities implemented in the Microsoft Excel program that are useful when planning experiments, especially those with more elaborate designs. Researchers familiar with Excel could create similar spreadsheets without much effort, but the utilities make the process rapid and simple, even for the experienced user. Time is a precious commodity, and the utilities will save time. They also help to avoid clerical and typing errors, or, when such errors occur, make them very easy to repair. Here and in later chapters (5, 7, 9, and 15), the utilities follow several conventions that are outlined in Box 4.2. Each utility for this chapter pertains to a design with a particular number of factors, and it is created with a specific number of levels of each factor, a number sufficient for most experiments. Methods to address a need for fewer or more levels are presented in Box 4.3.

One- and two-factor studies

The utility, **Design 1way**, allows the user to specify the names of 24 separate groups, perhaps several inbred strains of mice. Each name entered into one of the boxes on the first sheet then appears in the appropriate place on the full list of subjects for the study. As with all data entered in Excel, the name can be longer than allowed by the width of the box. Only part of a very long name may appear in the Full list sheet, but all the information will still be there. To expand the view to show the full names, create an empty spreadsheet, then select and copy all the cells from the Full list sheet to the new spreadsheet. Be sure to choose the option Values

---

4.2 CONVENTIONS FOR USE OF UTILITIES

Most utilities are *protected*, indicated by P in the file name. When protected, the user cannot change most cells because they are locked. Most formulas are also hidden from view.

A few utilities are not protected, so the user can make major changes in the number of groups.

The user can enter information only into certain cells. For character data, the cells are highlighted in some salient color. For numerical data, they are always *yellow*.

In most utilities, the Recalculation option should be set to Manual; the user must press F9 to see the results of recently added information.

It is suggested that the user create a folder on the local computer and copy utilities to it. When data are entered into a spreadsheet, it is then wise to save it under a new name to indicate its purpose and save it elsewhere.

4.3 ADAPTING A UTILITY TO DIFFERENT LEVELS

If the utility involves eight levels of factor A but you want to study only six, enter the real names for those six levels and then some extreme letters such as "zz" in the other cells. Copy the Full list to a blank spreadsheet (use the copy Values Only option), sort on factor A, and delete all those with zz. Do the same for any other factor that has too many levels in the prototype.

Likewise, if you want to study fewer than 10 mice per group, copy Full list to a blank sheet and delete the higher replications.

To work with more than the allowed number of levels, make two or more spreadsheets from the prototype, each with the desired names for each level, copy each Full list to a blank sheet, and append them to create one very large list. For example, if a study is to have 20 strains x 3 labs, make three copies of the utility **Design 2way**, entering eight strains in factor A in the first, eight in the second, and four in the third along with four cells having "zz." Enter "zz" in position four for factor B. Combine the three into one large spreadsheet, sort and purge all cases containing "zz."

Only when the small copy options box appears at the bottom right of the pasted material. The name can also include spaces and any character available in Excel, such as α or β. In this and all other Design spreadsheets, the Full list entails 10 mice per group. If fewer or more than 24 strains are anticipated, the approach described in Box 4.3 enables the user to create an appropriate list of subjects. Likewise, fewer or more than 10 mice per group can readily be accommodated with a few quick changes in the spreadsheets.

The utility, **Design 2way**, allows the user to specify names of eight levels for factor A and four for factor B— a total of 32 separate groups.

Three- and higher factor studies

The utility, **Design 3way**, entails eight levels of factor A, four of factor B, and three of factor C — a total of $8 \times 4 \times 3 = 96$ groups, a very large study indeed. Few experiments with mice ever exceed three independent and fully crossed study factors with so many groups. At 10 mice per group, the spreadsheet allows for 960 animals, more than most labs would care to examine in a single experiment. Note that the names for levels of factor A can only be entered into the first row of names. Wherever these are to be repeated, a simple "0" is shown. After a value is entered into a cell of factor A, it appears in the appropriate cell in the other matrices and in the Full list.

The utility, **Design 4way**, combines eight levels of factor A, four of factor B, three of factor C, and two of factor D — a total of $8 \times 4 \times 3 \times 2 = 192$ groups. This behemoth might be used when sex is added to the design as a control factor, for example.

The utility **Design 5way** extends the four-factor design by adding factor E with two levels, bringing the tab to 384 groups, more than most of us will ever encounter in a lifetime of research when study variables are considered. When two of the five factors are control variables, however, this design may be just what is needed for a major experiment. In this and the six-factor situations, the value of the utilities will become apparent.

An example is shown in Figure 4.6 of the Excel spreadsheet **Design Example 5way** where eight inbred strains are tested in four labs after injection of a drug or saline or without any injection. Factor D is shipment with two levels and Factor E is sex, female, and male.

The utility **Design 6way** adds yet another factor F with two levels, most likely another control variable. A study with three study variables and three control variables is quite realistic in many kinds of mouse research. It is highly unlikely that the lab will be able to test 10 mice for each of $8 \times 4 \times 3 \times 2 \times 2 \times 2 = 768$ groups. Even one mouse for each group may be expecting too much, and certain control variables will not be perfectly balanced in the final list of subjects for testing. Nevertheless, the Full list can serve as a convenient place to start the process of balancing and randomization, as described in Chapter 7 on logistics.

Breeding list

The utility **Breeding list** entails 16 combinations of a female genotype mated with a male genotype. This is convenient for implementing the full reciprocal crossbreeding design shown in Table 4.3. It can also be adapted to other breeding schemes involving fewer or more parental combinations.

Sample Size

Once the design and measuring instruments for a study are decided, a sample size must be chosen. A researcher cannot seek ethics approval, apply for a grant, or defend a thesis proposal unless a good reason can be given for testing a specific number of animals. It is not enough to say that other studies of the same general topic used 10 mice per group, so 10 should be used for this study. What if all those other studies were using too few or too many animals? The new study would then be the latest in a long line of weak experiments, if the investigator can persuade two or three committees to approve it anyway.

Many experiments done with lab animals do not estimate sample size correctly, instead relying on crude guesses about the proper numbers to purchase. Others simply divide the total funds available to purchase animals by the number of groups in the design and then by the dollar cost per animal, including shipping, which will yield the number of animals one can afford to test. If this is the method, an elegant research design with many elaborate control groups may be executed with strict discipline but then fail to discover anything of interest. It is far better to conduct a modest study with just a few carefully chosen treatment and control conditions using a sample size sufficient for the task than it is to launch a grandiose assault on the bastions of knowledge with too few troops.

The correct sample size to use in an experiment depends primarily on three things: the design of the study, the size of the treatment effects that one reasonably expects or hopes will occur, and the acceptable level of risk that effects of that size will indeed occur but the statistical tests will fail to detect. Methods to estimate sample size depend on statistical theory. Therefore,

Mouse Behavioral Testing. DOI: 10.1016/B978-0-12-375674-9.10005-9

a brief review of some of the basic principles of statistical inference is needed, just enough to understand the elements of a sample size calculation (Wahlsten, 2007).

UTILITIES TO DO THE CALCULATIONS

Computing sample size from statistical theory can be a major challenge because: (a) the theoretical distributions of values are complex and asymmetrical, and (b) a different formula needs to be used for each experimental design. Methods and tables provided by Cohen (1988, 1992) and Borenstein and colleagues (2001), several books on the topic (Bausell & Li, 2002; Kraemer & Thiemann, 1987; Murphy & Myors, 2004), as well as programs such as Gpower (Erdfelder, Faul, & Buchner, 1996; Faul, Erdfelder, Lang, & Buchner, 2007) or SamplePower (Borenstein, Rothstein, Cohen, Schoenfeld, & Berlin, 2000) can be helpful, but there are many situations where those approaches are not readily applied.

Instead, it is possible to make the calculations with the aid of a series of utilities created with Microsoft Excel, each adapted to a specific experimental design. Calculations of sample size in relation to power are based on a normal approximation to the non-central t distribution (Wahlsten, 1991) for some tests, and a normal approximation to the non-central F distribution (Severo & Zelen, 1960) for more complex designs. With the utilities provided in Chapter 5, the user needs to choose the right utility for the specific application and then enter appropriate information into the yellow boxes in the spreadsheet. All other cells in the spreadsheet are protected so that they cannot be modified from the keyboard, and calculations take place behind the scenes. The user must have a good understanding of the basic concepts involved in power analysis but does not need to know calculus or matrix algebra. This is much like following a recipe in the kitchen. A person does not need a background in organic chemistry to be a good cook, but some knowledge of *why* onions go into the pot before the tomatoes can help to avoid embarrassing and unpalatable errors.

76

POPULATION AND SAMPLE

A fundamental challenge in most studies is that the investigator would like to draw conclusions about an entire population of people or animals but has only limited information on a sample (Figure 5.1). Provided the sample is not biased and is large enough, this process can yield valid answers. What constitutes the population is sometimes obvious. Perhaps the investigator wants to learn something about the health of school children in the United States. That is a very large and geographically diverse population, and obtaining a representative sample will be difficult, but the limits of the population are clear enough. Results of the study of a sample should be pertinent to American children who were not in the study, but they

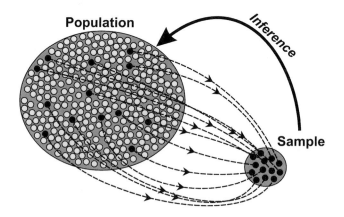

FIGURE 5.1

A sample consists of individuals drawn from a larger population. The members of the sample are tested and measured, and their data are used to draw conclusions about the population.

cannot be generalized to children in Canada, Mexico, Germany, or China because none of them were in the sample. In field studies of animals, there is usually a definable population, sometimes of uncertain and fluctuating size, and the task of sampling it adequately, without major biases, is daunting.

Two kinds of populations

What is the population in a study of laboratory animals? The animals are usually bred specifically for use in a research project. If a unique genetic strain of mice exists only in one lab, just enough animals will be bred to satisfy the needs of the investigators. In the case of a common strain such as inbred C57BL/6 mice, commercial suppliers usually have a large stock on hand because so many labs order them every week. In both situations, however, the population of mice in the statistical sense does not exist as furry critters needing to be fed. Instead, it is a conceptual or hypothetical population, the set of mice of a particular genotype that *could* be bred, should somebody care to do this. Given the proclivities of mice, the population is effectively infinite in size.

Thus, there are two kinds of populations. A *finite population* is made up of identifiable, countable individuals, whereas a *hypothetical population* is a virtually infinite set of individuals that could be produced under given conditions. A *sample* is then a subset of one of these populations. It is always a group of real individuals that can be treated and measured. The fundamental task of statistical inference is to draw valid conclusions about a population on the basis of information on a sample. If the sample is representative and reasonably large, the inferences will usually be close to the truth and the likely margin of error can be estimated.

There is a good reason to inject a tone of uncertainty into this discussion. The members of a population are never all the same. If they were, just a small sample would suffice and the method of assigning subjects to treatment groups would not matter. Instead, individuals differ from one another, and sometimes these differences can be very large. For example, in the 129S1/SvImJ and 129P1/ReJ strains of mice a substantial number of them have no corpus callosum connecting the cerebral hemispheres, while most seem to have an anatomically normal cerebrum (Wahlsten, Metten, & Crabbe, 2003b). In several inbred strains of mice, including C57BL/6, some show a consistent preference for using the right paw, whereas others are left-pawed (Biddle & Eales, 1999; Collins, 1975). On most behavioral tests, the members of a genetically homogeneous inbred strain yield substantially different scores, despite all efforts to control the local environment and conditions of testing. An important consequence arises when sampling from a population that has large differences among its constituents.

Specifying an entire population

If every member of the population could be measured, this would eliminate all the guess work. The mean of the population and the relative frequency of scores far above or below the mean would be known exactly. The true correlation between different measures and the equation of the straight line that best predicts one (Y) from another (X) could be computed with total confidence. Yes, this would be ideal, but researchers never have all the information.

Instead, the population mean and other parameters are represented by hypothetical values, and for this purpose Greek symbols are used. Several are shown in Box 5.1. For example, suppose that the mean score of the population on some test is expected to be $\mu = 100$ and the scores are distributed above and below the mean in a bell-shaped or normal distribution. The dispersion of scores above and below the mean is well summarized by the standard deviation σ. Suppose that $\sigma = 15$. Given the values $\mu = 100$ and $\sigma = 15$, it is then possible to enumerate every member of the hypothetical population. The utility **Normdist 10K** generates a population with N = 10,000 members that has exactly the mean and standard deviation specified by

5.1 SYMBOLS FOR POPULATION PARAMETERS AND SAMPLE STATISTICS

Descriptor	Population	Sample
# members	N	n
Mean	μ	M
Variance	σ^2	S^2
Standard deviation	σ	S
Effect size	δ	d
Correlation	ρ	r

5.2 USE OF STATISTICAL UTILITIES

All Excel utilities mentioned in this book are available at the website for the book. They are arranged by chapter for convenience. A detailed guide to using the utilities is provided for each chapter under the names **Guide_Chapter4**, **Guide_Chapter5**, etc. For utilities that compute power and sample size in Chapter 5, there is another spreadsheet in the Web site entitled **Comparisons** which compares results of calculations using the utilities to several other readily available software packages, as well as the tables from Cohen (1988). In most cases, the utilities provided for this book call for a sample size that agrees exactly or differs by only one or two subjects. They provide a major advantage over tables, because there is no need to interpolate between tabled values.

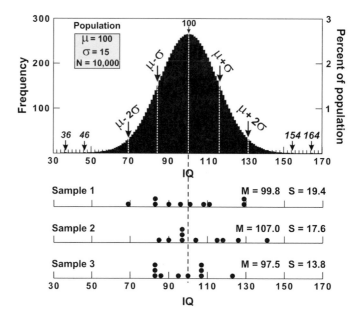

FIGURE 5.2

The frequency distribution shows a population of 10,000 individuals with true mean $\mu = 100$ and standard deviation $\sigma = 15$. Almost all (about 98%) individuals are within 2σ of the mean. Three samples of 10 individuals are then drawn randomly, and their sample means (M) and sample standard deviations (S) differ from each other as well as from μ and σ.

the user (Box 5.2). One such frequency distribution is shown in Figure 5.2. If the user would like to generate some other population, it is an easy matter to enter different values for μ and σ into the boxes and press F9 to calculate a new set of 10,000 scores. These can then be copied to another spreadsheet or a statistical program and portrayed in a graph.

TABLE 5.1 Computing Parameters from N = 10,000 Scores Using the Utility Normdist 10K when μ = 100 and σ = 15, Where Columns A to I and Rows 5 to 10,004 are Labels in Excel

	A	F	G	H	I
	Subject#	X	(X − μ)	(X − μ)²	Z = (X − μ)/σ
5	1	36	−64	4,096	−4.27
6 ...	2	46	−54	2,916	−3.60
4,897 ...	4,893	100	0	0	0.00
6,181 ...	6,177	104	4	16	0.267
10,003	9,999	154	54	2,916	3.60
10,004	10,000	164	64	4,096	4.27
	Totals:	1,000,000	0	2,252,066	0
	Total/10,000:	μ = 100		σ² = 225.2	

A few parameters can be computed to summarize the data. This is done in **Normdist 10K** in the columns to the right of the normally distributed X values. Table 5.1 shows some of the details, omitting the hordes of numbers in the large range of scores. Sums or totals of each column are computed at the bottom of the spreadsheet and can be reached quickly with the shortcut ctrl-end. The sum of all X scores or ΣX is used to find the mean of all N scores: $\mu = \Sigma X / N$. N can be found using the count() function, as shown in the bottom row of the spreadsheet. For a normal distribution, the mean μ will be exactly at its center because it is symmetrical. The next column shows the deviation of each score from the mean $(X − \mu)$. This value is of little use on its own, and its sum over all scores must be $\Sigma(X − \mu) = 0$. It is, however, a means to an important end. Square it to obtain the squared deviation $(X − \mu)^2$. This value will always be positive, and it will be much larger for X values that are farther from the mean. Its total for all scores $\Sigma(X − \mu)^2$ is a very important quantity in statistical work, the sum of squared deviations of X from the mean, which is abbreviated $SSX = \Sigma(X − \mu)^2$. The value of SSX depends on the degree of dispersion of X scores from the mean *and* the number of scores. Obviously, it will be much larger for 10,000 scores than 10 scores. Dividing SSX by N, however, removes its dependency on N and yields another very important quantity, the average squared deviation of X from its mean, which is known as the *variance* or σ^2 for a population, calculated as $\sigma^2 = [\Sigma(X − \mu)^2] / N$. Taking the square root of the variance yields the *standard deviation* σ. This value is a useful descriptor of the dispersion of scores around the mean, because almost all scores will occur in the range from two standard deviations below the mean to two standard deviations above it, or $\mu \pm 2\,\sigma$, as indicated in Figure 5.2. When σ is defined as 15.0, the variance should be 225.00, but it is 225.21 in this example because of small rounding errors in making X an integer. On the far right of the spreadsheet and table is the standard score Z for each X score, where Z is the number of standard deviation units X is from the mean μ, such that $Z = (X − \mu)/\sigma$.

Random choice of a sample

Now draw one score randomly from the population having μ = 100 and σ = 15 created by **Normdist 10K**. To avoid bias, a random number generator in Excel can be used to select the individuals for the sample. The Excel function *rand()* creates a random number between 0 and 1.0, such that every possible value between those limits is equally likely. Multiply this number by the population size N, then round up to obtain the subject number with no decimal places. The formula for this operation is Subject# = *roundup(10000 ∗ rand(),0)*. It is important to round up, because rounding down could yield the non-existent subject # 0. Once the subject number is determined, the corresponding value of the datum X can be found in the

5.3 FIVE RANDOMLY DRAWN SCORES

rand#	index	subject#	X
0.890204	8902.04	8903	118
0.234819	2348.19	2349	89
0.543443	5434.43	5435	102
0.680559	6805.59	6806	107
0.351122	3511.22	3512	94

spreadsheet. It is tedious to scroll down a list of 10,000 scores to find the right one, so using the Find function with subject number can save time. Five examples are shown in Box 5.3. Their sample mean of X scores is M = 102.0, which is a little larger than the population mean $\mu = 100$.

Choosing a random index for subject number and then looking up the X value in a table can be tedious when the hypothetical population has 10,000 individuals. Thankfully, it can be automated in the case of a normal distribution with a quick sleight of hand. First obtain a random number with *rand()*, then use this probability as the variable P of the Excel function *normsinv(P)* that finds the standard normal deviation Z with area P under the normal curve that is less than Z. This formula is Z = *normsinv(rand())*. Next, use the values of μ and σ to generate a random X that is normally distributed: X = μ + Zσ. To make this score a whole number, the function becomes X = *roundup((μ + σ*normsinv(rand())),0)*, where μ and σ are replaced by references to the cell numbers containing those values in the spreadsheet. The procedure can be made even more compact by using the Excel function *norminv(P,mean, stdev)*, where mean and standard deviation are replaced by cell references. The function returns a random normal value with the same measurement scale as the mean and standard deviation. The procedure is then X = *roundup(norminv(rand(),mean, stdev),0)*. These and other Excel functions are applied in the utilities, but they are hidden from view in order to keep things clear and simple for the user.

The procedure is implemented in the utility **1sample 1group** that draws 100 scores randomly from an infinite population having any specified mean and standard deviation. If a random sample of only n = 10 scores is desired, simply use the sample mean of the first 10 scores. The results for the next 90 can just be ignored. To generate a new random sample, simply press F9 Recalculate. Results for three independent samples of n = 10 individuals from a population with $\mu = 100$ and $\sigma = 15$ are shown in Figure 5.2 plotted on the same scale as the entire population. Note that in no case does the sample mean M equal the population mean $\mu = 100$. The first sample is very close, with M = 99.8, but this is clearly fortuitous because the next sample has M = 107.0, a much larger deviation from μ. In each case, the difference between the sample mean and the true mean arises from *sampling error*. Perhaps "error" is a bad term, because no mistake was made. The difference M − μ is an inevitable consequence of random choices of individuals to be in the samples and the fact that the members of the population differ widely.

The **1sample 1group** utility can demonstrate a fundamental property of a random sample: the larger the sample size, the closer the sample mean M will be to the true mean μ and the smaller the difference M − μ. This rule applies to the average of several independent samples, however. For a single sample, M − μ could be larger than M − μ for another sample that has a somewhat greater sample size. The outcome for any one sample depends on the luck of the draw. Only after drawing many such random samples does a pattern emerge in the data.

To see just how much the sample means will depart from the true mean μ when samples are large and small, run **1sample 1group** over and over many times. For example, the first

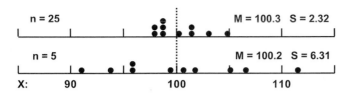

FIGURE 5.3
Ten samples are drawn randomly from the population in Figure 5.2. In one series of 10 samples, there are 25 individuals in each sample, whereas in the other series there are only 5. The frequency distributions show the scatter of sample means around the true mean of $\mu = 100$. In both cases, the mean of the sample means is very close to 100, but the dispersion of sample means is much greater when $n = 5$.

10 samples might have $n = 25$ and the second 10 might use $n = 5$. Results for this type of series are shown in Figure 5.3. The means of the two sets of 10 samples are almost identical (100.3 vs. 100.2), but the dispersion of means is considerably greater when n is smaller. When another series is run, the specifics will differ but the same general trend should be evident. The fact that both sets had means very close to 100.0 in this example is just a lucky happenstance. The means of 10 new means will almost certainly differ by a greater amount.

When first learning about sampling, it is a valuable exercise to do these kinds of experiments with sample size by repeating a run over and over to generate several independent samples. Before long, however, this will become tedious. A modification of the spreadsheet enables the user to obtain 100 random samples in a flash. The utility **100samples 1group** does this for a population with any desired mean and standard deviation. Suppose $\mu = 25$ and $\sigma = 5$. For each random sample, 50 scores are obtained and the sample mean is found for the first 5, 10, 15, 20, 25, and all 50 scores. To compare the dispersions of sample means for these six sample sizes, recalculate the utility with F9 six times (see Box 5.4) and copy each result to another spreadsheet. The 100 samples, each with a different n, can then be imported into a statistical program and graphed. Results are shown in Figure 5.4.

The copying operation can be done in Excel by first selecting or highlighting the desired values in a spreadsheet, then giving the Copy command (ctrl-c), then placing the cursor into a cell of the new spreadsheet and giving the Paste (ctrl-v) command. A small menu will appear at the bottom of the pasted material. Open it and choose Values Only. If this is not done, the formulas will also be copied to the new sheet, which will impede further calculations or import into a data analysis program. Another way is to right click on the highlighted material, choose the Paste Special option, then select Values.

From Figure 5.4 it should be obvious that larger samples have less dispersion around the population mean, whereas all sample sizes yield values centered around the true mean. For each distribution of sample means (M), the standard deviation of the sample means (S_M) is smaller for larger samples. This standard deviation of sample means is better known as the *standard error of the mean* or SEM. In theoretical terms, its expected value has a very simple relationship with the population standard deviation σ: $\text{SEM} = \sigma/\sqrt{n}$. Note that for the example in Figure 5.4, the true $\sigma = 5$, whereas for samples $n = 25$, $\text{SEM} = 5/5 = 1.0$.

5.4 INDEPENDENCE

The utility **1sample 1group** and several other utilities compute means of a random sample for several different sample sizes. Because a single run of the program generates values by finding the mean cumulatively, the sample means for different sample sizes in a single run most emphatically are *not* independent. To compare patterns for different sample sizes, it is mandatory that *different runs* of the program be done for each sample size.

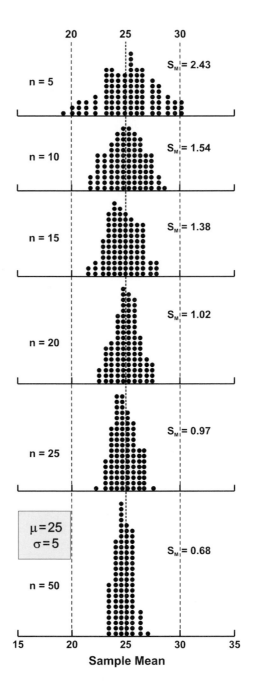

FIGURE 5.4

One hundred samples of various sizes drawn randomly from a population with μ = 25 and σ = 5. In every case, the mean of the sample means is close to 25, but those with larger sample size have most means much closer to the true mean. The standard deviation of the sample means (S_M) is also known as the standard error of the mean.

The sample value of $S_M = 0.97$ is very close to expectation. The utility states the theoretical value in the column next to the observed S_M.

In practice, the true mean μ and the true standard deviation σ are not known, and there is only one sample, so SEM must be estimated from the sample standard deviation S: SEM = S/sqrt (n). The SEM provides an excellent indicator of the precision of the estimate of the true mean on the basis of the sample mean, because, as a standard deviation, almost all values of sample M will occur in the range μ ± 2 SEM. This is why a bar chart of group means often is shown with a bracket to represent the standard error of the mean. It provides a convenient reminder of just how much the true mean might depart from the sample mean M, given the sample size n. Perhaps the brackets should be drawn at 2 SEM above and below the sample mean. By consent among scientists, however, the field has adopted the narrower range that emphasizes the most common result, rather than a wider range encompassing greater errors that are present in many studies.

Just as the sample mean M is subject to sampling error, so are the sample standard deviation S and variance S^2. The utility **100samples 1group** also does for the variance of a sample what it does for the mean. If results are compared for sample mean and variance, one noteworthy difference will be seen: the distribution of the sample mean tends to be close to a bell-shaped, normal distribution, whereas the distribution of variances is skewed to the right and has a lower limit of 0. If the individual scores are normally distributed, the variance will have a χ^2 distribution. By trying a few experiments with this utility, the user should come to understand that the standard error of the mean estimated from just one sample is subject to sampling error and could deviate by a considerable amount from the true value. Thus, those ubiquitous standard error brackets provide no guarantee that the true mean is that close to the sample mean.

At this point, several principles demonstrated with the utilities can be summarized:

1. The mean of a random sample is usually not equal to the true mean of the population.
2. Even for a rather large sample, the sample mean can deviate considerably from the true mean. Large deviations are less likely for a large sample size, but they are still possible.
3. When many samples of a particular size are drawn randomly from a population, the mean of those sample means tends to be close to the population mean, regardless of the sample size. Small samples do not bias the estimate of the mean; they just make it less precise.
4. The dispersion or variability of sample means around the true mean tends to be less when the sample is larger, and the standard deviation of the sample means is close to $\sigma/\sqrt{n}$. Hence, the precision of an estimate of the true mean depends strongly on the sample size.

Comparing two groups

Rarely is a researcher interested in just one group in an experiment. There is almost always an experimental treatment group and some kind of a control or comparison group, at the very minimum. A typical study compares the means of the two groups and uses this difference to help decide if the treatment really did anything worthy of note. A really large difference between their means is a good sign that the treatment worked, while a trivially small difference suggests the treatment did little or nothing. In the great majority of real experiments the difference is not exceptionally large and things are not so clear.

This situation can be addressed with another utility (**1sample 2groups**) that draws random samples from *two* separate populations with any desired mean and standard deviation, then finds the difference between the sample means, and does this for several different sample sizes. Experiments with this utility will be instructive, but the process can be accelerated using the utility **100samples 2groups** that generates a sampling distribution instantly. The results for individual samples are provided under the **diffs** tab.

First run the utility for a scenario where the group means are *identical* at $\mu = 25$ with $\sigma = 5$, such that the true difference (Δ) between the population means is $\Delta\mu = 0$. This would be obtained if an experimental treatment had absolutely no effect on the mice. The distributions of sample differences between the group means are shown in the left panel of Figure 5.5 for four sample sizes (n = 5, 10, 20, 50). Each frequency distribution represents 100 random samples. Next, run another series of tests with a difference of $\Delta\mu = 2.5$ between the group means, and then a further series with a difference of $\Delta\mu = 5.0$. The gray zone in each distribution shows the number of sample differences between the two sample means that are negative, such that the group with the smallest true mean has the largest sample mean.

83

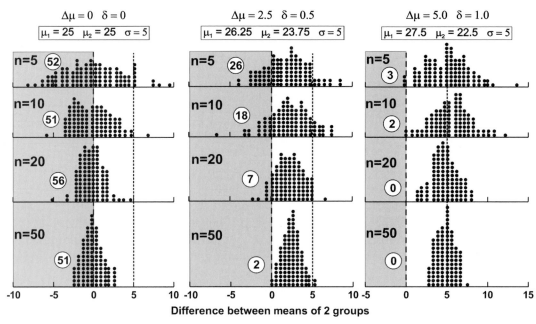

FIGURE 5.5

Random samples are drawn from two populations and the difference between the two sample means is found. This is done 100 times independently for sample sizes ranging from 5 to 50 under three models of true means. In the left panel, the means of the two populations are identical. Nevertheless, the differences are widely dispersed around 0 for small samples and half of them are in the gray zone where the difference is negative, regardless of sample size. In the middle panel where the populations differ by 2.5 units and express effect size $\delta = 0.5$, the number of samples in the gray zone is strongly influenced by sample size. In the right panel where the populations differ by 5.0 units and express effect size $\delta = 1.0$, very few samples are in the gray zone, regardless of sample size. Circled numbers are the number of samples with a difference less than 0. The heavier dashed line represents no difference between sample means and the finer dashed line indicates a difference of 5 units.

As expected, when $\Delta\mu = 0$, about half are >0 and half <0. When sample size is small, many differences are a considerable distance from 0. For the middle case when $\Delta\mu = 2.5$, the number of samples with a difference less than 0 depends very strongly on sample size. For a larger true difference $\Delta\mu = 5.0$, most samples show a positive difference, even when sample size is rather small. This figure shows quite vividly how the results in a sample depend strongly on two quantities: the true difference between the two group means and the sample size.

Comparing several groups

Sometimes a study involves more than two groups. For this situation, another utility (**1sample 8groups**) draws random samples from eight hypothetical populations having any specified means and standard deviations. It then finds the mean of the sample means as well as the standard deviation among those means. It does this for every possible sample size from 2 to 100 in each group. Thus, it is a relatively straightforward exercise to draw a series of samples when population means are quite different, and then repeat this for a series where the population means are identical. Looking at the pattern of eight means for a single run with one sample size, one can perceive how drastically or minimally they may depart from the true values. Figure 5.6 shows results for five random samples with n = 10 under two scenarios: eight inbred strains whose true means are identical at $\mu = 25$ units, each with $\sigma = 5$, and eight strains with means 1 unit apart for adjacent strains. There is considerable variation among the means when all strains have the same true mean. This variation would be less if a larger sample size were used. In the case where the strains really do differ considerably, there are many striking departures of strain rank orders from the true order of means. In this specific example, strain H, with the highest true mean, always exceeds strain A with the lowest mean, but it

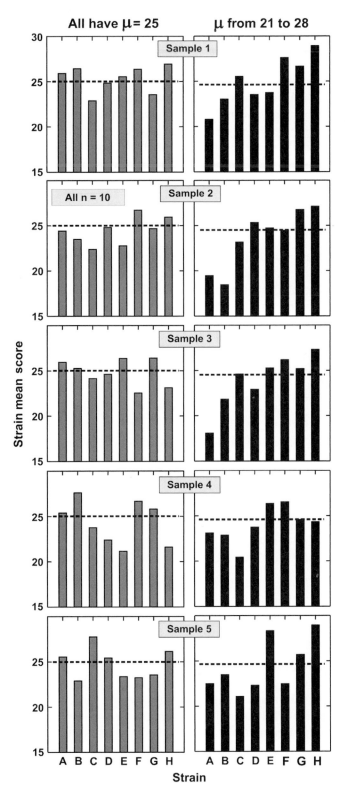

FIGURE 5.6

Random samples of n = 10 mice per strain are drawn from eight mouse strains. In the series on the left, the populations have identical true means of 25 units. Nevertheless, there is considerable variation among group means in five independent samples. In the second case on the right, true means of the eight groups are spaced one unit apart (21, 22, 23, 24, 25, 26, 27, 28, 29, 30), such that the mean of the eight strain means is 24.5 units. A general trend of increasing means can be perceived in the five samples, but there are several instances where the rank orders of strain means do not follow the order of the true means.

sometimes has a lower sample mean than strain E in the middle range. Again, with a larger sample size, rank orders of strain sample means would agree more closely with the true values.

By experimenting with various settings in these utilities and running them with different sample sizes, an appreciation should be gained of the profound influence of sample size on the departure of sample means from the true means of the populations. Clearly, larger samples generally provide better estimates, and variations in sample size have a major impact when the true differences between groups are not very large. The big question is how large samples really need to be for most purposes. This can be vaguely perceived by looking at many graphs created under a wide range of models, but there is an easier way to find the correct answer directly.

SIZE OF AN EFFECT

It makes good intuitive sense that a small treatment effect can only be detected with a high probability when using a large sample. A similar principle is valid in all areas of science. To peer further into the cosmos and see smaller objects, one needs a larger telescope. To visualize really small subcellular structures, a very large microscope is required, such as an electron microscope with a big magnet. To detect small trends of change in Earth's climate, measures from many widely dispersed geographic regions over many years are essential. To go beyond intuition inspired by graphic examples, a numerical index of the size of an effect that might reasonably occur in our experiments is needed. Only then can the required sample size be calculated.

What is an effect?

To decide on a good index of effect size, researchers must agree on what is meant by an effect. In experimental work with mice, this almost always refers to a change in some characteristic, some feature of the animals' behaviors or brains, that is caused by a specific treatment such as a mutation or a novel environment. Cause is a strong word. Proving that A causes B, and is not merely a correlate of B, or that both are caused by some third factor C, is no easy task. To infer causation, one must use an experimental design where some individuals receive the treatment, while others, who are comparable in every other way, do not. This is the essence of a good experiment: a treatment group and a control group differ in only one respect.

Great care needs to be taken so that all kinds of potential influences are equated through proper balancing and randomization of the study (Chapters 4 and 7). The magnitude of an effect cannot be demonstrated with just one group of individuals. The pre- versus post-treatment design is a good way to track changes in the behavior of an individual, but it cannot reveal what specifically is attributable to the treatment. After all, individuals change from one trial to the next or one day to the next because of habituation and other kinds of learning, lagging motivation, sensitization, drug tolerance, and so forth. To interpret a pre- versus post-treatment difference, it is important also to see two measures on individuals who did not receive the treatment in the period between the two tests. The conclusion is inescapable: the impact of a treatment is apparent from the difference between treated and control groups.

Index of effect size

The magnitude of an effect can be expressed either in units appropriate for the specific test or in relative units that compare one group of individuals to others. Perhaps a drug decreases latency to fall off the accelerating rotarod by 10 sec, or 24 h of food deprivation decreases body weight by 3 g. For experts intimately familiar with those phenotypes in mice, the original units of measurement will convey a good idea of the size of a treatment effect.

There are two shortcomings to the use of real units. First, considerable expertise in a phenotypic domain is needed to interpret most measures. Suppose an article states that an anxiolytic drug increases percentage of time spent in the open arms of an elevated plus maze by 5%. Is that a large or small amount? Only an expert would know. Furthermore, the judged size of the effect would depend on the baseline level of open-arm exploration in the particular lab. Second, data from many kinds of tests are incommensurate. Response latency in seconds and volume consumed in milliliters entail different measurement scales that cannot be transformed into one another.

5.5 N FOR POPULATION, n − 1 FOR SAMPLE DATA

The variance for all N scores in a population is $\sigma^2 = [\Sigma(X - \mu)^2]/N$, whereas for a sample of n scores it is $S^2 = [\Sigma(X - M)^2]/(n - 1)$, where M is the sample mean. Why the n − 1 instead of N? Recall that M is an estimate and will not be the same as μ. M is based on the same scores that are used to compute S^2, which tends to make S^2 a little smaller than σ^2. The bias is corrected by dividing the SS in the numerator by n − 1.

The solution to this conundrum is to compare the *difference between groups* to the *variation within groups*. The change caused by a treatment can be compared with the variation within a group of individuals who are treated the same way, and ubiquitous individual differences that are *not* caused by the treatment. A relative indicator does not depend on the measurement scale and can be used to compare a wide range of phenotypes. It has the added advantage of being closely related to several common kinds of statistical tests that are used to analyze data. The specific kind of indicator depends on the design of the study. An excellent discussion of different effect size indicators is provided by Nakagawa and Cuthill (2007).

Effect size for a study with two groups

When there is only a control and an experimental group or perhaps two independent groups such as males and females, the difference between groups 1 and 2 can be represented by the difference in their mean scores ($\Delta\mu = \mu_1 - \mu_2$ for a hypothetical population, $\Delta M = M_1 - M_2$ for a real sample). The variation within groups (Box 5.5) can be expressed as the standard deviation (σ for the population, S_1 and S_2 for the samples, usually (Box 5.6) combined into a pooled estimate for the two groups or S_{pooled}). A good indicator of effect size is then the *number of standard deviations by which group means differ*, or in the case of small effect sizes, the fraction of a standard deviation. The effect size indicator for a population is $\delta = \Delta\mu/\sigma$, whereas for two samples it is $d = \Delta M/S_{pooled}$. This index is widely understood in behavioral neuroscience. It is common in meta-analysis (described on page 90) and sample size estimation. For the examples shown in Figure 5.5 where $\sigma = 5$, the population effect size is $\delta = 0$ when $\Delta\mu = 0$, $\delta = 0.5$ when the groups differ by 2.5 units or one-half σ, and $\delta = 1.0$ when $\Delta\mu = 5.0$.

There are widely accepted standards for small, moderate, large, and very large effect sizes δ or d in mouse research (Figure 5.7). Generally speaking, an effect size of one standard deviation is considered large while $d = 0.5$ is a small effect (Wahlsten, 2007). The standards encountered

---

5.6 POOLED ESTIMATE OF VARIANCE

If the variance in group 1 is S_1^2 and in group 2 is S_2^2, and if these sample values estimate the same true variance σ^2, then the best estimate of σ^2 is obtained by pooling the sums of squares for the two groups using $SS_1 = (n_1 - 1)S_1^2$ and $SS_2 = (n_2 - 1)S_2^2$. Then $S^2_{pooled} = (SS_1 + SS_2)/(n_1 + n_2 - 2)$ and the pooled standard deviation S_{pooled} is the square root of the variance.

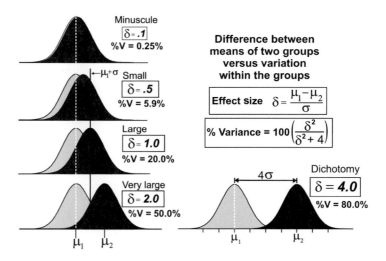

FIGURE 5.7

Standards for effect size δ that compares differences between two group means to the standard deviation of scores within the groups. For each case, the percentage of total variance in scores attributable to the difference between group means (%V) is shown.

in psychological studies of humans are somewhat lower because of the greater difficulty in controlling sources of within-group variance (Cohen, 1988, 1992). Very small effect sizes of δ = 0.1 are sometimes encountered in the literature. Gender differences in mathematical ability are even less, and some authors regard these as "trivial differences" (Hyde, Lindberg, Linn, Ellis, & Williams, 2008). It is now apparent that common genetic polymorphisms influencing human intelligence in the normal range must have effect sizes of δ <0.1 (Dreary, Spinath, & Bates, 2006). In the realm of mouse behavior genetics, such small effects would indeed be seen as trivial, but Figure 5.7 uses the less pejorative term "minuscule."

Effect size for more than two groups

If a study has J groups, the hypothetical means are μ_1, μ_2 ... μ_J, and the standard deviation within groups is σ_{within}. The magnitude of variation between the groups is then the standard deviation of the J group means, termed $\sigma_{between}$. Cohen (1988) has proposed effect size $f = \sigma_{between}/\sigma_{within}$, and his extensive tables provide guidance on choosing sample size for different values of effect size. The utilities employed here also use Cohen's f when there are more than two groups.

It is sometimes helpful to express matters in terms of variances (σ^2) rather than standard deviations (σ). Suppose there is a set of scores X_i for an entire population of N individuals distributed across J groups. When squaring each difference from the true mean to obtain $(X_i - \mu)^2$ and adding them, the result is the *sum of squared deviations from the mean*: SS = Σ $(X_i - \mu)^2$. The sum of squares for the entire population is SS_{total}. Now, perform a bit of algebraic magic. The group mean is μ_j for each individual in group j (where j ranges from 1 to J). Add and subtract this constant to $(X_i - \mu)$ and rearrange terms a little: $(X_i - \mu) + \mu_j - \mu_j = (X_i - \mu_j) + (\mu_j - \mu)$. Thus, the deviation of one individual's score X_i from the population mean can be divided into two components, its deviation from its group mean and the deviation of its group mean from the grand mean. Now square the expression $[(X_i - \mu_j) + (\mu_j - \mu)]^2$ and add it up over all N individuals. The result, after a bit more algebra, is very important: $SS_{total} = SS_{between} + SS_{within}$. When between- and within-group differences are independent, it can also be show that variances add to total variance: $\sigma^2_{total} = \sigma^2_{between} + \sigma^2_{within}$.

A handy indicator of effect size is the proportion of total variance attributable to the variation among group means: $\sigma^2_{between}/\sigma^2_{total}$. For the kinds of experiments typically done with lab

5.7 STANDARDS FOR EFFECT SIZE

Two Groups	δ	ω^2
Very small	0.25	1.5%
Small	0.50	6%
Medium	0.75	12%
Large	1.0	20%
Very large	1.5	36%
J Groups	**f**	ω^2
Very small	0.10	1%
Small	0.20	4%
Medium	0.35	11%
Large	0.50	20%
Very large	0.70	33%

mice, this ratio can be symbolized as ω^2. When there are more than two groups, it has a neat relationship with Cohen's f: $\omega^2 = f^2/(1 + f^2)$. For a study with just two groups, $\omega^2 = \delta^2/(\delta^2 + 4)$. Box 5.7 shows the relations among these indicators for small, medium, and large effect sizes. For either situation, a large effect is one where the difference between or among the groups accounts for about 20% of the total variance among the N individuals. An effect of that kind still leaves 80% of the total variance unexplained by the treatment conditions. Perhaps 20% does not seem like a big thing, but years of experience with research using animals teaches that many interesting kinds of treatments do not have spectacular effects, yet they are meaningful. If we took seriously only those effects that could account for more than 50% of the variance, our field would be dishonored by a crude array of sledgehammer treatments.

Finding effect size from published data

To determine the correct sample size for use in a future study, the likely or desired effect size must be estimated from prior information. The best guide is probably data from a preliminary study of the same phenotype of the same mouse genotypes receiving the same treatments in the same lab. Alternatively, a similar study may have been reported in the literature. It is possible that a published article cited values of d or ω^2 in a clear and concise manner, so that they can be adopted for planning the future study. If not, there are several ways to find these quantities from sample data.

1. When the article provides a table of means and standard deviations or SEMs for each group, these can be entered into the utility **Effect size from article**.
2. If the article shows a bar chart with standard error bars, then means, standard errors, or standard deviations can be measured directly from the graph and results entered into the utility **Effect size from article** (Box 5.8). This can sometimes be achieved by scanning, enlarging, and printing the graph, then measuring lengths with a ruler. If the article is available in pdf format, the measurement feature of Adobe Acrobat can be used to enter the lengths in the utility.
3. If all descriptive statistics are lacking in an article but there are several statistical tests reported, effect size can often be determined from even this sparse information using the utility **Effect size from article** when there is a t test on two groups or an F test on more than two.
4. Effect size is subject to sampling error, and the estimate of d for two groups from a single article may not be the best guide to the choice of δ for a sample size calculation. When several published articles are available that addressed the same general topic and

5.8 FINDING MEAN AND SEM FROM pdf

When a pdf file for an article is available, Adobe Acrobat can measure a chart directly or values can be plugged into **Effect size from article**. An example is shown in Figure 5.8 for Figure 5a from Singer, Hill, Nadeau, & Lander (2005). Steps are as follows:

1. Open the pdf file and zoom the graph as far as possible to enlarge it.
2. Go to Tools—Measuring—Distance tool to start the length measurement.
3. Place the cursor on one end of the Y axis scale and then double click the other end. Length in inches or centimeters will appear. Enter this value into left panel of the Scale ratio box.
4. In the right panel of Scale ratio, enter the number of units using the Custom option.
5. Proceed to measure all other points in the Y dimension. Their values in real units will be shown on the screen.
6. Use the Acrobat Tools—Select—Snapshot tool to copy the graph with measures to the clipboard and then paste this into PowerPoint or a photo editing program to save and print it.

employed similar measures, *meta-analysis* can be used to obtain a better estimate of δ than from a single instance (Borenstein, Hedges, Higgins, & Rothstein, 2009; see also www.meta-analysis.com). The **Effect size from article** utility also implements a meta-analysis for a series of studies that examined two independent groups. It yields not only the best overall estimate of δ but also a range of plausible values. The Q statistic assesses the extent of heterogeneity among the studies. If the P value for Q is quite small, this suggests that some of the studies were fundamentally different from the others. In such a case, rather than adopt the combined estimate of δ, the user is advised to explore why there is heterogeneity and then identify a subset of studies most like the one planned.

SIGNIFICANCE OF AN EFFECT

In scientific journals tests of statistical *significance* with an accompanying P value are often encountered. This word has created confusion and befuddled the minds of generations of students and professionals alike who struggled to comprehend the twisted logic of the statistician. The word is terribly misleading, because it appears to be an index of effect size. Doesn't the language suggest that a large effect will be highly significant and a very small effect be perhaps insignificant? Nevertheless, a very small effect can be highly significant and an apparently large effect can be non-significant, depending on the sample size. Significance in the technical sense is a probability, not an effect size.

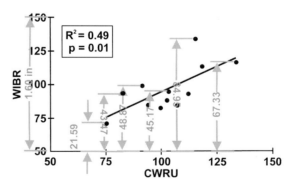

FIGURE 5.8

Measurements of group means in the Y dimension using the Adobe Acrobat Measure tool. Gray arrows were generated by Acrobat. The measure of the Y axis (1.60 in. equivalent to 100 units of Y) established the scaling factor. A series of measures in the X dimension was also done but is not shown in the example. R^2 and p values are from the original report based on strain means (black dots) observed at Case Western Reserve University and the Whitehead Institute for Brain Research. *(Based on Singer, Hill, Nadeau, & Lander, 2005. Redrawn with permission.)*

When sample size is small, the value of a sample mean can be a considerable distance from the true population mean (Figure 5.4). Likewise, with small samples the difference between means of two groups can be surprisingly large when their population means are identical (Figure 5.5). Unless very large sample sizes are studied, data analysis entails a certain amount of guess work. Formal statistical tests help to make these guesses in a way that keeps the rate of errors reasonably low. What kinds of errors are these?

Scientific research with mice compares reasonable hypotheses about the origins of specific results. If one group receives a treatment while the other group serves as a control, then one hypothesis is that the treatment has a real effect, and the effect causes the group means to differ. A logical complement to this hypothesis is that the treatment has *no* effect, and the difference between group means arises merely from sampling error. Representing the true group difference as $\Delta\mu$, the *Null hypothesis* of no effect is $H_0: \Delta\mu = 0$ or $\delta = 0$. Notice that the null is a very specific numerical hypothesis. The data that could occur when the null hypothesis is true are shown in Figures 5.4, 5.5, and 5.6.

Because a substantial difference in group means *could* occur when the null hypothesis is true, it would be foolish to reject it unless the difference in means is quite large. A standard is needed to aid the decision about the null. Statisticians base this decision on a probability: the probability of making a *false positive discovery*. Here the word "discovery" means the investigator concludes that a treatment had a real effect, hence something was discovered.

Statistical inference is a guessing game. Figure 5.9 shows the flow of logic in a test of hypotheses in a typical study with mice. It is not possible with conventional methods to prove directly that a treatment had an effect. Instead, the null hypothesis that there was *no* effect is evaluated. If it can safely be rejected, then the treatment must have had some effect. Obviously, before the data are collected, the researcher really does not know whether the null is true or the treatment will have an effect. After the data are in hand, something must be decided about the null: reject it as not credible or retain it as plausible. If the difference between control and treatment groups is fairly large, the null may be rejected. Most likely this will be the correct thing to do, but there is a risk that it is wrong; the difference *could* have arisen from the insidious presence of sampling error, a fortuitous assignment of exceptional mice to the two groups, and the null might actually be true. If so, the decision to report a discovery is a false positive and the error of inference is termed a *Type I error*. A false positive can occur only when the null hypothesis really is true. If there really is an effect, a Type I error cannot possibly be committed.

Statistical tests of significance such as the t test comparing two groups or the F test on several groups using the analysis of variance (ANOVA) provide convenient ways to compute the probability of a Type I error, given the data that were actually obtained. Methods for conducting diverse tests are explained in detail in many formal statistics courses and texts. For the present purpose, at the planning stage of a research project, it is sufficient to state that, in the parlance of mathematicians who invented the statistical tests in use today, the statistical *significance* of a group difference is the probability of committing a Type I error. The researcher has total control over the probability of a false positive. The acceptable risk of a Type I error is

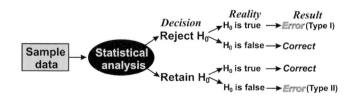

FIGURE 5.9

Logic of inference in a study where the null hypothesis, H_0, that a treatment has no effect is evaluated with a statistical test. Two kinds of errors can occur, depending on whether the null is actually true or false.

set before the data are collected. Furthermore, a Type I error is just as likely with a very large sample as it is with an embarrassingly small sample.

How low a risk is low enough? For many investigators, an acceptable risk is 1 chance in 20 of making a mistake. Symbolizing the probability of a false positive as Greek α, the criterion is $\alpha = 0.05$. As a conditional probability, this is $\alpha = P(\text{Reject null} \mid \text{null is true})$. The value of α determines how large a difference in sample means $\Delta M = M_1 - M_2$ there must be to reject the null that $\Delta\mu = 0$ or $\delta = 0$. The test statistic t has a very neat relation with effect size $d = \Delta M / S_{pooled}$ when there are n_1 and n_2 subjects in the two groups: $t = d \sqrt{[n_1 n_2 / (n_1 + n_2)]}$. The theoretical t distribution is centered around 0, and the shape of the distribution depends solely on a quantity known as degrees of freedom (df) that in turn depends on sample size, which in the case of two groups is $df = n_1 + n_2 - 2$.

A specific t distribution is portrayed in Figure 5.10 for a situation where both groups have a small sample size of $n = 10$ mice each and $df = 18$. In that case, the equation for t is simply $t = 2.24d$ or $d = 0.45t$. The equivalent sample d for various t values is shown in the graph. If the researcher has a good reason to expect that the mean of group 1 will exceed that of group 2 in a situation where the treatment works, only positive values of d and t need to be considered (one-tailed test). The value $t = 1.734$ has 5% of the area under the curve to the right of it, such that $P(t > 1.734) = 0.05$. That t value is equivalent to a sample $d = 0.775$, a moderately large effect according to the standard in Box 5.7. Thus, if the obtained d is greater than 0.77, the researcher will reject the null hypothesis and conclude that there was an effect. If there is no reason to expect one group to exceed the other, then the area under the curve needs to be divided in two, one portion on the upper and the other on the lower tail, which then requires

92

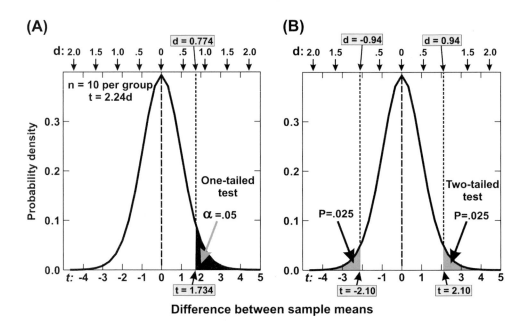

FIGURE 5.10

Two t distributions that portray the probability of observing various magnitudes of difference between means of two groups when the null hypothesis is true. (A) In the case on the left, only positive deviations from 0 are considered plausible, whereas (B) the case on the right divides probability between two kinds of deviations, + and −. The upper scale shows values of sample effect size d that correspond to values of the test statistic t when sample size is $n = 10$ per group. The critical value of d required to reject the null hypothesis is $d = 0.774$ for the one-tailed test in (A) and $+0.94$ or -0.94 for the two-tailed test in (B).

the t statistic to be greater than 2.10 or less than 2.10, or d to be beyond the limits ± 0.94 in order to reject the null.

Again it should be emphasized that *significance* of a group difference is not at all the same as *size* of a treatment effect. Suppose that instead of 10 mice per group, the researcher employed $n = 100$ per group. The equation for t would be $t = 7.07d$ and $d = 0.14t$. When $df = 98$, the critical value of t with 5% of the area under the curve to the right would be $t = 1.66$, which corresponds to $d = 0.23$, a very small effect. If the researcher found that $d = 0.24$, the null would be rejected and a "significant" treatment effect would be concluded.

False positives happen

The choice of Type I error probability or the false positive rate is very important. The higher the value of α, the more false positives will be detected when the null is true. False positives do occur and sometimes are published. A vivid example was presented recently when investigators used functional magnetic resonance imaging (fMRI) to assess the response of the brain of a dead salmon to visual stimuli. They applied a standard method of analyzing fMRI data to find especially active regions of a living brain. Remarkably, the brain of the dead salmon "responded" to a signal in a manner that would often prompt other researchers to conclude that a human subject could see a stimulus. The point of their demonstration was that criteria for detecting a noteworthy response by the brain are often too lax and give rise to a ridiculous number of false positive results. The original poster at the Human Brain Mapping Conference in June of 2008 is available at http://prefrontal.org/blog/2009/06/human-brain-mapping-2009-presentations/, and a fascinating story of the origins and aftermath of the study is available at http://prefrontal.org/blog/2009/09/the-story-behind-the-atlantic-salmon/.

The field of human behavioral genetics has witnessed a remarkable number of false positives. Several years ago, a report was published claiming that a polymorphism in the insulin-like growth factor receptor 2 gene (IGF2R) was related to human intelligence (Chorney et al., 1998). The report was heralded in the media as the first conclusive proof of a genetic influence on IQ. In Canada *The Globe and Mail* (Nov. 17, 1997) lauded this as "the first specific gene for human intelligence" and speculated that the finding "could settle the debate about whether genetics or education and lifestyle determine human intelligence." Other investigators were unable to replicate the finding, however, and the claim quietly descended into obscurity. More recently, five large-scale scans of the human genome for markers associated with IQ differences failed to replicate virtually every prior claim in the published literature of a specific gene effect on IQ in the normal range (Posthuma & de Geus, 2006).

Another widely cited study of interaction between stressful events in childhood and the severity of depression in adults in relation to genotype at the serotonin transporter (5-HTTLPR) gene (Caspi et al., 2003) has now been refuted by a meta-analysis of several studies, most of which failed to replicate the original finding (Risch et al., 2009).

In some fields of science, an inability to replicate a prominent finding may provoke speculation about cheating, fraud, or poor methodology. In a recent case, a study published in *Science* in the field of protein biochemistry was subsequently retracted by the authors after failure to repeat the results of an experiment (Zhang et al., 2004). Unproved accusations of data faking led to a threat of suicide and denial of tenure to one author (Service, 2009). If colleagues in behavioral genetics viewed failures to replicate in the same way, not many of us would survive or retain tenure to retirement age.

Whenever research is done with samples of animals that differ greatly in a population and researchers rely on statistical significance tests to guide decisions, they must live with the ever-present danger of false positive results. Some promising results will never be replicated. This could happen because of deliberate deception, but random sampling probably accounts for the vast majority of failures to replicate. This problem can be attenuated to some extent by

93

adopting a more stringent criterion for Type I error, such as $\alpha = 0.01$ or even 0.001. If the risk of a false positive conclusion is only 1 in 1,000, the literature will not harbor many such claims. Unfortunately, adopting $\alpha = 0.001$ sets an almost impossible standard for proving that an effect occurred. Consider the example with n = 10 mice in each of two groups. When $\alpha = 0.05$ with a one-tailed test, the results must indicate an effect size of d > 0.77 to justify a claim of a significant effect. With $\alpha = 0.001$, the t ratio would need to be 3.61 and the null would be rejected only if d > 1.62, a very large effect; more than twice as large as needed when $\alpha = 0.05$. This option can generate other kinds of errors, such as failing to detect an effect that is very real but not huge.

The more tests, the higher the risk of a false positive

Consider a realistic scenario where several independent research groups are studying the same phenomenon, perhaps working with the same gene knockout on tests of memory. If two groups both use $\alpha = 0.05$ to guide a decision, and if the null that the knockout has no effect is true, then the probability that both groups commit a Type I error is comfortably small, 0.05^2 or 1 in 400. The probability that one group does *not* make a Type I error is $1 - 0.05 = 0.95$, so the probability that *neither* group makes a Type I error is $0.95^2 = 0.902$ by the multiplication rule for independent events. By the rule of complementary events, the probability that one or both groups makes a Type I error is then $1 - 0.902$ or about 10%, double the rate for just one group. If α is the standard adopted in each lab and there are L labs working with the same mutation, the probability that at least one lab makes a Type I error is $1 - (1 - \alpha)^L$. In the highly competitive field of neuroscience, it would not be surprising to find five labs working with the same mutation, and the probability that at least one would "discover" a non-existent effect is 0.23. Thus, in fields of study where there are likely to be several intrepid knights in pursuit of the same scientific grail, it would be wise to adopt a value of α that is less than the ubiquitous 0.05. A good choice for work with mice is $\alpha = 0.01$.

An even more common situation occurs within a single study in one lab when mice are subjected to a battery of behavioral tests. If separate groups of mice are used for each test such that the tests are independent events, and there are 10 tests, then the probability of at least one false positive when $\alpha = 0.05$ is used for each test is $P(\geq 1 \text{ false positive}) = 1 - (1 - 0.05)^{10} = 0.40$, an uncomfortably high risk. One remedy is to use the Bonferroni adjustment of α, where the criterion for any one of T independent tests is $\alpha = 0.05/T$. That adjustment ensures that $P(\geq 1 \text{ false positive}) = 1 - (1 - 0.05/T)^T \approx 0.05$. If T = 10 tests, then $\alpha = 0.05/10 = 0.005$ would be used for each test. This adjustment is perhaps unduly conservative and makes it considerably more difficult to discover real effects, which drastically reduces the power of a test.

A more reasonable alternative has been proposed that seeks to limit the false discovery rate without recourse to the simplistic and overly cautious Bonferroni adjustment (Benjamini & Hochberg, 1995; Benjamini, Drai, Elmer, Kafkafi, & Golani, 2001). First, all of the observed P values for tests done in a single study are ranked from large to small. Suppose there are m such values and the index for each value is i, where i = 1 for the smallest P and i = m for the largest P. Starting with the smallest P value having i = 1, compare it to the critical value $P_{critical} = \alpha(i/m) = \alpha(1/m)$, the same as would be used for the Bonferroni adjustment. If it is significant, proceed to the next largest P value having i = 2 and compare its P to $P_{critical} = \alpha(2 / m)$. Continue moving down the list of P values until one is reached where its $P > P_{critical} = \alpha(i/m)$. It and all test results with larger P values are then considered non-significant. This method yields more significant effects than the Bonferroni procedure when there are real effects to be detected.

POWER OF A TEST

If the treatment had a real effect but the effect was not very large, the researcher might be misled by the data and decide to retain the null hypothesis, in which case a *Type II error* would be committed by making a *false negative* conclusion (Figure 5.9). This is the second way that nature can fool the researcher. The probability of making a Type II error is symbolized by β. If $\beta = P$ (Retain null | null is false), then $1 - \beta = P$(Reject null | null is false), which is the statistical *power* of the test. Power is the probability of detecting a real effect. The researcher would like this quantity to be high. Whereas Type I error is not affected in any way by sample size, Type II error and power are very strongly determined by sample size (Cohen, 1992; Wahlsten, 2007).

There are two steps in finding β and power. First, determine what value of d, termed $d_{critical}$, would be required to reject the null that $\delta = 0$ when a specific α level is used. For the example of a one-tailed test in Figure 5.9, that value is $d_{critical} = 0.774$. Second, propose a value of δ that would be able to be detected as a significant effect, and find the probability under that scenario that d would exceed $d_{critical} = 0.774$. If the sample d exceeds 0.774, the researcher would reject the null hypothesis that $\delta = 0$, and the probability of that happening is the power of the test (see Figure 5.11A). There is no way of computing power in general; it can only be done for a specific alternative hypothesis about the true value of δ. This is why the plausible value of δ is such an important issue when planning an experiment. For the example in Figure 5.11, this is done for the alternative hypotheses $\delta = 0.5$ and $\delta = 1.0$, small and large effects. Note that the value of β is very strongly determined by the choice of δ. For the small effect, $1 - \beta = P$(Reject $\delta = 0 \mid \delta = 0.5$) ≈ 0.27 is the area under the curve to the right of $d = 0.774$. For the large effect size, the comparable power is $1 - \beta = P$(Reject $\delta = 0 \mid \delta = 1.0$) ≈ 0.69. Thus, if the researcher is using 10 mice per group and chooses Type I error probability $\alpha = 0.05$, the probability of a Type II error when the treatment has a large effect size of one standard deviation ($\delta = 1.0$) is $\beta = 0.31$. Had the researcher instead opted for $\alpha = 0.01$, the power would be even lower.

In practice, the calculations of power and sample size are done *before* any data are collected. Hence, every pertinent parameter in the model is in Greek symbols: α, β, and δ, with $\delta = 0$ for the null and the alternative δ as some positive value of effect size that the researcher would like to be able to detect. Thankfully, the values of β and $1 - \beta$ do not need to be computed. Instead, the investigator stipulates what β and power are. The tricky part then becomes finding a sample size n that ensures β will have that value. The calculations can be challenging, and for that reason extensive tables or computer programs are employed to find n. Fortunately, once the user has a reasonably good intuitive grasp of the basic principles of statistical inference and can make informed choices of α, β, and δ, actually finding n can be easy.

FINDING THE CORRECT SAMPLE SIZE
Two independent groups

When comparing means of two groups, the null is that their true values are the same, while the alternative hypothesis is that the treatment had some non-zero effect size δ. In the utility **Samplesize 2 2×2 Jgroups**, the value of δ is entered into the yellow box and the Recalculate command is used. A table is then generated to show n for all reasonable combinations of α and β. The user must decide which pairing of α and β is best suited to his or her purpose. For example, entering $\delta = 0.75$, a medium sized effect, immediately shows that sample size should be 26 mice per group if Type I error probability is to be 0.01 and power is to be 90%. One might be tempted to make α as large as possible and β as high as possible to yield a modest sample size. This gambit might pass through a committee that contains no experts on statistical inference, something quite possible for a local animal ethics committee but not a grant review panel. Thus, the investigator might obtain permission to do the work but no funding to make it happen.

2 × 2 design, 4 groups

One of the most common designs in work with mice involves two genotypes subjected to either a treatment or a control condition (Box 5.9). The investigator may be interested in

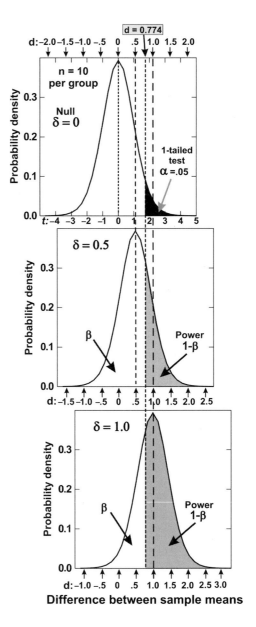

FIGURE 5.11

Distributions of sample effect size d under three population effect sizes δ for the difference between two independent groups, each having n = 10. (A) The top curve shows what is expected when the null hypothesis δ = 0 is true, as in Figure 5.10. That hypothesis will be rejected if the sample d exceeds 0.774. (B) When δ is actually 0.5 (a small effect), the probability that d > 0.774 and the null hypothesis is rejected is the power of the test, represented by the area shaded in gray. (C) Power is much greater when δ = 1.0, a large effect.

5.9 THE A×B FACTORIAL DESIGN WITH 2 LEVELS PER FACTOR

	Strain1	Strain 2
Control	A1B1	A2B1
Treated	A1B2	A2B2

whether the treatment has an effect when averaged over the two genotypes, but most likely the central question is whether the treatment has a larger effect on one genotype than the other. This entails a test of interaction. In the 2×2 sheet of the **Samplesize 2 2×2 Jgroups** utility, simply enter the hypothetical group means and the standard deviation that one expects.

The necessary sample size is determined for each main effect as well as the interaction. In most situations, the three values of n will differ. The user then chooses the sample size to detect the smallest effect that is of major interest in the study; the effect that will occasion the largest sample size. For many kinds of results, detecting the interaction will require the largest sample. Suppose the effect of the treatment is 0.5 standard deviation for one inbred strain and 1.0 standard deviation for the other. These effects are embodied in the four groups A1B1, A1B2, A2B1, and A2B2 if their means are 25, 30, 35, and 45, respectively, while standard deviation within groups is 10. If α is set at 0.05 and power is $1 - \beta = 0.80$, the A main effect will require 6 mice per group and the B main effect will require 13, whereas the interaction effect will require 101 mice per group. It is not at all surprising in the light of these figures that so many studies obtain data that appear in the graph to express an interaction while the statistical test says it is not significant.

J groups, omnibus test

In a one-way design with J independent groups, perhaps several inbred strains, the researcher might want to know whether there is any significant variation among the strain means. The null hypothesis is that all strains come from populations with identical means. The *omnibus* alternative hypothesis is that one or more of the populations have different means. This is a non-specific alternative to the null that is tested with an F ratio using the one-way analysis of variance. A significant F ratio might lead the researcher to do a post hoc test such as Newman—Keuls or Fisher's LSD test to compare every group to every other group. For the purpose of determining power and sample size for the omnibus test, the utility **Samplesize omni Jgroups** uses Cohen's effect size f as the index that is computed from the J population means entered by the user. Separate sheets are provided to find Cohen's f for a one-way design involving 3 to 10 groups, and another sheet can be altered to handle any case with 12 or more groups. After the means and σ are entered and f is determined, the user then transfers f and J values to the sheet with the power tab, enters them into the yellow boxes, and creates a chart with separate power curves for $\alpha = 0.05$, 0.01, and 0.001. The user can then print the graph and use a straight edge to find the proper n on the X axis or read the value of n in the table next to the chart. An example is shown in Figure 5.12 for a study of six inbred strains where the investigator anticipates the means will differ to an extent that Cohen's f = 0.25, a small effect, or 0.5, a moderately large effect.

J groups, linear contrast

Several of the more complex designs presented in Chapter 4 are best analyzed with the aid of linear contrasts among group means. This is emphatically true for the designs shown in Tables 4.2, 4.3, and 4.4. When a contrast is used to answer a specific question, it does not matter whether or not the omnibus F test is significant. A contrast aimed directly at an intricate pattern in the results might reveal a substantial effect while the overall, non-specific F test does not. The utility **Samplesize 2 2×2 Jgroups** provides a convenient way to compute sample size to detect a contrast (Wahlsten, 1991) by using separate sheets with designs having 3, 4, 5, 6, 8, 10, 12, and 16 groups. After entering the hypothetical group means and standard deviation, the user then enters the contrast coefficient (c_j) for each group. The sum of the coefficients must be zero ($\sum c_j = 0$). Sample size n is then shown for different values of α and power. The utility also computes Cohen's f, and that value can be transferred to the power sheet of **Samplesize omni Jgroups** to find n needed for the omnibus F test, should the investigator care about this.

The study of midazolam effects on rat anxiety (Pesold & Treit, 1994) described in Table 4.2 is ideally suited to a linear contrast. Only the group receiving midazolam in the septum without the antagonist drug is expected to show a difference in open-arm exploration. Based on the available data, it would be reasonable to expect the mean of that group to be 40 and the mean of all other groups to be 20, with a standard deviation within groups of 17, the value derived

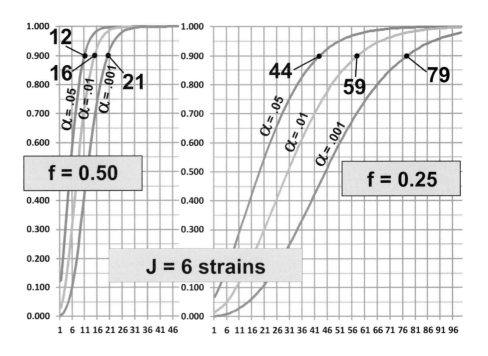

FIGURE 5.12
Power (Y axis) versus sample size (X axis) for two effect sizes involving six inbred strains of mice. Each chart shows power for α = 0.05, 0.01, and 0.001. Specific sample sizes to achieve power of 90% are highlighted for each effect size. Curves were generated by the utility **Samplesize omni Jgroups**.

from the MS_{within} for the original data. The contrast coefficients testing the specific prediction are −1 for the seven groups not expected to be affected and +7 for the other group. A sample size of n = 13 per group would be required to confer 90% power on the test when α = 0.01 is used. For the omnibus F test, to achieve the same power with the same value of Type I error, about 20 per group would be needed. Thus, using a planned contrast makes the study more efficient in the use of animals.

Consider the example for the 16 reciprocal hybrid cross design in Table 4.3. The first step is to propose a realistic model that embodies the effects one hopes to detect with a statistical analysis. Suppose the investigator uses the elevated plus maze to assess anxiety-related behavior and relies on percent time in the open arms of the maze as the best indicator. Suppose further that the baseline level of exploration averaged over all groups is 30% time in the open arms with a standard deviation within groups of 8%. The following model is used to determine the true group means in Table 5.2 by adding the effects for each cross to the baseline 30%. The + and − values in the model are in units of percent time; they are not contrast coefficients.

1. Mitochondria: From strain A increases score 2 units, from B decreases score 2 units.
2. Y chromosome: With inbred mother, A is −2, B is +2; with F_1 mother, A is −4, B +4.
3. Autosomes: Additive:100% A is +4, 75% A is +2, 50% A is 0, 25% A is −2, 0 A is −4.
4. Maternal environment: A is −2, B is −4, F_1 is +6

The model involves certain effects that cannot be detected with contrasts because of the nature of the design. For example, there are no groups that differ only in the A vs B maternal environments; they are confounded with the source of the mitochondria.

TABLE 5.2 Contrast Coefficients, Means, and n for Reciprocal Cross Experiment Shown in Table 4.3[a]

#	♀	♂	Mean (μ)	Mito-chondria	Y Chrom	Y on Inbred vs F_1	Auto-somes	Auto-som. on Inbred vs F_1	Mat. Env. Inbred vs F_1
1	A	A	32	0	0	0	1	−1	0
2	B	B	22	0	0	0	−1	1	0
3	A	B	32	0	0	0	−1	1	0
4	B	A	22	0	0	0	1	−1	0
5	A	A × B	34	0	−1	−1	1	−1	0
6	A	B × A	30	0	1	1	−1	1	−1
7	A × B	A	36	1	0	0	1	1	1
8	B × A	A	32	−1	0	0	1	1	0
9	B	A × B	24	0	−1	−1	1	−1	−1
10	B	B × A	20	0	1	1	−1	1	0
11	A × B	B	40	1	0	0	−1	−1	0
12	B × A	B	36	−1	0	0	−1	−1	1
13	A × B	A × B	42	1	−1	1	1	1	0
14	A × B	B × A	34	1	1	−1	−1	−1	0
15	B × A	A × B	38	−1	−1	1	1	1	0
16	B × A	B × A	30	−1	1	−1	−1	−1	0
			n:	15	8	52	27	#DIV/0!	7

[a]In F_1 hybrid cross (e.g., A × B), strain of female parent is given first, as shown in Table 4.3. Numerical model used to generate group means is stipulated in the text. Standard deviation of scores within each group is set at $\sigma = 8$. Sample size n is shown for $\alpha = 0.05$, one-tailed test, and power of 80%, as calculated by the utility **Samplesize 2 2×2 Jgroups**.

Next, linear contrasts need to be posited that ask the specific questions of interest to the researcher. Several of these are shown in Table 5.2, using +1 and −1 for the coefficients c_j. The specific questions posed by each contrast are:

1. Do mitochondria from strain A increase scores more than B? The specific groups to be compared are (7 vs 8) (11 vs 12) (13 vs 15) (14 vs 16). Those with A mitochondria are +1.
2. Does a Y chromosome from strain A increase scores more than B? Specific groups to compare are (5 vs 6) (9 vs 10) (13 vs 14) (15 vs 16), with those from A receiving +1.
3. Is the Y chromosome effect larger when the mother is an F_1 hybrid than an inbred? This question about interaction compares [(5 vs 6) (9 vs 10)] vs [(13 vs 14) (15 vs 16)].
4. When two groups differ by 25% in the proportion of autosomal genes from strain A, do those with more A alleles score higher? This contrast compares pairs of groups, each of which differs by 25% A: (1 vs 6) (2 vs 9) (3 vs 5) (4 vs 10) (7 vs 14) (8 vs 16) (11 vs 13) (12 vs 15).
5. Is the A strain autosomal effect larger when the mother is an F_1 hybrid than an inbred? This compares [(1 vs 6) (2 vs 9) (3 vs 5) (4 vs 10)] vs [(7 vs 14) (8 vs 16) (11 vs 13) (12 vs 15)].
6. Do mice with an F_1 hybrid mother show less anxiety-related behavior than those with an inbred mother? The comparison is (6 vs 7) (9 vs 12).

Finally, after group means and the standard deviation are entered into the utility, the contrast coefficients for one contrast are entered into the 16 boxes and F9 Recalculate prompts the computer to display the proper n for each value of α and power. Results of this exercise are shown in Table 5.2. The contrast value for effects of strain A autosomes on an inbred versus an F_1 hybrid background is 0, which yields a divide by 0 error (#DIV/0!) in Excel and can be ignored. The model did not specify such an interaction, so the contrast could not detect it. For the other contrasts, the required sample size ranges from 7 to 52 mice per group. The researcher must decide which among these effects are really worthy of an adequate sample size. If the interaction of the Y effect with maternal environment is not central to the goals of the study, there is no need to use so many mice. If the overall effect of A autosomes is established by evidence from previous studies, it is not very important to seek replication. Reciprocal hybrid crosses have the unique capability of revealing non-Mendelian effects, which is probably why the design was chosen. Among those effects, the mitochondrial effect appears to be the smallest of major interest, so the proper sample size for the study is n = 15 mice per group. With that sample size, the power to detect Y chromosome and maternal environment effects will exceed 80%.

The method of contrasts has great flexibility and can be adapted to many kinds of sophisticated experiments. The design of the experiment determines what the contrast coefficients should be and how many contrasts are appropriate. The null hypothesis is always that the true contrast effect is 0. The alternative hypothesis to be tested with desired power is embodied in the specific values of group means derived from a model of things that determine the behavior. For the purpose of estimating sample size, the model should represent the types of results that might reasonably occur in the experiment when all the effects to be tested are present. The size of each effect should be something that the researcher would like to be able to detect, if it does occur in the planned study.

Two-way factorial design

When there are more than two genotypes in a study, each tested under two or more treatment conditions, perhaps the most important question posed by the research is whether effects of the treatments depend on genotype or if the effects of the two factors act independently and algebraically are additive (Wahlsten, 1990). Suppose eight inbred strains of mice are tested simultaneously in three labs using the accelerating rotarod. Plausible strain means are shown in Figure 5.13 based on data from the author's lab. If the aim is to determine whether a study can detect rather small lab effects and a strain by lab interaction, a model must be proposed in which such effects exist. A power analysis then compares that model to the null model where there are no genuine strain, lab, or interaction effects and groups show different sample means solely because of sampling error. The model posits no lab effect at all for the BALB strain and a very small effect for FVB, small effects for five other strains, and a large lab difference for C57. Wherever there is a lab effect, Lab 3 always yields the longest and Lab 1 the shortest fall latency. If Lab 3 yielded the shortest latency for certain strains, this would reduce the average effect of labs across all strains while increasing the magnitude of the interaction effect.

(A)

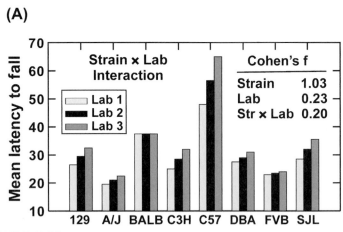

(B)

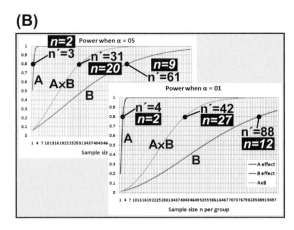

FIGURE 5.13

(A) A model of means of eight inbred strains tested on the accelerating rotarod in three labs. There is a large strain effect, a small lab effect, and a strain × lab interaction. The null model (not shown) asserts that the mean is 32.3 in all groups. It will be rejected if an ANOVA finds a significant F ratio. (B) The right panel shows power versus sample size n' for the three effects when α = 0.05 and 0.01. Effect sizes were determined from the means using the utility **Cohens f J×3 design** and then power curves were generated by the utility **Samplesize J×K Cohens f**. Insert n' values into boxes in the utility to find the correol n for each group.

The analysis of variance partitions variations among the 24 group means into three components; main effect of strain averaged over labs, the main effect of labs averaged over strains, and the interaction deviations. The partition is performed for the 8×3 factorial design in the utility **Cohens f J×3 design** that has three rows for the levels B1, B2, and B3. Once the user has entered the 24 means and the standard deviation, pressing the F9 key produces the values of f for each effect. The spreadsheet also shows how interaction is determined. In the first box where means are entered, the marginal means for each strain and each lab are stated right where one might expect — in the margins of the table. This is why they are called marginal means, not because they are barely large enough to warrant attention. The marginal means are then used to determine what each group mean would be if the two factors are additive. Consider the BALB strain in Lab 3. The mean of BALB over all labs is 37.5 and the mean of Lab 3 over all strains is 35.0. The grand mean of all 24 groups is 32.2 ("grand" not because it is wonderful but because it takes account of every group in the study). Thus, the average effect of the BALB genotype is $37.5 - 32.2 = 5.3$ units, and the average effect of Lab 3 is $35 - 32.2 = 2.8$ units. If strain and lab combine additively, then the mean of BALB in Lab 3 should be simply the grand mean plus the two effects: $32.2 + 5.3 + 2.8 = 40.3$. The actual mean, however, is 37.5, and this value deviates from 40.3 by -2.8 units, which is the interaction deviation. Interaction effect size is found by squaring and adding all those deviations across the 24 groups. As shown in Figure 5.13, the effect size f is much larger for the strain effect than the lab or interaction effects.

Knowing the number of levels of each factor and the effect sizes present in the model, these numbers are then inserted into a separate utility, **Samplesize J×K Cohens f**, that plots power curves for the two main effects and the interaction effect in separate charts for $\alpha = 0.05$ and 0.01, as shown in Figure 5.13B. The investigator is virtually certain to obtain a significant strain difference, but a fairly large sample will be needed to find the relatively small lab and interaction effects. The power to detect the interaction effect is somewhat higher than for the lab effect that actually has a slightly larger $f = 0.23$ because the degrees of freedom for the interaction effect are much greater (14 vs 2).

One virtue of these utilities is that the investigator can easily try several different models with various kinds of interactions simply by changing a few means in the 8×3 table in **Cohens f J×3 design**. The general method for finding power and sample size for the J×K factorial design is extended across a family of utilities that makes it convenient to plan any study from a 2×2 to an 8×8 design. For example, if one factor has just two levels, the utility **Cohens f J×2 design** can handle every design with two levels of one factor using separate sheets for the 8×2, 7×2, 6×2, 5×2, 4×2, and 3×2 designs. Separate utilities are provided for J×2, J×3, J×4, J×5, J×6, J×7, and J×8 designs. Because the routines to find power for each sample size from 2 to 100 are rather bulky and require six separate sheets, one for each kind of effect and each value of Type I error, the one utility **Samplesize J×K Cohens f** provides this service to all of the others that compute Cohen's f. It can be used equally well to make power curves for a multigroup effect that is estimated from published data using the **Effect size from article** utility.

Designs with more than two factors

The basic methods for finding main effects and interaction deviations with two factors can be extended to more complex designs. Cohen (1988) described the general approach to finding effect size f for higher order interactions, and that description quite adequately demonstrates how onerous a task it can be. This is a situation where spreadsheet-based utilities would be very helpful. At the same time, it is evident that a very large number of utilities would be needed to address the extraordinary number of possible design variants. For example, one might like to use a $8 \times 3 \times 2$ design or a $4 \times 3 \times 2 \times 2$ design, or even a $6 \times 3 \times 2 \times 2 \times 2 \times 2$ monstrosity. As will be shown in Chapter 7, it is common practice to balance the order of testing when there are three study factors and three control factors. If an investigator would like to know whether a control factor has a significant interaction effect involving a behavior, a power calculation could aid in planning a study. A general approach is presented here that can determine power even for a five-way interaction effect using a convenient utility. The entire process cannot be automated easily as it is for a one-way design, but the rich harvest of information about a complex design can make the extra effort worthwhile.

This method was devised to investigate claims about the consequences of standardizing behavioral test conditions and the local environment across different laboratories (van der Staay & Steckler, 2002; Wahlsten, 2001; Würbel, 2002). The issue will be discussed at greater length in Chapter 15 on laboratory environments. Recently, Richter, Garner, and Würbel (2009) proposed that standardization will actually decrease the likelihood of replicating test results in different labs, a claim that at first reading seems counterintuitive. They further proposed that results will be more replicable across labs if each lab subjects mice to more than one housing condition and tests animals at more than one age. Specifically, they advocated "heterogenizing" age and cage conditions within each lab by adding age with two levels and caging with two levels (standard and enriched) to whatever design is used to study effects of different treatments on several mouse strains. The author therefore desired to conduct a power and sample size analysis for a rather large study with five fully crossed factors and evaluate several models with and without interactions.

The initial work was done with a familiar task, the accelerating rotarod test of motor coordination, using eight high-priority inbred strains from the Mouse Phenome Project injected with ethanol or saline. Based on prior experience with simultaneous testing in more than one lab (Crabbe et al., 1999; Rustay et al., 2003a), a model was devised where the entire study is run independently in three different labs. The design is thus 8 strains × 3 labs × 2 treatments × 2 ages × 2 cage conditions. The example also allows for half of the mice to be male and half female but does not allow any interactions with sex. Thus, it is a six-way factorial design with no six-way interaction, just five-way. For most people in this field, a five-way interaction will be quite sufficient for any conceivable power analysis.

The first step is to construct the design in the form of a spreadsheet. This is done in Chapter 4 using the **Design 6way** utility that embodies an $8 \times 4 \times 3 \times 2 \times 2 \times 2$ design having 768 independent groups, a number of groups that exceeds the number of animal subjects in most experiments. Reducing the scope to $8 \times 3 \times 2 \times 2 \times 2 \times 2 = 384$ groups still makes the study rather daunting, but the work will be divided among three labs, so that any one lab needs to test mice in 128 groups — a large but not impossible number if the sample size per group does not need to be too large. This study will be close to the upper limit of what most researchers would care to contemplate.

The utility **Simulation 6 factor experiment** has 192 cells into which hypothetical means are typed by the user, plus two cells for male and female effects and one box for the standard deviation within groups (σ). A portion of the spreadsheet in Figure 5.14 shows group means for a model that entails a five-way interaction. A bar chart (Figure 5.15) helps to demonstrate several effects. The model embodies a large strain effect as well as a large ethanol effect (shorter fall latencies) and a modest lab effect in which means are generally lowest for Lab 1 and highest for Lab 3. Careful inspection reveals that the ethanol effect depends on strain as well as lab, such that there is a strain × lab × ethanol interaction, perhaps not a large interaction but a real one nevertheless. Looking even more closely, it can be seen that the age and cage effects, although not large, differ across strain, lab, and treatment in an irregular pattern. For certain strains in certain labs, there is no age or cage effect of any kind, but which strain shows that lack of age and cage influence differs across labs and treatment condition. Enriched cages (condition C2) usually increase fall latency and older mice usually stay on the rod longer, but not for every strain in all labs. The sex difference is not allowed to interact with the other factors. The spreadsheet then generates two normally distributed replicates for each of 384 groups using the hypothetical group mean and σ, resulting in 256 mice for each lab and 768 for the entire study. This entails considerable computation, but it is only the beginning.

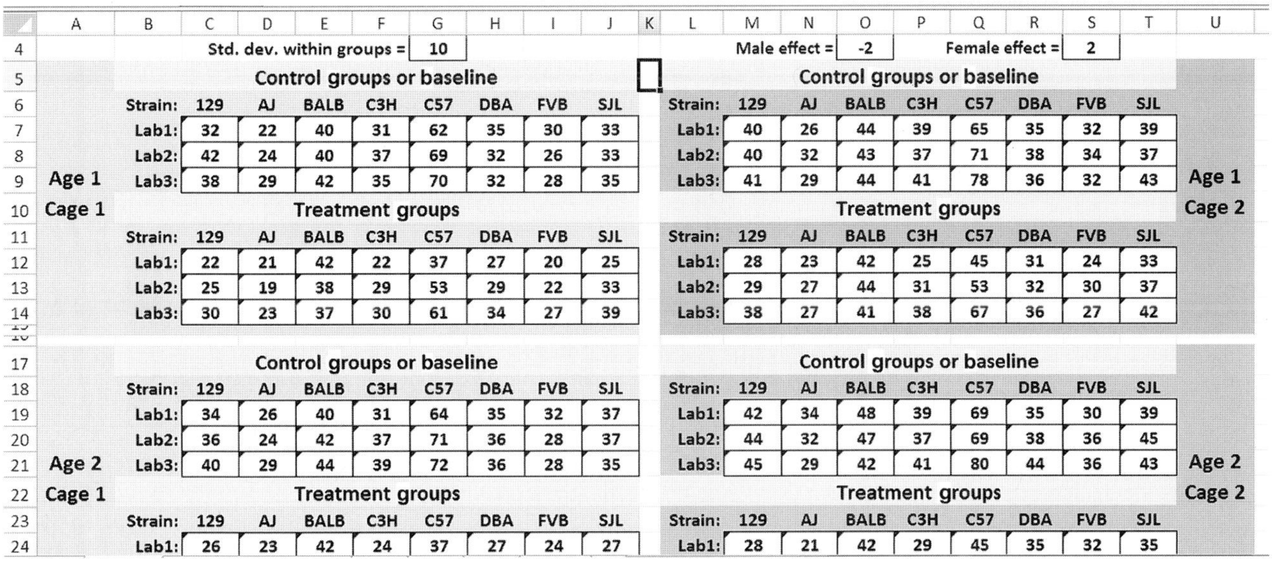

	129	AJ	BALB	C3H	C57	DBA	FVB	SJL		129	AJ	BALB	C3H	C57	DBA	FVB	SJL	
Std. dev. within groups = 10									**Male effect =** -2				**Female effect =** 2					
Control groups or baseline									**Control groups or baseline**									
Strain:	129	AJ	BALB	C3H	C57	DBA	FVB	SJL	Strain:	129	AJ	BALB	C3H	C57	DBA	FVB	SJL	
Lab1:	32	22	40	31	62	35	30	33	Lab1:	40	26	44	39	65	35	32	39	
Lab2:	42	24	40	37	69	32	26	33	Lab2:	40	32	43	37	71	38	34	37	
Age 1 Cage 1 Lab3:	38	29	42	35	70	32	28	35	Lab3:	41	29	44	41	78	36	32	43	Age 1 Cage 2
Treatment groups									**Treatment groups**									
Strain:	129	AJ	BALB	C3H	C57	DBA	FVB	SJL	Strain:	129	AJ	BALB	C3H	C57	DBA	FVB	SJL	
Lab1:	22	21	42	22	37	27	20	25	Lab1:	28	23	42	25	45	31	24	33	
Lab2:	25	19	38	29	53	29	22	33	Lab2:	29	27	44	31	53	32	30	37	
Lab3:	30	23	37	30	61	34	27	39	Lab3:	38	27	41	38	67	36	27	42	
Control groups or baseline									**Control groups or baseline**									
Strain:	129	AJ	BALB	C3H	C57	DBA	FVB	SJL	Strain:	129	AJ	BALB	C3H	C57	DBA	FVB	SJL	
Lab1:	34	26	40	31	64	35	32	37	Lab1:	42	34	48	39	69	35	30	39	
Lab2:	36	24	42	37	71	36	28	37	Lab2:	44	32	47	37	69	38	36	45	
Age 2 Cage 1 Lab3:	40	29	44	39	72	36	28	35	Lab3:	45	29	42	41	80	44	36	43	Age 2 Cage 2
Treatment groups									**Treatment groups**									
Strain:	129	AJ	BALB	C3H	C57	DBA	FVB	SJL	Strain:	129	AJ	BALB	C3H	C57	DBA	FVB	SJL	
Lab1:	26	23	42	24	37	27	24	27	Lab1:	28	21	42	29	45	35	32	35	

FIGURE 5.14

Top portion of the first sheet (captured from the computer screen) in the utility **Simulation 6 factor experiment** where population means are entered into $8 \times 3 \times 2 \times 2 \times 2 = 192$ cells plus two cells for a sex difference and one for the standard deviation within groups. The spreadsheet allows the user to create a model with up to a five-way interaction but no interaction with the sex factor. The specific means in the example do in fact express a 5-way interaction as well as four 4-way interactions, five 3-way interactions, five 2-way interactions, and all five main effects. The interaction effects are easier to perceive in the bar chart in Figure 5.15. Results of 100 independent simulations analyzed with ANOVA are provided in Table 5.3.

The utility repeats the entire process to generate 100 independent simulations of the entire experiment and formats them in a convenient form for reading into a data analysis program such as SPSS or SYSTAT. Thus, the key to the method for complex designs is simulation based on a model of true group means in a population.

The next step is simple enough but requires more technician or student time. An ANOVA with five factors must be done for each of the 100 runs of the simulation and then results for all main effects and interactions compiled in a spreadsheet. This requires about two hours of work by someone adept at statistical analysis and Excel. If the sex effect is to be included in the ANOVA, be prepared to spend a lot of time staring at a blank computer screen while the transistors hum and heat up the office. In the work to generate results shown in Table 5.3 using SYSTAT on the author's home computer, a 4-factor ANOVA ran in 1 sec and a 5-factor ANOVA required only 5 sec, whereas a 6-factor analysis that included sex required 1 min 25 sec. By pooling males and females in two replications, the first pass through the process will have $n = 4$ mice per group. The most salient statistic is the P value or achieved significance level for each effect in the ANOVA table. If the investigator plans to use $\alpha = 0.05$ as the significance criterion in the eventual study, then he or she wants to know how many of the 100 runs of the simulation yield a P value less than 0.05. Because there are 100 runs, this quantity is the power of the test of each particular effect. This will be an approximation to power because random sampling is employed to generate the data, but there is never any need to estimate power to the nearest percent. Nobody cares whether power is 78%, 80%, or 83%; all are reasonably close to 80%. The results of 100 runs based on the model in Figure 5.14 are summarized in Table 5.3 for every effect in the ANOVA. In a disturbing number of instances the value of $1 - \beta$ is only slightly higher than α, denoting a startling risk of Type II error. For effects that definitely were not present in the population means, Type I errors occurred in 1 to 9% of runs, with an average over 10 null effects of 4.3%, a value reasonably close to an α level of 5%. Should the user desire a more precise estimate, the simulations could be repeated and results combined.

It is not at all difficult to adapt the **Simulation 6 factor experiment** utility to a study with three, four, or five factors. For example, if the investigator does not care to employ mice of different

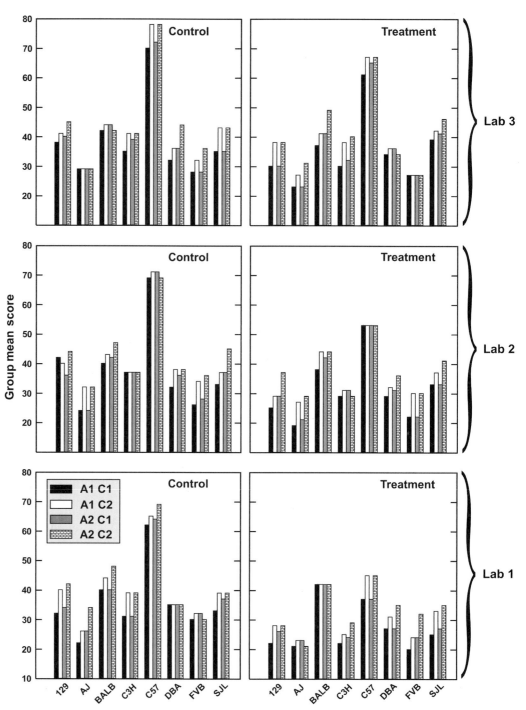

FIGURE 5.15

Group means for the five-factor design stipulated in Figure 5.14. Main effects of all factors are readily perceived, as are most of the two-way interactions. The three-way interaction of strain × lab × treatment is expressed in a way that alcohol effects are greatest in Lab 1 for several but not all strains. For example, there is no consistent alcohol effect in any lab on BALB mice, whereas C57 mice are most impaired in Lab 1. It can also be seen that the patterns of age and cage effects differ by strain and treatment in a way that is not the same in the three labs, which expresses a complex five-way interaction effect.

TABLE 5.3 Number of P Values <0.05 in 100 Runs of a Simulation of a Strain (S) × Lab (L) × Treatment (T) × Age (A) × Caging (C) Factorial Experiment. This Number Approximates Percent Statistical Power to Detect an Effect with a Five-Factor Analysis of Variance.[a]

Main Effects	#	2-way Interaction	#	3-way Interaction	#	4-way Interaction	#	5-way Interaction	#
S	100	SL	97	SLT	21	SLTA	22	SLTAC	3
T	100	ST	100	SLA	6	SLTC	36		
L	100	SA	3	SLC	24	SLAC	3		
A	61	SC	10	STA	3	STAC	7		
C	100	LT	71	STC	6	LTAC	5		
		LA	7	SAC	4				
		LC	9	LTA	2				
		TA	1	LTC	3				
		TC	5	LAC	2				
		AC	4	TAC	5				

[a]The analysis was done with n = 4 observations in each cell, 576 degrees of freedom for the within-groups variance. Effects shown in bold typeface were definitely present in the model shown in Figure 5.15, whereas those in italics were absent in the model. Two effects (AC, STC) were present in the model but so small that they could not possibly be detected. Results were generated by the utility **Simulation 6 factor experiment.**

ages housed in different kinds of cages, the means for just one age—cage condition in the spreadsheet can be copied to the other three conditions, which forces those factors to have zero effects. There will then be 16 mice in each strain—lab—treatment group, 8 of each sex. If the user plans to study mice of only one sex, enter 0 into the boxes for males and females. Because data are randomly generated, sample size can easily be reduced by half or more by selecting a subset of cases using a function in a statistical program. For example, sample size can be reduced by half by selecting just mice in replication #1 for analysis. Sample size can be doubled readily by running the simulation twice and then appending results of the second to the bottom of the spreadsheet with results of the first. For interaction effects in Table 5.3 where power ranges from 20 to 40%, one can be confident that doubling the sample size to n = 8 per group would enhance power substantially. Especially if the user is interested in only a few specific interaction effects, it would not take long to try n = 4, 6, and 8 because most of the effects in the ANOVA output can be ignored.

Ethics Approval

107

In most jurisdictions a research proposal must be approved by an animal ethics committee before the work can begin or the proposal can be submitted to a funding agency. The process of applying for approval is governed by extensive rules and regulations established by each country, province, state, or canton. The International Council for Laboratory Animal Science (ICLAS) has established a working group to promote the harmonization of animal use guidelines transnationally, and consensus on several principles has been reached (Demers et al., 2006). However, the investigator is now required to abide by procedures of the local institution, governments, and granting agencies that fund the work. Each institution provides a wealth of information on policies and procedures, and this is the place to start when getting ready for ethics approval.

In this chapter, several aspects of ethics review are discussed: the general principles, historical changes in policies, and special issues that arise in work with mice. Examples are provided from the experience of the author who has worked with several species of research animals and served on animal ethics committees, including work as the Vice-Chair of the University of Waterloo Animal Care Committee and Chair of the Institutional Animal Care and Use Committee (IACUC) at the University of North Carolina Greensboro (UNCG).

GOOD ETHICS AND GOOD SCIENCE

In the vast majority of research on mouse behavior, there is no fundamental conflict between high ethical standards and good science. The best behavioral data come from healthy mice living in relatively low stress environments subjected to tests that are well adapted for them.

Nevertheless, there are specific situations that require some difficult value judgments. These judgments are informed by the broad ethical principle of work with animals: there should be

6.1 SOME PRINCIPLES OF RESEARCH WITH LAB ANIMALS; FROM ICLAS WORKING GROUP ON HARMONIZATION

- Death or severe pain should be avoided as end points.
- Earliest possible end point should be used that is consistent with the scientific objectives.
- Seek to minimize any pain or distress … while meeting scientific objectives.
- Duration of studies involving pain and distress should be kept to a minimum.
- Staff must be adequately trained and competent in recognition of species-specific behavior … and signs of pain, stress, and moribundity.
- Animals should be monitored by means of behavioral, physiological and/or clinical signs … to permit timely termination of the experiment.

Source: *From Demers et al., 2006.*

no unnecessary pain or suffering. An elaboration of this principle is provided in Box 6.1. How the principles are applied depends on the particulars of the experiment. Suppose the researcher wants to learn more about the genetic aspects of the pain response. To do this, the mouse must experience some pain. Ethical conduct requires that the experimenter who administers and evaluates the pain must have a high level of expertise and know specifically how to recognize symptoms of pain in mice (Box 6.2). Uneducated squeamishness, uninformed anthropo-morphism, or just plain ignorance in the ways of mice should yield to the opinions of experts on these matters.

Consider an example from the 1970s at the University of Waterloo. A biochemist applied for permission to administer microcystin to mice in order to assay the level of toxin in a sample. The toxin occurs in blue-green algae blooms affecting public water supplies (see www.who.int) and even some dietary supplements (Gilroy, Kauffman, Hall, Huang, & Chu, 2000). An extract from the algae was to be injected into the animal's peritoneal cavity. The study aimed to monitor changes in blood chemistry, not behavior. When asked if the substance would cause pain, the chemist said it did not because the animals they had tested soon became quiet, rather than hopping wildly around the cage squealing in agony. When we observed the mice in his lab, however, it was evident that they were suffering greatly. They were lying prone and motionless on the cage floor with their abdomens indented. Such behavior is not normal for mice. Clearly, the chemist knew almost nothing about mice and was not qualified to conduct the research. The situation was potentially dangerous to humans as well, because his proposed assay could not detect signs of mild but real pain and distress caused by much lower levels of the toxin.

6.2 SYMPTOMS OF PAIN IN MICE

Reduced food and water intake; weight loss
Piloerection and hunched appearance
Increased respiration rate
Reduced vibrissae movement
Biting at the source of pain
Writhing of abdomen
Hunched posture away from light
Quiet and unresponsive (severe pain)
Hypothermia

Source: *Canadian Council on Animal Care Guide, Volume 1, page 191.*

A high level of expertise is needed not just to measure pain (Mogil, 2009) but also to minimize the amount of pain that must be endured to achieve the study goals. In almost all research, that level of pain will not be intense and will not need to last very long.

THE ERA BEFORE REGULATION

As recently as the 1950s and 1960s in most jurisdictions, there were few or no laws specifically governing use of animals in research or limiting what could be done with animals in the name of science. There were broad spectrum laws in many countries prohibiting cruelty to animals but they were rarely applied to what was widely viewed as legitimate scientific work. A few examples, some quite upsetting, illustrate some of the old ways that thankfully have been abandoned. The current situation at most institutions is far better for the mammals that work and reside in modern laboratories than it was four decades ago.

The author began work with animals in graduate school at Yale University. During a tour of the psychology department rat colony on the first day, it was noticed that one rat had a little blood around its nostrils, a sign of chronic respiratory disease that was rampant in many rat labs. The animal care technician said he would take care of it right away. This created a good impression. Yes, put the poor beast on an antibiotic right away. But, no, the man lifted the animal by its tail, spun in around twice in a large circle, slammed it against a table top, then tossed it in the garbage pail! That same day, the new student was given his departmental service task: killing all the rats used each week by famous experts on the psychology of learning in animals. Dozens of rats were dumped into a large garbage can connected to a natural gas outlet that caused slow asphyxiation. Meanwhile, the dog colony was filthy, and clogged drains overflowed with every heavy rain. To prevent the dogs from disturbing people in nearby labs by barking, their vocal cords were cut out by poorly trained personnel and they recovered without the comfort of analgesic drugs. There were no inspections of conditions of the animals by external authorities.

A number of practices were common at the time that have now have been abandoned or suppressed. Some researchers in the United States and Russia subjected dogs to traumatic levels of electric shock far in excess of what was needed to motivate performance (Solomon & Wynne, 1953). Solomon and Corbit (1974) later reviewed this work: "The dog appeared to be terrified during the first few shocks. It screeched and thrashed about … ." In some studies of brain function, cats were subjected to brain surgery while lightly anesthetized and paralyzed by succinylcholine or curare (Hubel & Wiesel, 1959), a drug that blocks contraction of skeletal muscles but does not blunt sensation or consciousness (Smith, Brown, Toman, & Goodman, 1947), so that they were drowsy but awake during the procedure. Mouse fights were sometimes allowed to proceed to the stage of severe wounding (Collins, 1970) or even death.

THE ERA OF REGULATED RESEARCH WITH ANIMALS

The dawn of institutional animal ethics rules in mouse research came at different times in different places (Box 6.3). One of the first laws that conferred real powers to regulate use of research animals was passed by the Province of Ontario in Canada in 1971. It required all labs using animals to register with the government and authorized an appointed inspector to enter any establishment where animals were being used in research. The inspector could make recommendations to a university administration but could also shut down a particularly inhumane project.

That happened the first time that the newly appointed inspector, a veterinarian at the Ontario Veterinary College in Guelph, toured labs at the University of Waterloo. He entered a room where the professor was conducting a study of overpopulation in mice. A large plastic cage had been inoculated with two freely breeding pairs of mice, and a surplus of food and water was present at all times. When a litter was born, the experimenter counted the number of pups but did not intervene. Before long the cage was literally half full of mice and grossly polluted with

6.3 CHRONOLOGY OF GOVERNANCE CONCERNING USE OF LABORATORY ANIMALS

1956 International Committee on Laboratory Animals established (no powers)
1966 USA: Laboratory Animal Welfare Act passed to regulate only dogs and cats crossing state lines
1968: Canadian Council on Animal Care established (voluntary compliance)
1971 Canada: Province of Ontario passes Animals for Research Act (inspector with considerable power)
1972 Canada: Province of Alberta proclaims regulation 33-72 for research animals
1972 USA: NIH Guide for the Care and Use of Lab Animals
1985 USA: NIH Authorization Act establishes local IACUCs.
1985 USA: Animal Welfare Act passed
1986 UK: Animals (Scientific Procedures) Act passed
1986: European Commission Directive 86/609 on animals in research is proclaimed
2004: First animal welfare regulation in China

feces and urine. As new litters were born, the pups were carried away by other mice and often eaten alive. Processes that seemed to limit population growth in rats (Calhoun, 1962; Ramsden & Adams, 2009; Wynne-Edwards, 1965) clearly were not working with those lab mice. The inspector was aghast. The study was halted that same day and the mice were killed humanely. Almost every person in the facility applauded his decisive action.

The Canadian Council on Animal Care (CCAC) conducts annual inspections of all research facilities and in its early days found many unethical practices. During one such inspection at the University of Waterloo, the author accompanied the CCAC panel through the labs. In one room an undergraduate student was subjecting rats to forced exercise by placing them in a deep tub filled with water. The study aimed to document changes in the blood and muscles caused by prolonged exercise. It was late in the afternoon and many rats had been tested that day without a change of water. The liquid had become opaque with fecal matter and had many fecal boli floating on the surface. One panel member asked the young lady how she knew when a rat was getting tired. She said that was easy: the animal would dive to the bottom of the tank and disappear from view for several seconds while it rested on the bottom. Asked if any animals ever drowned during the test, she said, "Oh yes, some of them never come back up." The professor in charge of the work was directed to alter his protocol immediately. Had he not done so, federal research funds could have been withheld from the entire university.

In both of these examples, no worthy scientific purpose was served by the ongoing procedures that offended almost everyone's sense of ethical practices. Most researchers welcomed the scrutiny from outside the university.

Much later (1985–1986), local ethics committees and extensive government regulations were established in the United States, the UK, and the European Union. The bureaucratic apparatus for the oversight of animal use in research grew rapidly and detailed guidelines on many aspects of animal use were published. Major organizations involved in this work are named in Box 6.4.

ETHICAL STATUS OF MICE AND RATS

The scope of ethical principles across diverse species is a topic of current debate. Some jurisdictions give special consideration to homeothermic (warm-blooded) birds and mammals but not poikilothermic (cold-blooded) reptiles and fish. Others exalt all vertebrates but not insects or mollusks. These debates move away from the topic of this book — mice. The main question here is whether mice or rats warrant more or less protection than mammals such as dogs, cats,

6.4 ORGANIZATIONS PROVIDING INFORMATION ON ETHICS IN ANIMAL RESEARCH

USA
American Psychological Association
Society for Neuroscience
Organization for Laboratory Animal Welfare
Canada
Canadian Council on Animal Care
EU
Federation of European Neuroscience Societies
European Science Foundation
International
Association for Assessment and Accreditation of Laboratory Animal Care International (provides information on many countries, including Japan, Korea, China)
International Council for Laboratory Animal Science

or monkeys. Many people have sentimental attachments to pet dogs and cats, whereas rats and mice are often seen as pests or vermin by the general public. Sentiment, however, is not a defensible basis for deciding what is ethical. In law, all mammals deserve the same protection in research settings, even the smaller rodents. Certain species of primates may be given higher status and greater protection because they are more similar to humans than mice or rats, but this does not depreciate the level of consideration that should be given to mice.

The dual role of mice in society gives rise to an interesting conflict of ethical principles. In an insightful commentary, Herzog (1988) noted that three separate ethical standards were applied to mice in the same building at his university. The "good" mice used in research were protected by numerous regulations and a duly constituted ethics committee. The "bad" mice were former lab mice that escaped and ran freely around the building. They had no protection at all and were to be exterminated by any possible means. The "food" mice were captive, live lab mice fed to predators such as snakes and owls and were often bred for that purpose. Their fate was to be eaten alive without any anesthetic or analgesic. Thus, things could be done to feral and food mice that would be abhorrent to an ethics committee reviewing an ethics protocol.

Sometimes the same government that throws a cloak of protection over lab mice and rats funds savage attacks on the same species when they infest a farm, food store, or dwelling. Epling (1989) drew attention to the province of Alberta in Canada where there are no wild rats. It is not that rats eschew Alberta. On the contrary, they would be more than willing to live in the rich granaries on those prairies. For that reason, the government created the Alberta Rat Patrol (Figure 6.1), a roving squad of exterminators that patrols the borders with Montana and Saskatchewan looking for recent intruders (see http://www.youtube.com/ entry on Rat Busters about the Alberta Rat Patrol). If any are found, they are gassed, shot, stomped with a boot, or trapped by a variety of lethal means.

Mice outside of laboratory cages fare little better. A visit to any local hardware store reveals a wide array of poisons and traps to use on mice. Warfarin is popular, despite the slow death caused by internal bleeding. Sticky glue traps kill by slow starvation and dehydration, as does the snap trap when it catches only a tail or leg, which is a common event. Mice as food can be legally purchased from a pet store and then fed live to a pet snake.

Clearly there is a double standard. Researchers are required to negotiate many time-consuming regulatory hoops and hurdles to win the privilege of working with lab mice, while mice outside the laboratory cage have no protection. Apparently humans want high-quality data from lab

FIGURE 6.1
Alberta Rat Patrol poster. *(Reprinted with permission of the Provincial Archives of Alberta; see https://sales.ccs.alberta.ca/paa/ store/select.aspx?item=199.)*

mice, and they are therefore protected from stress, carefully nurtured, and even pampered. On the other hand, feral and wild mice eat our food without permission and spread disease and filth, and humans benefit by getting rid of them however they can. If lab mice were to be stressed and tormented the way feral mice often are, the data from them would be worthless.

THE FUNDAMENTAL ETHICAL PRINCIPLE OF ANIMAL RESEARCH

Perusing the rules and regulations for research with animals it is sometimes difficult to find a clear statement of what is considered ethical and what is not. Procedures for approval are usually explained in great detail, but a vast degree of discretion is conferred on local committees, and few examples from real life in the lab are described. In the United States, every person, from the novice student to a professor with long experience, must take multiple choice examinations via an Internet site maintained by the Laboratory Animal Training Association (LATA; www.latanet.com), the Collaborative Institutional Training Initiative (CITI; www. citiprogram.org), or some equivalent body before being allowed to touch a live animal. The Base LATA module sets forth three principles: (1) animals should be well cared for, (2) use should be scientifically justified, and (3) pain and distress should be limited to that which is absolutely unavoidable. These can be condensed into just one phrase: there should be *no unnecessary pain or suffering*. This differs from the mantra of ethical practice with human research subjects, which states *do no harm*. If an animal research project is considered crucial for attaining knowledge that could be of use to humans, some degree of discomfort, stress, or pain may be permissible. Especially in biomedical research on disease processes, mice may be infected with foreign genes, viruses, or cancer cells to create models that may be valuable in the search for a cure for human disease. Just as humans exploit animals for food, they also exploit them for scientific research.

Thus, if a project involves some noteworthy suffering, it should be demonstrably necessary to do the work with mice because it cannot be done with humans. If the project is an exploration of basic behavioral processes with no clear application to human welfare in the near future, the degree of permissible suffering is substantially lower. In addition to the question of whether the research is necessary, there is the second question about whether the *degree* of intended suffering is really necessary. The investigator should strive to inflict the minimum level of stress or pain required to reach an important scientific goal. Any kind of suffering beyond that level is unethical and should be prohibited.

While the general principle of no unnecessary suffering is clear enough, its proper application to specific studies often is not easy to judge. There is no formula that can be applied across a wide range of procedures. In most jurisdictions, a committee is given powers to decide what is acceptable. Consequently, a project may receive approval at one institution but not another, or in one country but not another.

Questions of ethics are often considered by grant review panels and even referees of papers submitted to a journal for publication. Most granting agencies and journals require that a study has undergone some kind of local ethics review and been approved by a duly constituted committee before submission. Such approval does not end all further scrutiny, however. A grant review panel could refuse to fund a study on ethical grounds or a referee could object to a submitted paper, despite local approval.

For example, the journal *Genes, Brain and Behavior* instructs prospective authors:

> When experimental animals are used, the methods section must indicate that adequate measures were taken to minimize pain or discomfort. Experiments should be carried out in accordance with the Guidelines laid down by the National Institute of Health (NIH) in the USA regarding the care and use of animals for experimental procedures or with the European Communities Council Directive of 24 November 1986 (86/609/EEC). All studies using … animal subjects should include an explicit statement … identifying the review and approval committee for each study, if applicable. (http://www.wiley.com/bw/journal.asp?ref=1601-1848).

Journals such as *Behavioral Brain Research* published by Elsevier have a similar policy:

> "Reports of animal experiments must state that the 'Principles of laboratory animal care' (NIH publication No. 86-23, revised 1985) were followed, as well as specific national laws (e.g., the current version of the German Law on the Protection of Animals) where applicable." (www.elsevier.com/locate/bbr).

THE 3RS

In the United States and several other countries, regulatory bodies have embraced the principles of the "3Rs" first proclaimed in 1959 by Russell and Burch (1959). These principles asked researchers to continually seek to *refine, replace,* and *reduce* the use of live animals in research (Guillen, 2010).

Replacement

This means *replacing* live animals with tissue culture, computer simulation, or other techniques. In the study of mouse behavior, this option generally does not exist. Behavior by its very nature is an emergent property of a whole organism living in a fairly complex environment that includes other members of the same species. The laws of behavior are known only in a broad outline and are not established well enough to make quantitative predictions about behavior that do not need to be verified using live mice.

Reduction

At first glance, *reduction* of the number of mice used in a study would seem to be a good thing both financially and ethically. If the procedures impose pain or stress, then the fewer animals made to suffer, the better. Nevertheless, it is a bad principle that adds nothing useful to the fundamental principle and could compromise the quality of data if applied incorrectly by non-experts sitting on ethics committees. As discussed in Chapter 5, using too few animals tends to prevent the researcher from detecting real effects of a treatment, which in turn could waste the lives of all the mice in a study, including those in a benign control condition. Ethically, good research should use the *correct* number of mice. In many kinds of experiments, especially those evaluating interactions, this entails a larger number of mice than is customarily used. If an experiment is worth doing at all, it is worth doing the right way with the correct number of mice.

Most fields of science progress from an early stage where investigators measure things with really large effects to a more advanced stage where a sophisticated understanding of natural laws leads to experiments designed to detect rather small effects. For example, Galileo's proof that planets circle the sun relied on facts that were visually obvious when aided by a telescope, whereas Kepler's demonstration that planets move in ellipses rather than perfect circles relied on many precise measurements over many months and years of the locations of celestial objects. In the study of animal behavior, especially work on genetic influences, our science now has a serious interest in relatively small effects of a quantitative trait loci (QTL) (Flint, 2003; see Chapter 2), and much larger samples are required to detect such effects.

Some applications for ethics approval ask the investigator to list the databases of published studies that were searched over what time period with what keywords. The purpose of this exercise in an ethics review is to discourage duplication of studies, and it implies that duplication would violate the reduction mantra. Nobody disputes the importance of doing a thorough literature review before commencing any study of mouse behavior. Nevertheless, the no-duplication rule is fundamentally flawed because it denies the critical importance of replicating experimental results in other laboratories. Local ethics committees usually lack the requisite expertise to pass judgment on the state of the science in a small subfield and decide whether enough replication has been done, and the applicant almost always knows his or her own field far better than any local committee member. Grant adjudication panels, on the other hand, are well qualified to judge this matter. The author has never encountered a case where a research proposal was turned back because of duplication. This issue arises mainly in undergraduate labs that use live animals, not in research by professionals.

Refinement

The third R, *refinement*, is different. Refinement means we should seek ways to reduce the levels of pain or stress required by a test, shorten the length of a trial or the period of time spent in the mouse colony, and obtain data that have greater validity as well as precision. The idea is to elevate the quality of mouse life in the lab while obtaining higher quality data that meet the goals of the study. This principle is fully consistent with the edict against unnecessary suffering, and its realization enhances the scientific level of work with mice. Chapter 13 addresses task refinement in great detail. The imperative to seek task refinement does not add anything to the basic principle; it is just a more explicit statement of that principle.

Thus, while the basic principle of ethical practice in animal research, no unnecessary suffering, does not conflict with good science, some of the elaborations or additions to the principle (replace, reduce) tend to work against good scientific practice in the study of mouse behavior in particular. Refinement is a laudable objective but adds nothing of value. The three Rs have become a flag that commands the expression of loyalty to those who run the animal ethics establishment in many countries; researchers are compelled to salute it.

CASE STUDIES

Abstract principles are applied to mice by fallible humans. Two examples highlight some of the complexities and challenges. These should help to guide the researcher in planning experiments. They also illustrate why no simple formula or universally applied regulation is able to address all situations.

Death as an end point

As pointed out in Box 6.1, death as an end point is to be avoided, but that does not mean it is prohibited in all circumstances. Biomedical research in particular often causes disease in mice in order to discover cures. Cancer research frequently involves inoculation of healthy mice with malignant cells that grow into large tumors. Behavioral studies are sometimes done with mice to assess how the growth of the tumor is altered by social interactions, isolation rearing (Palermo-Neto, Fonseca, Quinteiro-Filho, Correia, & Sakai, 2008; Trainor, Sweeney, & Cardiff, 2009), or stress (Fernandez, Rezola, Moreno, & Sanchez, 2008), and the influence of the cancer on cage mates may be studied as well (Tomiyoshi et al., 2009). The central question for an ethics committee is whether death is really necessary to achieve the goals of the study. In most studies, it is not. For example, one might monitor the growth of a tumor by measuring its diameter and terminating the experiment long before the animal's quality of life becomes severely impaired. In a study where mice are infected with melanoma cells, metastatic foci in lung and other tissue can be counted after animals have been humanely killed (Fernandez et al., 2008).

Nevertheless, studies with mice are still being published that entail death as an end point. A recent study done in the United Arab Emirates included both tumor volume and percent survival in a study of *Salmonella* infection effects on a highly tumorigenic melanoma cell infection of mice (al-Ramadi et al., 2009). A study from Brazil infected mice with highly aggressive Ehrlich tumor cells and reported tumor cell numbers as well as Kaplan–Meier survival curves for control animals living in social groups and those housed individually (Palermo-Neto et al., 2008).

These studies point to a very difficult conflict of ethical principles that can only be decided by a committee of our wisest peers. By including death as an end point, it might be possible to detect smaller effects of a treatment than if only tumor volume or cell number were measured well before the animal became seriously ill and at risk for dying. If so, then fewer animals would need to be studied to detect an effect of any particular size. Which is best: fewer animals subjected to more suffering, including death in some cases, or more animals spared the pain of terminal cancer? There is no easy answer to this question.

Death as an end point is inherent in another popular field of mouse research, longevity and aging, where mice are housed in the usual low-stress lab conditions until they die from one cause or another. In many instances the cause of death will be some kind of cancer or a degenerative disease. One common experiment seeks ways to prolong life either through caloric restriction (Mair & Dillin, 2008) or by altering certain genes related to insulin and insulin-like growth factor signaling (Selman et al., 2009). Prolonging life is seen by many people as a very desirable goal, and achieving this in mice can be viewed as an enhancement in their lives. Nevertheless, the experience of an animal approaching the end of its life without veterinary help or amelioration of pain often involves substantial suffering. Researchers cannot intervene to treat maladies without compromising the goals of the study, and they cannot euthanize a clearly ill animal because days to death is the dependent measure.

In genetic studies where the focus is mouse behavior rather than biological disease, there is no behavioral test that requires death as an end point. There are only two situations where death of a mouse is at all likely. Water mazes and the forced swim test (Chapter 3) require mice to swim for several minutes, and there is a risk of drowning. The C58/J strain in particular sinks below the ripples on its first trial in a water tank (Wahlsten et al., 2005). Whenever mice are placed in water, it is imperative that the experimenter be present and watch them closely, then rescue any animal that sinks and remove it from the study. Provided this is done quickly, there will be no need for mouth-to-mouse resuscitation. The other situation is a test of fighting among males. The dominant mouse may wound and even kill the vanquished opponent if they are left undisturbed for too long. Hence, there is a need in these kinds of studies for continuous observation of the bout and intervention when one animal begins to wound the other.

Food deprivation

Ethics committees sometimes want to impose strict limits on how long a mouse can go without being fed. It is important to know whether there is a firm scientific basis for some specific time limit or if people are judging on the basis of a time without eating that would make them uncomfortable. There are two situations where food deprivation is used. In the first, hunger motivates a mouse to explore its environment and learn to perform a task to obtain a food reward. Satiated animals usually perform poorly or inconsistently. Thus, some degree of hunger is necessary for this kind of research. The second situation is the study of aging, where it is now clear that moderate caloric restriction prolongs life in many species (Koubova & Guarente, 2003; Longo & Fabrizio, 2002; Mair & Dillin, 2008).

If some level of starvation is essential, how much is enough or too much? A reasonable standard can be established based on objective criteria (Chapter 11). In almost all situations in behavioral research, the requisite duration of starvation will be far less than a dangerous level that imperils health or survival. C57BL/6J mice can survive for eight days with no food (Coleman, 1979). As shown in Chapter 11, 48 h deprivation is readily tolerated by most mouse strains without noteworthy effects on health. That duration is generally not required to motivate performance, however. The critical information in making an ethical judgment in this situation is the duration of starvation required to obtain consistent performance on a task by almost all mice. There is no good reason to starve them for a longer time than that. If the researcher uses 6 or even 12 h without food, many mice will not be well motivated and performance will be highly variable, and the entire study may be worthless. Conducting a worthless study transgresses the fundamental ethical principle. A full 24 h of starvation yields good performance on many tests of learning in mice.

The next question is whether 24 h of starvation compromises the health and welfare of the animals. There are two useful indicators of health status in a study that uses starvation. Overall activity of the animal is important. If it is too seriously depleted of energy reserves, it will tend to be lethargic and inattentive to the task at hand. Food deprivation and hunger generally increase levels of exploratory activity in the short run. After all, hungry mice are motivated to search actively for food. If deprivation is so severe that activity levels decline, this suggests it is too severe. Another good indicator is the speed with which an animal recovers from deprivation. This is evident in body weight gain and food intake. On both accounts, 24 h deprivation appears to do little or no lasting harm.

It is also important to consider the baseline to be used in a study involving hunger. Common practice in the mouse lab is to provide free access to rich food at all times, something that all but the most spoiled children do not enjoy. This contrasts strongly with the intermittent availability of food in the life of wild *Mus* species. Free feeding may seem kind, but over the longer term it is not, and it does not yield the best model of animals in the untreated condition (Martin, Ji, Maudsley, & Mattson, 2010). For many strains of mice, especially C57BL/6 and NOD/LtJ, it leads to obesity and diabetes (Surwit, Kuhn, Cochrane, McCubbin, & Feinglos, 1988). Restricting food to as much as 50% of the free feeding consumption actually improves the health and prolongs the life of mice (Selman et al., 2009). Thus, a reasonable argument can be made that some degree of food deprivation is good for mice and therefore ethically acceptable. The only real benefit from free feeding mice in a colony room is that it saves technician time; there is no need to weigh the mice and dispense food daily, just keep that hopper full.

CATEGORIES OF INVASIVENESS AND SEVERITY

In 1991 the CCAC established categories of invasiveness for procedures done with research animals, and every approved protocol since that time has been assigned to one of them. In 2009 the Expert working group on severity classification issued its final report to the European Commission (EC) that may soon become policy in most of Europe. In the United States there are no comparable guidelines on categories of procedures. The CCAC and EC categories are very similar (Table 6.1), except that the EC Mild category is split into B and C in Canada and the EC Severe category spans the Canadian D and E categories. Each policy statement provides specific examples to guide researchers and committees in rating invasiveness or severity.

Both jurisdictions make note of certain procedures involved in neurobehavioral studies. The CCAC assigns intraperitoneal injection of non-toxic substances to category B, minor surgeries such as biopsies to C, and major surgery with general anesthesia to D. Food deprivation similar to what an animal would experience in nature is B, somewhat longer periods of deprivation that cause only minor distress are C. Studies of fighting behavior or those involving noxious stimuli from which escape is impossible are assigned to D.

The EC report emphasizes that the final assignment to a severity category must be done on a case-by-case basis taking into account several factors. Mild procedures include intraperitoneal injection, ear or tail biopsies, short-term social deprivation, and brief exposure to mildly

TABLE 6.1 Categories of Invasiveness or Severity of Procedures

Canadian Council on Animal Care: Categories of Invasiveness (1991)		European Commission: Final report of Expert Working Group on Severity Classification, July 2009	
A	Invertebrates; tissue or cell culture	Non-recovery	Done under general anesthesia with no recovery
B	Little or no discomfort or stress	Mild	Short-term mild pain or distress; no significant impairment of well-being or general condition
C	Minor stress or pain of short duration	Moderate	Short-term moderate or long-lasting mild pain, suffering, or distress; moderate impairment of well-being
D	Moderate to severe distress or discomfort	Severe	Severe short-term or long-lasting moderate pain, suffering. or distress; severe impairment of well-being
E	Severe pain near, at, or above tolerance threshold		

Sources: *http://www.ccac.ca/en/CCAC_Programs/Guidelines_Policies/POLICIES/CATEG.HTM, http://ec.europa.eu/environment/chemicals/lab_animals/pdf/report_ewg.pdf.*

noxious stimuli that the animal can successfully avoid. Moderate includes surgery under general anesthesia (category D in Canada), food deprivation for up to 48 h in rats, and administration of a moderately noxious stimulus that the animal cannot escape or avoid. Severe procedures impair an animal's condition over a longer period of time or cause long-lasting moderate pain. Electric shock from which the animal cannot escape is considered severe, as is forced swim or exercise done to exhaustion. The EC guidelines explicitly deal with genetic mutations, grouping those that cause mild, moderate, or severe impairment under the respective severity categories. They also set a lower threshold below which procedures are not rated as mild distress, including food deprivation for less than 24 h in rats, mere observation of behavior, and open field testing.

Changes in the EC severity levels have been proposed by a joint committee of the European Science Foundation and European Medical Research Councils (ESF-EMRC;http://www.esf.org/ research-areas/medical-sciences/activities/science-policy/animal-protection-in-biomedical-research.html). Several explicit examples are recommended for various severity grades, including the addition to Moderate of "induction of anxiety in animal models." Whether this includes tasks such as the light−dark box or elevated plus maze that assess anxiety or is meant to apply to treatments that induce high anxiety is not entirely clear. Given that the ESF-EMRC position includes open field, staircase, and labyrinth tests under Mild, it appears that the light−dark box would qualify for a similar rating, but this is not made explicit.

CATEGORIES FOR BEHAVIORAL TESTS USED WITH MICE

It may be helpful to researchers working with mice to have a more general classification of behavioral tests according to invasiveness or severity class. To date, no organization has offered such a listing. The author would like to propose the scheme shown in Table 6.2. The categories were devised by first ranking tests in terms of level of stress or pain, then finding criteria that divided tests into groups. The categories were established without regard to the official invasiveness or severity categories in Table 6.1 and are not intended to substitute for those categories fixed by law and regulation. Instead, they demonstrate how a more finely graded scale

TABLE 6.2 Categories of Severity for Mouse Behavioral Tests Arranged From Least to Most

Category	Test
1. Unobtrusive observation, no stress; voluntary performance	2-bottle preference test in home cage; burrow and nest construction; IntelliCage; observation in home cage; observation from distance; running wheel
2. Free exploration with minimal handling	Elevated plus maze; hole board; light−dark box; open field
3. Very mild stress with handling; short duration	Air puff inhibitory avoidance; balance beam; body weight on scale; grip strength; intraperitoneal injection of saline
4. Mild stress with handling; repeated trials	Accelerating rotarod; Barnes maze; hot plate and tail flick analgesia; maze learning after 24 h deprivation; startle reflex and prepulse inhibition; water escape learning
5. Moderate stress, short duration	Active avoidance with low intensity electric shock; conditioned place preference using shock; mating behavior when female not receptive
6. Moderate stress, intermediate duration	Fighting among males, stopped before wounding; forced swim test (6 min duration)
7. Severe stress of short to intermediate duration	Fighting to severe wounding; inescapable intense shock; shuttle shock avoidance learning
Off the scale; torture; beyond level at which pain can be tolerated	Being eaten alive by fish, owl, or snake; glue trap; snap trap; warfarin poisoning

can be used when working with specific tests for mice. Categories 1 and 2 correspond approximately with level B in the CCAC system, 3 and 4 are similar to CCAC level C, 5 and 6 are similar to D, and 7 is E. In terms of the draft EC severity categories, Mild would include 1, 2, 3 and 4 in Table 6.2, while Moderate would be 5 and 6, and Severe would be 7. Things done to feral or food mice are shown as Off the scale.

Researchers and ethics committees might contest some of the ratings in Table 6.2. For example, the draft EC severity scale includes the forced swim test in the Severe category. This does not seem appropriate. For one thing, the mice are not forced to swim (Porsolt, Brossard, Hautbois, & Roux, 2007). They are placed in a small tank of water and allowed to swim freely for 6 min without any kind of interference. The dependent measure is the time they are immobile. Some remain active during the entire period, whereas others stop swimming and float. Some strains of mice will commence passive floating after a fairly short time in the tank, long before there is exhaustion. A period of 6 min spent swimming is far less than what a mouse might do on a running wheel, where it is common for one animal to travel the equivalent of 10 km in one night. Floating behavior complicates the interpretation of the test score, but it does not indicate a severe level of discomfort. The test would belong in the Severe category only if the experimenter allowed an animal to sink and commence drowning. Immediate rescue would preclude severe suffering, provided that the animal was not placed back into the tank for the remainder of the trial after it had become waterlogged. It would be a simple matter to count the time remaining in a test when rescue was needed as time immobile, because the animal certainly was not swimming actively after sinking below the ripples.

Shuttle or two-way avoidance learning belongs in Category 7 rather than 5, because mice are typically very poor at this task and receive many consecutive shocks, sometimes of long duration (Carran, Yeudall, & Royce, 1964; King & Mavromatis, 1956). The task involves a major conflict because an animal is required on one trial to run back into the chamber where it received shock on the previous trial (Figure 3.3). If, on the other hand, a mouse always receives shock in one distinct chamber and must run in one direction into another distinct chamber to escape or avoid the shock, learning can be quite rapid and only a few shocks of very short duration will be received (Wahlsten, 1972a,b), which places that task in Category 5.

THE FUTURE

Current trends in the regulation of animal research are clear. In almost all countries, the extent of regulation is increasing and the scope of things permitted in the name of science is shrinking. Hoops are getting smaller, hurdles are getting higher. Central authorities are striving to impose more explicit rules that leave less to the discretion of local animal ethics committees. Ethics regulation has moved from a voluntary system administered by peers in the research community to a legislated system run by professional ethics bureaucrats who do not do research, and in many cases have never done research with animals. Once there is a bureaucracy in place, evolution of the ethics system tends to follow a predictable course dictated by two fundamental rules of bureaucracy. First, rules and regulations are never rescinded; once empowered, regulators never back down. Second, regulators can demonstrate their worth only by creating more regulations; a stable, steady state suggests they are not doing their jobs. This generates conflict because researchers naturally want to spend most of their time in the lab or at the computer doing research and resent time wasted on counterproductive paperwork.

As the bureaucracy for enforcing the law becomes more detached from the researchers in their labs, the regulators take on the role of animal ethics police who approach the job with an inherent distrust of the researcher. In many jurisdictions there are now moves to implement longer application forms with more details and longer so-called "training" modules that must be passed before someone is allowed to work with animals. Ethics committees are permitting less flexibility in implementing approved protocols and demanding formal amendments be filed for even very minor alterations of a protocol. In the United States most universities and large research institutions now have a "compliance" office with a mandate to enforce the rules. At one American university the compliance officer has begun to search for and read journal articles published by researchers at her university to see if anyone has done something that was not approved in the original protocol. The traditional principle of collegiality and mutual respect in many institutions is being replaced by mutual suspicion and distrust.

While the regulatory frameworks are becoming more complex and onerous, there is very little serious discussion outside local ethics committees of what should be considered permissible or unethical in work with mice. It is very difficult to find intelligent discourse about specific protocols in the bureaucratic maze of Internet Web pages. This situation offers an opportunity for the mouse researcher because, in the final analysis, the judgment of what is ethical is the thing that really matters. The content is foremost, while the structure or form is secondary.

There is much to be gained if experts in the testing of mouse behavior take some initiative in discussing what is and is not good practice in their field. Nothing puts the spotlight on the most important questions better than informed discussion of real experiments done with live mice. We need to develop our own guidelines and try to achieve a consensus within the community of researchers about what is good and bad practice. This more than anything will help to move the ethics review process toward greater reliance on peer review.

Logistics

After choosing mice, tests, research design, and sample size, and having passed the ethics review and obtained an approved protocol number, it is time to plan the execution of the experiment in full detail. The objective of this chapter is to generate a complete list of every animal to be tested, its group membership and treatment conditions, and its order in the sequence of testing on specific days. The list is used to make a stack of data sheets, and this stack can then be given to the technicians who will actually do the testing, if one is fortunate to work in a lab where there are full-time professionals to run the tests from 9 a.m. to 5 p.m. every weekday. If not, this list is for you. It is usually helpful to create the list in the form of a spreadsheet that can be used for entering data as they are collected.

THE TOTAL EXPERIMENT

First on the to-do list is a spreadsheet containing the total experiment with all group conditions and every animal in the study. This task is an application of design principles introduced in Chapter 4. The user can adapt utilities from that chapter to create the initial spreadsheet. The order in this list should be systematic. The design of the study determines the number of groups and the power analysis determines sample size, so the total number of rows in the spreadsheet with one row per animal is groups × number per group, plus one row at the top for the variable names. Animal number within a group might be thought of as a replication number. If a study entails 12 groups with 10 animals per group, then the full 12-group design has 120 animals in 10 replications.

It is wise to place the variables representing groups in the left columns of the spreadsheet and the variable representing animal number within a group to the right. It makes good sense to place the group variables that are determined prior to the study, e.g., strain and sex, in the leftmost columns and those representing treatment conditions to the right of them. These are not fixed rules, just suggestions that make later steps easier.

Four examples are provided in Excel files (shown in bold type).

One-factor design

The utility for this is **20strains Design**: 20 strains with 25 females of each strain are to be evaluated for breeding and pup care performance. Each will be mated with an experienced and fertile stud male of her genetic strain, but the male will be in her cage only until mating has occurred. This will be a large study with 500 mice but a very simple design. Test order is an issue only because the lab does not have 500 breeding cages and mated females should be alone in the cage at the time of birth and nursing in this particular study. Thus, only a fraction of the total experiment can be actively reproducing at any one time. The stud males can each impregnate many females and therefore can be ordered in one shipment at the outset of the study, but females will probably need to be delivered in several shipments. Shipment is not an experimental treatment as such, but it could have an impact if something peculiar happens to one shipment and it should be recorded in the data and given at least a passing glance during the analysis.

Two-, three- and four-factor designs

One utility is **3 × 2 Design**: 3 genotypes and 2 sexes with 12 mice of each group yields a total of 72 animals. No sex effect is expected and equal numbers of males and females are to be tested just to make sure there is no imbalance in the results arising from unexpected sex differences. Each animal will receive four brief behavioral tests in a single test session. After all mice have been tested and euthanized, brains will be removed and studied anatomically. For a three-factor design use **4 × 2 × 3 Design.** In this design 4 strains and both sexes receive one of three treatments: unoperated control, sham lesion, and electrolytic lesion of the dentate

122

(A)

	A	B	C	D	E
1	Pre-sort	Strain	Sex	Treatment	#
2	1	129S1/SvImJ	Female	Control	1
3	2	129S1/SvImJ	Female	Control	2
4	3	129S1/SvImJ	Female	Control	3
5	...	...	...	...	...
6	...	...	...	...	...
7	84	BALB/cByJ	Male	Lesion	3
8	85	BALB/cByJ	Male	Lesion	4
9	86	BALB/cByJ	Male	Lesion	5
10	...	...	...	...	...
11	...	...	...	...	...
12	150	C57BL/6J	Female	Sham	6
13	151	C57BL/6J	Female	Sham	7
14	152	C57BL/6J	Female	Sham	8
15	153	C57BL/6J	Female	Sham	9
16	...	...	...	...	...
17	...	...	...	...	...
18	195	DBA/2J	Male	Lesion	6
19	196	DBA/2J	Male	Lesion	7
20	...	...	...	...	...

(B)

Ord	Sqd	Female	Male
A1	1	Control	Sham
	2	Sham	Lesion
	3	Lesion	Control
A2	1	Control	Lesion
	2	Sham	Control
	3	Lesion	Sham
B1	1	Control	Lesion
	2	Lesion	Sham
	3	Sham	Control
B2	1	Control	Sham
	2	Lesion	Control
	3	Sham	Lesion

Ord	Sqd	Female	Male
C1	1	Sham	Lesion
	2	Control	Sham
	3	Lesion	Control
C2	1	Sham	Control
	2	Control	Lesion
	3	Lesion	Sham
D1	1	Sham	Control
	2	Lesion	Sham
	3	Control	Lesion
D2	1	Sham	Lesion
	2	Lesion	Control
	3	Control	Sham

Ord	Sqd	Female	Male
E1	1	Lesion	Control
	2	Control	Sham
	3	Sham	Lesion
E2	1	Lesion	Sham
	2	Control	Lesion
	3	Sham	Control
F1	1	Lesion	Sham
	2	Sham	Control
	3	Control	Lesion
F2	1	Lesion	Control
	2	Sham	Lesion
	3	Control	Sham

FIGURE 7.1
(A) Portions of the spreadsheet **4 × 2 × 3 Design** that shows groups listed in a systematic manner prior to balancing the order of testing. (B) Twelve possible orders of testing (Ord) for male and female mice subjected to three treatments, with the stipulation that in each group of three test squads (Sqd), there will always be one of each sex in each treatment condition. In the first four orders, the female in squad 1 is always a control, in the next four she is in the sham group, and in the last four she gets a lesion. After the test order is completely balanced, order within a squad will be randomized so that a female is not always tested first.

gyrus of the hippocampus. Two weeks after surgery, each animal is given two behavioral tests run in one day. A partial list is shown in Figure 7.1A.

A four-factor design is implemented in the utility **8 × 2 × 2 × 2 Design**: 8 strains and both sexes are reared in either an impoverished or enriched cage environment for 4 weeks, then subjected to a battery of 4 tests over 8 days, with half of each group receiving an ethanol injection shortly before each day's tests and the other half receiving a saline injection. The 8 × 2 × 2 × 2 design has 64 conditions with 8 mice in each for a total of 512.

After the total spreadsheet has been constructed, it might not be a bad idea to insert a column at the far left to number every mouse from 1 to N, the total sample size of the experiment. This can make it easy to sort and return the spreadsheet to its original state, if needed. Once all the balancing and randomization have been done, this column can be deleted and the spreadsheet saved under a name that includes the word "Final."

THE PRINCIPLE OF BALANCING AND RANDOMIZATION

In deciding the order of testing the various groups and individuals, the objective is to *avoid any kind of bias or confound* of test order with group or treatment condition. Animals receiving a particular treatment should not be concentrated in the earlier or later weeks or days of a study, nor should they be consistently tested in the morning rather than the afternoon, or be first or last to be tested in a squad. However the total sample is divided into smaller blocks for testing, equal numbers of animals in each condition should be included in each block, and there should be no regular ordering of groups within a block. Whether this is best achieved with balancing or randomization depends on the study design and other logistical constraints.

Balancing is a systematic ordering of groups within a block to guarantee equal numbers of each treatment condition, strain, and sex in the block. *Randomization* assigns the animals to test order using random numbers, and randomness could result in a clustering of one kind of group early or late in the order purely by chance. For most studies of substantial scope, it is usually necessary to employ both balancing and randomization in order to achieve an unbiased order for testing. There is no one correct way to do this. Creativity comes into play here, and the researcher can have some fun working out the fine details in a clever way. Two different ordering schemes are equally good if neither one entails any kind of noteworthy bias in the order.

What kinds of factors should be targeted by balancing and randomization? Everything imaginable that might possibly have an influence should be taken into account, even if there is no firm basis in the published literature to prove the importance of a factor. One of the great merits of a perfectly balanced and randomized test order is that it even controls for unsensed, unexpected influences. The order cannot prevent the intrusion of extraneous effects into the data, but it can ensure those effects are not concentrated in any particular group.

THE TOTAL SAMPLE DIVIDED INTO SMALLER UNITS

Only rarely will it be possible to run the entire experiment in a few hours of one day. Usually the testing is spread over several weeks or even months. The total sample is received in the lab in several shipments, then broken down into even smaller units for testing. Several kinds of units are typically involved.

Shipment is several boxes of animals that arrive at the lab on a truck at the same time after a journey in the belly of an airplane or a larger transport truck. Unless the study is very large, it is generally a good idea to receive equal numbers of animals from all genotypes and both sexes in the same shipment. Sometimes the supplier is not able to meet this condition for practical reasons, and the final test order may need to be altered at short notice to fit what actually arrives in each box. It is always wise to have each shipment sent at the same age.

Cage is just that, a housing arrangement where all mice are of the same genotype, age, and sex. They are assigned usually 3 to 6 to a cage by an animal care technician as the shipment is unpacked. Within a cage, animals will be individually identified by tail marking with a felt pen, ear punching, a little tattoo, or a tiny radio frequency chip (RFID). Which mouse gets mark number 1, 2, or 3 is usually not done either systematically or randomly. Instead, it is haphazard. By all means do not assign the mouse with tail number 1 to the control condition, tail number 2 to treatment A, and so forth. There is a tendency for the animal care technician to grab whichever animal is easiest to capture and give it tail #1, continuing until that one last elusive mouse is trapped, pinned, and marked. A cage usually has fewer animals than conditions in the study and therefore sometimes cannot be used in any simple way to balance the testing. However, the cage must be taken into account. It is obviously a bad idea to make all animals in one cage be the controls and all those in the next cage be in the same experimental treatment condition, unless a treatment is to be applied to the whole cage; for example, an experiment on enriched living conditions.

As animals are unpacked, placed into cages, and individually marked, they can be assigned a formal ID code that remains with them for the entire time in the lab. This ID code is for tracking purposes only, and it can be assigned in a systematic order by shipment, strain, and sex. Trying to balance or randomize the ID code can add to the confusion and generate errors. The ID code should be inserted into the spreadsheet for the study and included in the final data set used for analysis.

Batch is a broad division of a shipment into two or more portions that are sometimes needed because all animals in a shipment cannot be tested in the same week. One batch will begin testing first, while those in the next batch remain in the colony room for another week or two awaiting their turn. Shipping is expensive, and fewer shipments will probably cost less. At the same time, it is not good to have large differences in age at testing. If mice are to be tested at 8 weeks of age, after they have achieved reproductive maturity, little harm will be done when a shipment is divided into three batches that are then tested at 8, 9, and 10 weeks of age. Little change over that range of ages is expected for mice. For rapidly developing animals younger than 6 weeks, however, a one week difference in age can involve a substantial difference in phenotype. Soon after birth, even a one day difference can result in a major change in brain and behavior (Wahlsten, 1974).

Age difference may be confounded by an additional factor: acclimatization to the lab. Animals should be given at least a week and probably more to get used to the new housing conditions and animal care personnel. If they arrive in the lab at 7 weeks of age and testing starts at 8 weeks, the three batches will have a range of 1, 2, and 3 weeks of acclimatization time. If instead they arrive at 5 weeks, the range will be 3 to 5 weeks of acclimatization, which is probably better than 1 to 3 weeks. It must be admitted that there is little experimental evidence that one range is clearly better than another. Many logistical choices are made on the basis of experience, compromise, and intuition from a researcher who tries to "think like a mouse." Some researchers are better at this than others. Numbers of shipments and division into batches also depends on the number of cages available in the animal colony. Whatever the constraints that lead to division of a shipment into batches, the test order must be carefully constructed so the batch is not in any way confounded with group across the entire experiment.

Session divides the working day into time periods, between which the experimenter takes a break for lunch or coffee. All animals to be tested in a session are placed into holding cages and removed from the colony room at about the same time, then transported to the holding area in or near the testing room. It is common practice to house animals individually in a clean cage prior to testing. At the end of a session, all animals are returned to their home cages in the colony room. A session should not exceed one or two hours, because it is generally not a good idea to isolate a mouse for very long. Depending on the scope of a study, the session may provide a convenient basis for balancing the study, so that animals from all groups are included in a single session. As with batch, session should not be confounded with treatment group.

Squad is the smallest logistical unit in a behavioral study. It consists of a small number of mice in individual cages that are tested closely in succession, ranging perhaps from four to eight animals. Typically the experimenter does not take a break while putting animals in one squad through their paces. It is common practice to bring all cages in one squad from the holding area to a location close to the test apparatus so that the animals can be efficiently handled. A squad usually has so few animals that all groups cannot be represented in one squad. Order of testing within a squad often can simply be randomized rather than trying to find some elaborate balancing scheme. The order of testing within a squad should never be done systematically by group. Mice and rats leave odors wherever they go, and they are very sensitive to odors from other animals. Apparatus is usually cleaned after each animal to reduce odor trails, but telltale odors cannot be totally eliminated. Males and females can smell each others' presence, and some treatments can alter odors. Randomizing test order within a squad ensures that lingering odors do not bias the results of the experiment as a whole. Because odors tend to fade and mix with former odors, this factor is an issue mainly within a squad, especially for the animal tested immediately after the previous mouse.

The physical locations of *colony, holding,* and *testing* areas depend on the specific lab (see Figure 1.3). A colony room may contain animals assigned to different experiments or even in the employ of different investigators, and in a large facility the schedule of daily activities by animal care personnel in the colony room is often beyond the control of the investigator. The holding and testing areas, on the other hand, should be quiet and free from unscheduled intrusions by anyone other than the person who is doing the behavioral testing. Whether the holding and testing areas are in the same room or adjacent rooms depends on local conditions and competition for space.

After completing the spreadsheet for the total experiment as previously described, it is a good idea to add several new columns. ID code should be inserted at the very left of the spreadsheet. To the right of the major group variables, include the variables to represent shipment, batch, session, squad, cage number, and individual mark number. Because cage number and mouse number are usually decided at the time of unpacking, it is convenient to list them after shipment but before batch, session, and squad.

Things can get a bit complicated in all but the smallest study. It is suggested that two variables be put at the left of the spreadsheet, one for the order of rows in the total experiment *prior* to any balancing and randomization, and another for the final testing order *after* it has been perfectly balanced and randomized. Having these two lists of N numbers can help to check and double-check the ordering of animals for testing by sorting and resorting.

THROUGHPUT

The required number of shipments, batches, sessions, and squads depends on how many behavioral tests will be given to each animal and the duration of those tests. The time to test one animal determines *throughput* — the number of animals that can be comfortably tested in one work day by one person. Suppose two tests are given consecutively to an animal: one trial lasting 5 min on the elevated plus maze and then one 10 min trial in the open field 30 min later. The experimenter must have enough time to clean the apparatus after each animal and set up everything for the next animal. It is reasonable to expect that things done outside of the trial will require about 5 min. Time devoted to working actively with one animal would then amount to about $5 + 10 + 5 = 20$ min. If the open field test is done with automated apparatus that monitors the animal using photocells or video tracking, the experimenter can do other things during that 10 min trial, such as clean the elevated plus maze and run the 5 min trial on that maze for the next animal. Thus, it might be reasonable to expect four animals to be tested by one person in one hour. This tight schedule could not fill a normal eight hour workday because there must be time allowed for breaks between sessions and moving animals to and from holding cages in the colony room as well as transport from colony to holding area. Six hours of active data collection should be attainable, therefore one technician should be able to test 24 animals in one day.

Such a crude estimate of throughput needs to be evaluated with real testing of live animals by using spares or veterans of another study that are recruited for the new purpose. Preliminary testing is especially important when trials are of variable length, depending on how well the animal performs. This happens in studies that involve learning of complex tasks. There is always some kind of upper limit set for the time allowed to complete one trial. Just how many animals approach that limit will depend on their genotypes and the nature of treatments, especially when they are likely to impair learning. As learning progresses over days, trials will become shorter, and the extra time between squads can be used for data entry or other housekeeping chores.

Recent work in the author's lab involving alcohol effects on tests of motor coordination showed that one experienced technician could process 32 mice per day, but 48 per day was just too many and required the person to work overtime. If higher throughput is really necessary, it is best to employ two people and carefully coordinate their work when there is only one piece of any particular apparatus available.

Just how long a trial should be is a topic addressed in Chapter 12. Generally speaking, longer trials yield more reliable data while reducing throughput. It is not easy to find the optimum balance between the two. When using a behavioral test where trial length has been more or less standardized in other labs, one can adopt their parameters. Because greater reliability of a test tends to yield lower within-group variation, the power and sample size calculation discussed in Chapter 5 should be based on prior information where similar task parameters were used to collect the data.

PARTITIONING THE WORK DAY

Having decided on the tests and trial durations, it is good practice to construct a detailed time schedule to make sure everything can fit within the allotted time. Only then can a correct number of mice per squad and number of sessions be calculated. For this purpose, a spreadsheet is useful (see **Throughput schedules**). Let one row be the smallest unit of time, which in the case of a test with a 5 min trial is 5 min. The first column can be animal #1, the second animal #2, and so forth. For the example with the elevated plus maze (EPM) followed 30 min later by the open field (OF), running six squads end to end with a 5 min break between squads extends the work day past 8 hours and allows no time at the end for general clean up and organizing. This *Long Day* schedule would be fine for a lonely graduate student but not a technician with a family.

The *Shorter Day* schedule divides the 24 animals into 3 squads of 8 and starts the EPM test of animal #5 during the OF test of animal #1. A bit of practice is required to coordinate the action smoothly. Although the length of a trial should be kept rigidly to the schedule, the time spent at rest in the holding cage can be *approximately* 30 min, and this flexibility should make it possible to test two mice at the same time. This schedule fits nicely into a work day with 7.5 hours, allowing time to get everything set up in the morning and cleaned at the end of the day. Thus, three squads of eight is a good choice.

A more complicated example for 20 inbred strains subjected to saline or ethanol injections and evaluated on seven tests over several days is described in three Excel files. **Alcohol Timing** shows the throughput analysis and test scheduling, **Alcohol Jax order** shows how mice were distributed across four shipments, and **Alcohol Test order Ship1** shows the balancing and randomization for mice in shipment #1. These elaborate spreadsheets were completed with the assistance of Elizabeth Munn at the University of Windsor.

THE BALANCING ACT

Next, the researcher must deal with the specific design of the study. Suppose a three-factor experiment is conducted as previously described, where four strains of mice, both males and females, are given three treatments (control, sham lesion, hippocampal lesion). There are $4 \times 2 \times 3 = 24$ groups. This is perfect for the given throughput of 24 mice per day. One full replication of the experiment can be tested each day. The first step is to sort the spreadsheet by animal #, which places all mice with #1 in the first 24 rows. Be sure to save this under a new file name.

Simply randomizing the order of testing the 24 mice within a day could be done, but this would allow for the possibility of having all of one strain concentrated in the first squad. Balancing requires that each squad of eight mice includes one of each strain as well as four males and four females. There should also be two or three mice from each treatment group. Therefore, rearrange the mice to give two of each strain in the first squad, one of each sex. Then try to assign treatment to two of each strain so the male and female in one squad never have the same treatment. Furthermore, if the female in one squad gets a lesion, make sure the male in the next squad also gets a lesion. This sounds easy enough but can strain the brain in practice. There is no generally recognized method to achieve the desired result for any possible design. Here are two approaches.

Enumeration

In the spreadsheet, all 12 possible enumerations of squad membership for males and females receiving three treatments are shown for one strain. Each of these satisfies the conditions that in each squad there will be one of each sex and they will always receive different treatments; and that for the three squads in one order, each sex will receive each of the three treatments. Figure 7.1B shows the possible orders (Ord) A1 to F2, each with three squads (Sqd). Do not employ sequence A1 for all four strains because there would be four control females in the first squad of eight. This would not be a serious error, but the researcher can do better with a bit of effort. Without regard to strain, there are six sex—treatment combinations, and a squad of eight must have two instances of two of these combinations. Try to choose the combination of sequences for the four strains in a way that minimizes the repetition of any one combination in a squad as well as between squads.

If a sequence for strain 129 is chosen from sequence group A or B, there will be a control female in squad 1. Then choose sequences from groups C or D and then E or F to avoid another control female in squad 1. Squad 1 must have a duplicate of one of the six conditions; for the first replication, let there be two control females. If A1 is assigned to 129, perhaps B2 can be used for BALB, then D1 for C57, and F1 for DBA. The spreadsheet ($4 \times 2 \times 3$ **Balanced**) shows in **Sheet 1** that there are a few more duplications of sex—treatment combinations in a squad than desired. This is not a serious shortcoming, but it is inelegant.

Latin square

When six objects are to be arranged in different orders and the objects cannot communicate with each other or exert an influence on those adjacent in the order, the Latin square can be used (Table 7.1). Objects can be shifted to the right on each step or to the left on each step, but left and right should not be mixed in the same balancing operation. In the square for six objects, objects are shifted to the right by one position for each order. For six objects, there are six orders, and each object occurs only once in any one place in the sequence. For the example with six sex—treatment combinations, the squad numbers can be shifted to the left by one step, as shown in the spreadsheet ($4 \times 2 \times 3$ **Balanced**, **Sheet 2**). This yields three unique sequences. Then append these three end to end in three different sequences and assign them to

TABLE 7.1 Latin Square to Determine Order of Six Tests

	1st	2nd	3rd	4th	5th	6th
Order 1	A	B	C	D	E	F
Order 2	B	C	D	E	F	A
Order 3	C	D	E	F	A	B
Order 4	D	E	F	A	B	C
Order 5	E	F	A	B	C	D
Order 6	F	A	B	C	D	E

the first replication. The fourth strain must get a duplicate sequence, perhaps 123123, the same as strain 129. This method yields two duplicate sex–treatment combinations in every squad, a 129-DBA duplication in each case. Now, across replications, assign the sequences so that different pairs of strains are duplicated in a replication. In the spreadsheet the duplicated strains are highlighted in yellow. The nine replications have five duplications for two strains and four for the other two. It cannot be perfectly balanced when there are four strains and nine replications.

There are probably many other ways to balance the sequences. An expert in combinatorics or someone who enjoys working with puzzles may devise a more elegant solution. A criterion for an acceptable level of balancing can be proposed: if there are no biases in the end result and no human could possibly determine how one made the sequences just by looking at the end product, it is good enough for use with lab animals. It is not necessary to adopt extreme measures to achieve perfection.

RANDOMIZATION TO THE RESCUE

Assign the nine sequences to the nine replications and then sort on replication and squad. The method has achieved a major goal: every squad has two of each strain and four of each sex, and there are never more than three or less than two of any treatment. Nevertheless, there is an obvious problem — the four strains are always listed in the same order within a squad. Once a decent balance within a squad has been achieved, rearrange the order within a squad without changing that balance. At this point, any systematic scheme would overload the synapses, so it is time for randomization.

In a separate column to one side of the data, insert the equation = rand() into each cell by copying down the column (4 × 2 × 3 **Balanced, Sheet 3**) . This Excel function generates a random number between 0 and 1.0. Note that all the numbers will be recalculated when any change is made in one of them, unless automatic recalculation is turned off. For clarity, it is recommended that automatic recalculation be turned off, so that a recalculate command must be given when ready by clicking on the Recalculate symbol in the bottom task bar of Excel or going to the Formula-Calculation menu and selecting Calculate Now or pressing F9. Copy and paste the random numbers into a column to the immediate right of the data, using the Values Only option for the copy operation so that the formulae are not copied. Sort the columns of data by replication, squad, and random number. Voila! The sequence within a squad has lost all semblance of bias or consistency.

Now there is a finely balanced and randomized list for the experiment. Copy it to a new spreadsheet for use in the study. It is probably wise to add a variable to indicate test order within a squad. After the data are collected, it will then be easy to check whether being in the first squad of the day or early in the test order within a squad made any difference in the outcome of the tests. One hopes they did not, but if they did, this source of variation ends up in the within-groups variance and does not bias the test of treatment effects. Why? Because the test order sequence has been carefully balanced and randomized.

Finally, add a variable to indicate order in the entire experiment for each animal after balancing and randomization. This makes it easy to return the list to the final order after the next step or after double-checking other aspects of the list.

SHIPMENTS, CAGES, TAIL MARKS, AND ID NUMBERS

For the study with a $4 \times 2 \times 3$ design and nine animals per cell, it might be convenient to receive the animals in two shipments and have the second shipment acclimatizing to the routine of the colony room while the first shipment is tested. Nine mice per group cannot be divided into equal halves. It would make sense to order five for each cell in each shipment to provide a spare in the event that something goes wrong. Perhaps an operation on an animal in the first shipment was not successful or a control animal died with no explanation. In a large study, those things happen, and it is wise to plan for them. The tenth day of the experiment can then be devoted to filling the gaps in the data. For an experiment involving surgery, this may require doing one extra surgery per cell ahead of time. The researcher may then decide to test the extra animals on the tenth day, even if none in a group perished or went awry. Better to have one more than the minimum sample size than to fall below the quantity stipulated by the power calculation.

The design has $4 \times 2 \times 3 = 24$ cells and each cell contains 5 animals in one shipment, so there will be 15 animals of each strain and sex in the shipment, 5 of which will receive each treatment. This is a convenient number for arranging animals in cages. If they are housed three per cage, there can be one control, sham, and lesion animal in each cage.

When the animals arrive in their shipping containers, with all 15 in one container being the same sex and strain, they can be quickly distributed across the five housing cages using any kind of system; just grab and lift as they are caught. It is common practice to identify the animals in a cage before they are assigned to a treatment condition. In the present study, the two steps can be done as one. Suppose animals will be tail marked with one, two, or three bands of a felt pen. Catch the first and make it #1, the second #2, and the third #3. The balancing done earlier ensures that these tail numbers will be distributed evenly across treatment groups. All with tail #1 will be in squad 1. To the extent that the first captured mice are tamer than the more elusive ones, squad 1 will tend to be a little more tame than later squads. This minor bias will not be confounded in any way with the experimental group in the study and is probably not a bad way to start the work day.

Because balancing and randomization have already established the test order, there is nothing further to be gained by randomizing what strain goes in what cage number. Rats and mice cannot read and do not care. Better to keep the unpacking step as simple as possible. Just number the cages sequentially as shown in the spreadsheet. Likewise, assign tail numbers #1, #2, and #3 sequentially.

At this stage of the work, it is time to assign a unique subject ID code that identifies each animal. It should be on the cage tag, the data sheet, and computer files of data. One method is to start the code with a letter, followed by at least four digits. The author has used A for Alberta, W for Windsor, and now G for UNC Greensboro. You could use your own initials if you anticipate moving often. The ID codes can be assigned in the order mice are unpacked and written directly on the cage tags at that time.

The spreadsheet $4 \times 2 \times 3$ **Final** entails three sheets with the final specifications of the entire study. **Sheet 1** has the final testing order, taken directly from $4 \times 2 \times 3$ **Balanced (Sheet 4)**, with random numbers omitted and order within squad added. **Sheet 2** shows the subject ID codes, cage assignments, and squad membership for all mice. This sheet can be used during unpacking and tail marking. **Sheet 3** is the final test order by replication, squad, and order within squad, showing ID codes and cage numbers for each animal. This is the final sheet that will be used to coordinate the daily testing throughout the study.

CAGES IN THE COLONY ROOM

One caveat applies to cage numbers. It is good to number in sequence as they are unpacked, but the cages should not be placed on the rack in order by cage number. Otherwise, all mice of strain 129 would be on the top shelf, closest to the ceiling lights, which can alter vision (Greenman et al., 1982). It is good practice to distribute the cages of different strains and sexes across the rack so there is no preponderance of any group in any location on the rack. Be sure to have the designated cage number written in large numbers on the cage tag and explain to animal care personnel the distinction between cage number and location.

With a clear numbering system for cages and tails, it should be an easy matter for a technician to enter the colony room with a cart and collect the eight mice for the next squad to be tested. Putting a note on the tag about strain and sex in the cage serves as a double-check on animal identity in a squad.

THE DATA SHEET

The spreadsheet is the basis for ordering mice for testing, but it is not the thing to use for daily work. Instead, the work day should be organized as a stack of data sheets, one per animal, in the order specified by the spreadsheet. If there are to be 3 squads of 8 animals each, there will be 24 data sheets for that day, with each group of 8 joined by a paper clip or in a notebook in the correct order. It is good practice to prepare *all* data sheets, 240 of them in the case of the $4 \times 2 \times 3$ design with 9 animals per cell, before the experiment commences.

The data sheet should have an explicit location for writing each item of information given in the spreadsheet, plus a place for the actual date and time of testing, the name of the person who conducts the test, and all of the data to be collected. If some of the data are collected by computer, then the data sheet should have a place to record the file name for that animal. It is most efficient if places for each datum are set in the order most convenient for typing into the spreadsheet at day's end.

It is also good practice to provide plenty of space for notes on animal behavior and unusual events that might influence test results. These may not be entered into the spreadsheet, but they could prove helpful if the data for an animal are found during statistical analysis to be out of the range for others of that group or strain.

FINAL CONSULTATION

The person in charge of organizing the experiment may be in a position of authority over the technicians or students who will do the actual testing. Authority does not always equate with being correct about everything. It is a very good idea to review the entire plan for the study, including the data sheets and testing schedule, with technical staff before day 1 of testing begins. Presumably the staff were consulted during the planning stages. Nevertheless, the last consultation is perhaps the most important. Experienced technicians and keen-eyed students may notice something that the PI forgot to consider or a blip from a slip of the cursor along the way. Better by far to discover the flaw before testing begins, even if it means some embarrassment. A little flick to the ego is nothing compared with the abject humiliation that follows if a big flaw is found in the middle of the fifth replication of a study with six replications or is found during the data analysis stage. Especially when a study entails several groups and order of testing is balanced across several conditions that are not part of the formal design, things can get complicated and devilish mistakes can be elusive.

OTHER EXAMPLES

The one-factor design, involving 20 strains with 25 females of each strain evaluated for breeding and pup care performance, is enumerated in **20strains Design.** Mice are ordered from Jackson

Laboratoiem in five shipments with four females of each strain in a shipment, and then housed four per cage until mating. The stud males arrived earlier and are now ready for breeding. Number of shipments was decided by the number of cages and space available in the colony room for the study. Each shipment will be mated in two batches about two weeks apart, so that the lab is not completely swamped by dozens of new births in a few days. It is a good idea to balance across rack position in a study where females will be bred in the same colony room, probably on an adjacent rack. Reproduction is known to be very sensitive to odors, lighting, and noise. The spreadsheet **20strains Shipping** assigns animal ID codes and colony housing cages, distributing the various strains across the cage rack so there is no systematic bias in which strain is located where. The method used to do this is given on the fourth sheet *Notes on methods*. It is very important there is no bias in the location of the mating cages; hence the assignment to mating cage numbers is done for the 40 females within a replication and batch using random numbers, as shown in **20strains Mating**. The final list of mating cages for use in the lab appears on the third sheet of **20strains Final**. When the time comes, mating cages 1 through 40 for the first batch are then simply arranged in that order along the shelves of the rack. The next batch of cages 41 through 80 continues along the same shelf and probably extends to a new rack, depending on rack capacity.

The two-factor design with three genotypes and two sexes is specified in **3 × 2 Design**. There is a knockout (KO) strain with a targeted mutation originally created in the 129S1/SvImJ strain and then backcrossed onto C57BL/6J. Those two inbred strains are then included in the study as controls. The study will involve 12 replications, all from a single shipment of mice. There will be six mice in a squad, one from each group, and therefore all in the same replication. Order of testing is randomized within a squad. The spreadsheet **3 × 2 Final** shows the method for determining the final test order with brief notes on the method in the fourth sheet *Notes*. The final spreadsheet does not address the issue of how many mice per cage, which animal is in which cage, or how to assign tail numbers. The reader may wish to add those features to the spreadsheet.

MORE INTRICATE DESIGNS

Larger designs that can be applied with mouse research are presented in advanced texts on experimental design that are often the basis for an entire graduate course lasting one term or semester (Box, Hunter, & Hunter, 2005; Hinkelman & Kempthorne, 2008). Should the need arise, the user will then find it necessary to apply the principles of balancing and randomization to the specific design. No matter how intricate the design, the basic reasons and ways to create a test order that lacks any substantial biases or confounds are essentially the same in almost all work done with mice. Every study done in the mouse lab will have shipments, sessions, batches, squads, and cages.

Getting Ready for Testing

With a design, sample size, and detailed order of testing carefully prepared in advance and permission granted to start work, there is just one ingredient missing: the mice. It cannot be overemphasized that mice should not be procured until everything is ready for action, including all the apparatus and test protocols as well as technical staff trained in the procedures. That idyllic period between sending the order for mice and testing the first animal is not a good time for refining tests, getting the "bugs" out of apparatus, or learning how to give an injection with a hypodermic needle — unless one enjoys cavorting near the threshold of panic. Several steps are involved in getting mice to the point where testing can begin.

ORDERING MICE

When mice are purchased from a commercial supplier, timing of the order is critical. It is imperative that the supplier is consulted long in advance of the study. In fact, it is wise to contact the supplier about availability of specific mice at the design stage of the study. Some of the rarer genotypes are available only as frozen embryos, and several months are required to generate live mice from them that are ready for shipping. Other strains that exist as live mice may be few in number and two or three generations may be needed to expand the numbers enough to fulfill the needs of a large experiment. At a minimum of three months per generation, half a year might be needed to fill the order. Even this estimate may not be accurate, because the fertility of some inbred strains is marginal at best.

Web sites for reputable mouse breeders give many details about pricing and availability, but it is still essential to contact a live person at the facility to obtain a realistic time estimate to fill a specific order. Keep in mind that many laboratories across the country are doing experiments in the same time period and often will be ordering the same strains of mice. One large order from a major lab can quickly exhaust the supply of even the most common strains. The author's experience has been that Jackson Laboratory tries to maintain a surplus of the strains in highest demand, such that a small order can almost always be filled on short notice, but there is no guarantee about numbers and timing. Yes, buying mice is much like buying shrimp at the market; if somebody throwing a large wedding reception was there just an hour before you, there may be no more at the store until the shrimp boats return to the dock. You will need to wait, but at least your new stock will be fresh.

If the experiment entails several genotypes, arranging an order can be a logistical challenge. Staff at the supplier will have many years of experience with this kind of order and can be very helpful, but the researcher must be very specific about numbers and timing. At any one time in a breeding facility, there will be a certain number of females of a given strain pregnant and their expected birth dates will be known. There will also be a fixed number of live pups nursing, and their weaning dates will be known. In effect, when an order is placed, some of these nascent mice will be reserved for your order. When a study involves many genotypes, with some rare strains, things can get complicated because good research design calls for all genotypes to be shipped at the same time. The general rule is that more complicated study designs require longer advance notice to the breeder.

No list of commercial suppliers is provided here because these will usually be specific to one's country. Local animal care personnel will know of several breeders. Alternatively, the new investigator can find suppliers from the methods sections of recent articles or an Internet search.

SHIPPING

Reputable suppliers either ship the mice in company trucks or deal with a shipper that knows how to transport live animals. Climate control during transit is essential. The researcher hopes the journey will be smooth and free of stress for the mice, but this never happens. Boxes of mice get tossed and even dropped from time to time, they bounce when a truck strikes a pothole, and they slide to and fro when brakes are applied suddenly. The noise levels in the cargo hold of a jet airplane far exceed the pristine calm of the breeding room. Any way it is measured, the transit from supplier to lab is a radical change in environment that is full of stress. Mice often arrive in a disheveled state.

Consequently, it is essential to allow a considerable time for mice to acclimatize to their new surroundings. Two weeks, enough time for at least two cage changes, is a bare minimum for this purpose. The animals need to adapt to the new psychological environment, including new odors and animal handlers, and they also need to become accustomed to new food, water, and bedding material as well as cages. Very few research labs duplicate the conditions at commercial breeders. For example, mice at the Jackson Laboratory are maintained on acidified water with a pH of 2 that inhibits the growth of bacteria, whereas many labs use local tap water.

Given the rigors of shipping and the substantial change in environment in the new lab, it would not be surprising to find that behavior on tests of anxiety would be different for mice purchased from a commercial supplier versus those bred locally. Shipping was included as a study factor in a large experiment that was done simultaneously in three labs (Crabbe et al., 1999). Eight genotypes of mice were either shipped from the supplier or bred locally, and everything was timed carefully so that mice were nearly the same age in both conditions on the day testing commenced. The absurdly difficult logistics and travails of making the arrangements have been described in detail (Wahlsten et al., 2003). To the surprise and even disappointment of the harried researchers, there was no significant effect of shipping versus local breeding (Figure 8.1).

One study cannot rule out shipping effects for other phenotypes and genotypes in different labs and years. Shipping is a factor that deserves more attention because it often differs between laboratories, some of which breed locally while others order from afar. At the present time, there is no evidence to suggest that either option is clearly better than the other. The local situation will usually dictate the choice.

Shipping effects may also be present when a large study involves more than one shipment of mice that arrive in the lab in different months. Many things can change in a few weeks at the supplier, in transit, and at the local lab. Shipment is usually included in the research design as a control factor. Provided the study has been properly balanced for shipment effects (Chapter 7), shipment can be assessed in preliminary analyses. If there is a main effect of

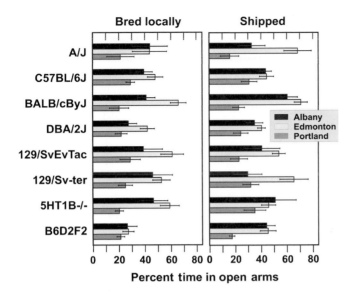

FIGURE 8.1

Comparison of performance (mean scores ± standard error of the mean) in the elevated plus maze for eight mouse genotypes tested simultaneously in three labs after being bred locally or shipped from the suppliers. (*Based on data from Crabbe, Wahlsten, & Dudek, 1999*). Mice in both conditions were the same age at testing. Challenges of conducting this kind of experiment are discussed by Wahlsten et al. (2003).

shipment but no interactions of shipment with study factors (Figure 8.2), interpretation of the results will not be complicated by the use of several shipments. When there is a main effect of shipment, the analysis may be more sensitive to treatment effects if shipment is included as a factor; otherwise, variance arising from shipment effects inflates the within-group variance, reducing power to detect treatment effects.

135

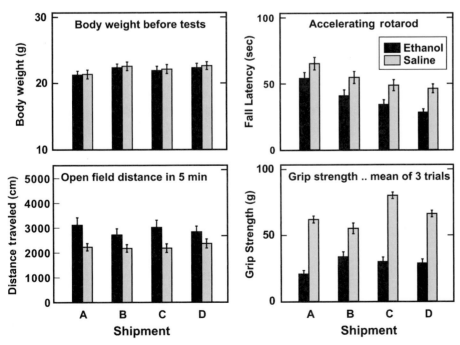

FIGURE 8.2

Mean scores ± standard error of the mean of four phenotypes averaged over 20 inbred strains of mice from the Jackson Laboratory that were delivered to the author's lab at the University of Windsor in four shipments. Half of the mice in the study received an injection of ethanol before certain tests and half received saline. There was a noteworthy decline in fall latencies on the accelerating rotarod across shipments, whereas other measures changed little with shipment.

Although there is no fixed rule about the age when mice should be shipped, it is generally good practice to bring them to the lab while relatively young, perhaps five or six weeks of age, and then allow a fairly lengthy acclimatization to the new surroundings. Mice younger than five weeks tend to be hyperreactive and prone to sound-induced seizures (Henry, 1984; Henry & Bowman, 1970), and they should not be shipped during that sensitive period. Most commercial suppliers charge more for older mice; hence it may also be cheaper to buy them young and house them in the lab, provided the per diem charge for care is not excessive.

Shipping between labs is fairly straightforward within a country. Many labs save the sturdy shipping containers that come from good breeders and reuse them for collaborative work. Crossing international boundaries can be a challenge because of customs and quarantine requirements for live animals. The author knows of instances where valuable knockout mice were covertly carried across a border in the investigator's pocket. This audacious action is not recommended. Mice will not trigger a metal detector, but they do have a peculiar odor and sometimes complain loudly. A loose mouse on an international flight could create mayhem.

UNPACKING

The shipment should be closely tracked from its origin, so that the lab can be ready for action. It is good practice to unpack a shipment of mice as soon as possible after they arrive at the lab. Perhaps the greatest peril in transit is excessive heat that can cause dehydration. The sooner the animals are placed into air-conditioned comfort in their new environs, the better.

Most reputable suppliers ship mice in very good condition, complete with a certificate stating they are free from a long list of potential infections and parasites. Good shipping containers will keep most vile infestations out of the box and away from the mice. The outside of the shipping container, however, will become host to a wide variety of pestilence. Never, ever should a shipping container be brought into a clean colony room. The best option is to unpack the mice in a separate "dirty" room or, if not available, then in the hallway on a "dirty" cart placed just outside the colony room. Two people must be involved in the operation, one to work on the "dirty" side and the other to handle the clean mice. A clear line needs to be drawn between the "dirty" sector and the clean zone. Separate carts can help to make the distinction clear.

The basic operation is quite simple. The dirty worker opens the container by touching only its exterior, then removes the lid and stands to one side. Then the clean worker picks up each mouse by the tail and quickly transfers it to a clean cage, never touching anything in the dirty zone. As a precaution against a mouse that jumps out of the container, a clean piece of paper under the box may keep the escapee clean for a few moments until it can be captured by the clean tech. Once all mice have been transferred, the dirty worker removes the used shipping container and brings the next one for processing.

MARKING FOR INDIVIDUAL IDENTIFICATION

Several methods are effective for identification of individual mice within a cage (Deacon, 2006). Each has certain advantages and liabilities. Which mouse within a cage is assigned mark #1, #2, and so forth, is not important because balancing and randomization (Chapter 7) preclude any confounds with mouse number. Consequently, it does not matter whether individual identities are established soon after arrival in the lab or shortly before the start of testing. A very useful description of several methods of marking and other procedures with lab mice is available from the National Human Genome Research Institute Office of Laboratory Animal Medicine at NIH (http://oacu.od.nih.gov/additional/biomethod.htm).

Tail marking with a waterproof felt pen is quite effective, even on black mice, using one to four bands plus no mark to identify five mice in one cage. The principal drawback is that the marks disappear in a few days as the skin cells slough off, and marks must be reinforced frequently, more than once per week. This method is useful mainly for short-term experiments when the marks are assigned just before the first measurement.

Ear punching is done with a special tool that makes a small hole near the outer margin of the ear. This is a form of tissue biopsy, although the patch of ear is usually discarded. Provided the instrument is kept very sharp and sterilized between mice, there is little pain or bleeding and minimal risk of infection. The critical skill for the technician is to hold the mouse properly so that it is immobilized briefly. A system for identifying more than 10 unique individuals is easy to devise by using different locations along the left and right ears. The hole from an ear punch should last many weeks.

Microchips are now available that can be injected under the loose skin on the nape of the neck and their codes are read with a radio frequency transponder from a distance. Radio frequency identification (RFID) is used to provide a unique signature for pets and livestock, and miniature devices are available for work with mice. This method is used in the IntelliCage (www.newbehavior.com) to determine which specific animal has arrived at a feeding station. The main disadvantage is that the chip is implanted under anesthesia because of the large size of the needle, but the period of anesthesia can be very brief. The RFID reader is costly but can be used many times, so that the cost per animal is low. The microchip has two potential advantages over tail and ear marks, methods that establish identity only within a cage. First, if there is a major disaster that leads to mice being spilled from many cages, every animal can be returned to its proper place. Second, the ID number is unique to the individual even across many studies and can be attached to tissue samples and computer files of data to ensure the right mouse is included in every analysis.

Metal ear tags can be used with mice but are close to the lower size limit for this technique. The principal drawback is that mice sometimes scratch at and remove them or tear them off during vigorous activity. If this happens to two mice in one cage, unique identity is lost.

Tattooing can be done with a miniature version of the multi-needle tattoo machine. The author has tried this and found that mice do not care to have large needles jabbed into their tails. Tattooing hurts. Considerable skill is required to make more than one mark.

Toe clipping causes pain and bleeding as well as risk of infection and is no longer used.

HOUSING

Many features of life in the colony room can have a substantial impact on the mice and their behaviors during testing. Certain things such as ambient noise can be important, but very difficult to control in an existing building (Turner, Parrish, Hughes, Toth, & Caspary, 2005). As discussed in Chapter 4, the position of a cage on the shelf can influence vision (Greenman et al., 1982). These and other aspects of the lab environment will be discussed further in Chapter 15. Three features of the colony room warrant further comment here.

Group versus individual housing

Mice are social animals, and it is best to house them in small groups of the same sex. This option is also more economical than individual housing. Most mouse cages can accommodate three to five animals comfortably. At first glance it may seem that keeping just one animal in a cage results in greater control and uniformity of environmental conditions. A large research literature, however, indicates that housing a mouse in isolation makes it more difficult to handle, increases the tendency of males to fight with each other when they are later placed in the same cage (Hood & Cairns, 1989; Lagerspetz & Lagerspetz, 1971), and has substantial and detrimental effects on many behaviors (Brain, 1975; Martin & Brown, 2010). Isolation can also modify effects of experimental treatments such as stress (Arndt et al., 2009) and drugs (Koike et al., 2009). Group housing is not the ideal solution in all situations. The consequences of cohabitation for male mice of some genotypes can be devastating; in a recent study half the males were killed by other males (Deacon, 2006). Isolation housing is practiced primarily when group-housed males fight and must be separated.

Cage enrichment

Mice can be housed in a simple plastic cage with nothing but bedding material to dig in and use for nesting. It is not difficult to add plastic objects to the cage for the animals to manipulate, and a small running wheel can fit into many cages too. It is evident that enrichment of the living cage can provide some benefits to the mice. High levels of enrichment alter gene expression (Rampon et al., 2000a), improve motor coordination and complex learning (Rampon et al., 2000b; van Pracig, Christie, Sejnowski, & Gage, 1999; van Pracig, Shubert, Zhao, & Gage, 2005; Wainwright, Huang, Bulman-Fleming, Levesque, & McCutcheon, 1994; Wainwright, Levesque, Krempulec, Bulman-Fleming, & McCutcheon, 1993), and create a more complex brain (Benefiel & Greenough, 1998; Greenough, Black, Klintsova, Bates, & Weiler, 1999; Kempermann, Gast, & Gage, 2002; Mohammed et al., 2002). Insufficient information is available concerning how this practice may alter a wide variety of behavioral processes (Benefiel, Dong, & Greenough, 2005) or how much enrichment of a mouse cage is needed to create a noteworthy improvement (Bailey, Rustay, & Crawley, 2006). There is a very sparse literature on the effects of modest levels of enrichment within the confines of the typical mouse cage, which is the kind of enrichment most likely to be implemented in most labs. It has been suggested that cage enrichment will not compromise many tests (van de Weerd, Van Loo, van Zutphen, Koolhaas, & Baumans, 1997; Wolfer et al., 2004); nevertheless, there is evidence that a modest degree of enrichment can influence behavior in a strain-dependent manner (Tucci et al., 2006; van de Weerd, Baumans, Koolhaas, & van Zutphen, 1994) and alter the effects of certain drugs (Solinas, Thiriet, El, Lardeux, & Jaber, 2009).

Light–dark cycle

Every lab makes a difficult choice about the light–dark cycle. Mice are nocturnal mammals with a circadian rhythm strongly entrained by light (Takahashi, 1995). They show relatively short periods of activity during the light phase. Levels of expression of many genes change with the time of day (Panda et al., 2002). Inbred strains differ substantially in the way they apportion activity between light and dark periods (Mouse Phenome Database entries for Mogil2, Seburn1; http://phenome.jax.org; Tankersley, Irizarry, Flanders, & Rabold, 2002). Although a good argument can be made for maintaining a behavioral test facility on a reversed light–dark cycle, animal care personnel prefer to do their jobs in the light phase. Perusal of the literature on mouse behavior genetics indicates that some labs use a reversed cycle and others do not.

GOING TO SCHOOL: TEST DAY

After the mice have been unpacked and marked, and given time to get used to their new cages, food, water, and animal handlers, it is time to begin mouse school and take some behavioral tests. While the animals are acclimatizing, the experimenters can be arranging apparatus and data sheets, double-checking the testing schedules, and running a few practice animals through the entire procedure to ensure there are no unpleasant surprises on the first day of formal testing. There are many ways to arrange events on the test day, and most ways are good enough, provided the researchers adhere to a few basic guidelines.

Consistency across days within a single study and between studies in a single lab is very important. Consistency means doing the same things at the same times each day, performed by the same people, if possible, especially when more than one person is doing the testing, a detailed, written schedule for each day is advisable. The alternative is to allow each technician to organize the test day. This could create a confound between the day of testing and the effects of seemingly minor procedural differences that could influence results. Once a detailed procedure is adopted in one lab, it can be applied to subsequent studies in the same lab, enhancing replicability of results within the lab.

Run two or three test animals at the start of each day to make sure all equipment is functioning properly and supplies are adequate. If there is any kind of problem with apparatus, it is far better to discover this with a practice mouse. A little practice can also help to get the experimenter into the groove for the day's work.

Experimenter effects need to be taken into account when planning the test day and week. The person who administers tests that involve considerable handling can have a major influence on results. In one large set of data on pain sensitivity collected over several years in one lab, the largest single influence on test scores was the experimenter (Chesler et al., 2002). Because of this, it is best practice to balance testing throughout the day and week with respect to experimenter. If test order within a day is properly balanced, there is no harm in having one person administer the morning tests while someone else does the afternoon. Experimenter is then a control variable (Chapter 4), and it will be confounded with time of day. If, on the other hand, a battery of tests is to be given over several days and different tests are given each day, it would be unwise to have one person do the testing on Monday and another on Tuesday, because this would create a confound between experimenter and the kind of test.

An example from a recent study in the author's lab is shown in Figure 8.3. Mice from eight inbred strains were tested in the open field before and after an ethanol injection. Half of the animals were tested by one experimenter and half by another, with test order balanced carefully for experimenter. There was no difference between experimenters before the injection, which is not at all surprising because the open field test entails only very brief handling of the mouse at the start of a trial. After the injection, results depended strongly on which person gave the injection. Only injections given by experimenter MB yielded ethanol activation of activity for certain strains.

Schedule adequate rest periods for the experimenter. Periodic breaks from testing serve to forestall fatigue and lapses of attention. The arrangement of squads and sessions makes it easy to schedule breaks. Breaks provide a buffer period that can absorb small emergencies that lengthen a session. One emergency is an escaped mouse that must be pursued through

139

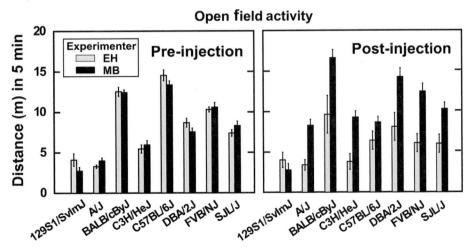

FIGURE 8.3
Experimenter effects (mean scores ± standard error of the mean) on open field activity after an intraperitoneal injection of ethanol in the author's lab at the University of North Carolina at Greensboro. Eight inbred strains were tested by two experimenters, and the experimenter was carefully balanced across squad and time of day in the study design. The experimenter effect on activity prior to injection was negligible.

a room full of apparatus and furniture. Minor apparatus malfunction and computer glitches happen in almost every study. If every minute of the test day is scheduled with essential events, any small emergency will lengthen the test day and may cause major inconvenience for the experimenter.

Coordinate testing with cage changing and routine animal care. A cage with fresh bedding and nest material is a major event in the weekly life of a lab mouse (Gray & Hurst, 1995). Invariably the animals become very active and work to arrange the new cage in a unique way. Sometimes they wrestle and seek to re-establish some kind of dominance relationship within the cage. Disturbance of a cage to install a clean water bottle or even to count the mice can also provoke exploration and alter activity rhythms. If at all possible, it is best practice to have cages and water bottles changed *after* the daily tests are completed. In a large animal facility, this will not be possible for all studies in progress at the same time, so the next best option is to have changes done at the same time of day and week. It is then wise to make a note on the data sheet for mice that were tested during a period of cage changing. Presuming that the order of testing has been properly balanced and randomized, disturbance of the mice by animal care personnel should not bias the results of a study, but they could increase the variance in test scores within a group, which in turn would require testing of larger samples.

A very good method for cage changing requires the person who gives the behavioral test to the mice to also give them a clean cage after the day's session is done. After the last mouse in the cage has completed testing for that day, all cage mates can be quickly moved to a clean cage. Thus, no mouse in the study would ever have a clean cage shortly before testing. Cage changing would occur at different times during the day for various mice, but this is a mere inconvenience, not something that affects the results of a study. For this approach to work, a good level of cooperation with the animal care staff is required.

Coordinate testing with the light cycle. Even though light cycle effects on behavior often are not large, it is good practice to minimize them by testing squads at fixed times of day with respect to the onset and offset of light. If lights come on in the morning to make the work of the animal care personnel easier, all studies in the lab are best run in the light phase. If the lab is on a reversed light cycle, consistency across studies in that lab is desirable.

One option involving only minor inconvenience is to have the lights go off at 1200 hours (noon) and come on at midnight. Cage changing and routine maintenance could be done in the morning while the lights are on. Behavioral testing could then be done entirely during the dark phase in the afternoon. Alternatively, there could be a morning test session during the light phase and an afternoon session during the dark phase, making light phase a control variable. This would not change the logistics, provided the entire order of testing within a day has already been balanced. If there are light cycle effects on the behavior or treatments being studied, dividing the sample between light phases will tend to increase the variance within groups. If the effects are small and do not interact with treatments, testing in both light and dark will increase the generality of the results.

There is surprisingly little evidence that the light cycle affects results of behavioral tests, despite abundant evidence for an entrained circadian rhythm in many physiological functions. One likely reason is that most studies of mouse behavior employ an acclimatizing period of at least 30 min prior to the actual test that awakens a sleeping animal from its slumber. An example from the author's laboratory is shown in Figure 8.4 for a study of ethanol effects on motor coordination of eight inbred strains. All mice were tested at the same chronological time of day, but half were housed in a colony room where lights went on at 0600 (6:00 a.m.), while the other half were in an identical adjacent room having a reversed light cycle (Figure 1.3).

Light cycle effects on behavioral tests might be more apparent if the test were given shortly after the mouse is removed from its home cage. If so, the result would not be highly relevant to routine work in the mouse lab because behavioral tests are never done that way.

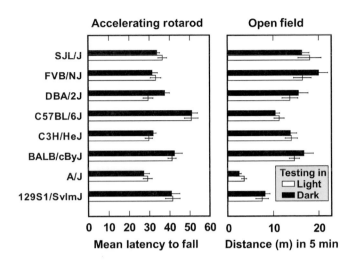

FIGURE 8.4
Mean scores ± standard error of the mean on two tests for eight inbred strains of mice that were maintained under opposite light—dark schedules in adjacent colony rooms at the University of Windsor (see Figure 1.3). All mice were tested under the same low level of room illumination (5 lux) after a 30 min acclimatization period in the test room. Light—dark cycle had no discernable effect on any measure in the study.

Emergencies in the building can inflict damage to the test schedule as well as to results. In some buildings, fire alarms are surprisingly frequent in the absence of any real fire. Routine fire drills also occur in some institutions. If an alarm is signaled by a loud bell or horn, this could damage the hearing of many kinds of mice and even induce seizures in those near weaning age (Henry, 1984). The effects can be minimized by using flashing lights as the alarm signal, in which case there must be alarm lights inside each test room. Other kinds of emergencies that may require evacuation of the facility most likely will result in data loss. If a major portion of a replication of a study is lost, it is probably best to repeat the entire replication.

Preparation for the next day is best done at the end of the previous day. Apparatus needs to be cleaned thoroughly before substances deposited by the mice dry. If a cleaner with strong odor is used at the end of the day when mice are no longer in the test room, odors should be gone by the next day. End of day is also a good time to ensure there are adequate supplies for the tests on the next day. Everything should be done to avoid any need for urgent action at the start of the next day.

Notes and entries in the test log should be written at the end of each test day while events of the day are still fresh. There are usually entries on data sheets when each mouse is tested, but more general activities and unusual occurrences need to be written in a log, along with thoughts about ways to improve the tests in future studies.

Prelude to Data Analysis

143

Having spent months collecting data, the time has finally come to analyze all those facts and transform the work into a published article. For many people, this is like making sausage. They enjoy consuming the end product but do not care to know how it is made. But somebody must do it, and all agree the maker should know the process very well.

The first thing to be done is to put piles of data scribbled onto sheets of paper and stored in voluminous computer files into a compact form that a data analysis program can ingest. For this purpose, one large *spreadsheet* with a rectangular matrix of rows and columns is created. Even this first step can go awry and cost time and frustration. One serious mistake here, and every other painstaking step along the way may need to be repeated after discovering and repairing the glitch. All those lovely graphs and elaborate tables will be tossed into the trash bin, while you crumple to the floor sobbing.

A little tip can prevent really big disasters. Everyone knows how important it is to save work fairly often, in case the program crashes, to back up new work at day's end, and to keep the backup file in a location separate from the computer used to create it. Yes, do all these things, but they will not protect against one of the worst mistakes when creating a spreadsheet; performing an operation on the spreadsheet file and saving it under the same name will erase some of the previous information. When making a change in the file, the little Undo arrow can be used to reverse it, but Undo will not fix a mistake that has been saved under the same file name. So, whenever doing major things like sorting data or making global changes to data in many locations in the spreadsheet, save a copy under one name before the change and another name after it. For example, the name might be Mysheet1 12dec09.xlsx before and Mysheet2 12dec09.xlsx after the change. With this numbering scheme, it is always possible to know which is the most recent version. Making the current date part of the file name is also wise. Do not rely solely on the creation date displayed by Windows, because this sometimes changes when the file is simply copied to a new location and the dates and times on different computers may not be set the same way. Once the investigator is fully confident that the

Mouse Behavioral Testing. DOI: 10.1016/B978-0-12-375674-9.10009-6

spreadsheet is in its final, correct form, all those preliminary versions can be cut and copied to a folder named Preliminary to get them out of sight.

KNOW YOUR OBJECT

It is important to recognize what the *object of measurement* is in any study. This is the individual thing that is measured, be it a live mouse, a slice of brain tissue, a burrow dug into the earth, or even a concerned citizen who is interviewed. Any study involves measures or ratings of several such objects. Hopefully the study was designed so that the measure of one object was independent of the other objects; that is, the measure obtained for one did not then determine what value occurred for the other objects. In the interests of efficiency and a desire to learn as much as possible, several kinds of measures were probably collected for each object. For example, one may have weighed it, noted its coat color and sex, administered a few quick behavioral tests, then did unspeakable things to determine what it ate for breakfast, as well as how much protein, fat, and environmental toxins it contains.

CONTENTS OF ONE ROW OF DATA

Usually the best layout of a spreadsheet is to place all information about *one object in one row* across the page. The different values across the row are the variables. The next row then contains the corresponding information for the next object. If the study involved 237 separate samples, then there should be 237 rows of data, plus at least one row at the top of the spreadsheet that gives the name of each variable. In statistical programs, the row for an object is sometimes referred to as a case or record, and there may be a number from 1 to 237 assigned to each case by the program.

It really does not matter which object is typed into the first, second, or last row of the spreadsheet. The statistical program will not care about this. One may later decide to sort the data in order to place all objects with a certain characteristic into adjacent rows so that the numbers can be scrutinized in a search for patterns in the results.

THE ID CODE

What about the order of variables across the row? With one exception, there are no fixed rules on what to put where in a row, but guidelines are suggested to make later data analysis more convenient. One variable in particular should always be placed into the very *first column* of every row, the object ID code. There is a serious, all important, overpowering need to be sure each object has a unique *identity code* that connects the data in its row in the spreadsheet to the raw observations written onto the data sheet or recorded in a computer file. If objects are entered into the spreadsheet in the order specified by the ID codes, it will be easy to sort the sheet this way and that, then sort again by ID code to put it back into the original order.

There are many schemes for ID codes, and a new investigator will probably want to use the same scheme employed by the boss or lab director. It is suggested that the ID begin with a letter or perhaps two letters, such as a person's initials, followed by the object number. One tip: think ahead to how many objects could be studied in graduate work or even an entire career. Surely more than 1000 objects could be studied. If one's initials are CK, then code number CK0158 would be for the 158th object that was measured. Why the 0? In sorting, some programs might not arrange CK45, CK158 and CK3421 in the desired order, so use CK0045, CK0158, and CK3421. Then all will sort nicely.

ORDER OF VARIABLES ACROSS A ROW

Concerning everything else across a row, there are two general kinds of variables. Whatever the focus of a study — toxins, nutrients, drugs or genetic differences — some measurements might be affected by those things and others could not be affected because they have their features determined prior to the experiment. It is generally good practice to place independent variables that describe more or less fixed aspects of the object toward the left side of the row. Those variables often have the same values for all objects in a study group. They could include, for example, species, strain, sex, age, color, geographic location, sampling date, or experimental treatment condition. Then come the dependent variables representing aspects of the objects that were measured and that might be affected by treatment conditions. They could include body weight or length, brain chemical composition, behavioral activity, or reactions evoked by odors in the air.

If measurements were taken of the same object on different days or at several times during one day, it makes later analysis easier if all measurements of one kind are entered into adjacent columns. Suppose core body temperature and food consumption were measured every hour for 6 h. Make the order of variables BODY1, BODY2, BODY3, BODY4, BODY5, BODY6, FOOD1, FOOD2, FOOD3, FOOD4, FOOD5, FOOD6; rather than BODY1, FOOD1, BODY2, FOOD2, and so forth. This is not a requirement; it is just for convenience. Once deep into a complex statistical analysis, however, it soon becomes evident why the former order is better than the latter.

NAMING VARIABLES

Naming variables is also important. Be concise but not too cryptic. If possible, use an acronym that provides a hint about the nature of the measurement, but try not to make it too long. No blanks or punctuation in the variable name, please. Some programs allow this but others will protest and refuse to work.

Two further recommendations really help over the long term when several studies are separated by months or even years. First, be sure to place a complete, detailed *definition of each variable name on a separate page* of the spreadsheet where it can be easily checked when needed. Unless someone has a photographic memory now and forever, some of those abbreviations will not be as meaningful when looking back at the data after months or years spent working on something else. Second, if at all possible, use exactly the *same variable name for the same kind of measure in different studies*. Not only does this ease the load on cluttered neural circuits, but it makes it easier to combine data sets from separate studies to compare results over different times and places. It might be wise for a large laboratory with several students and technicians to establish a standard list of variable names posted in the office. New recruits can then be inducted into the naming ritual of the local lab cult.

THE VALUE OF USING THE RIGHT VALUES

The actual values entered for each variable of an object will, for most measurements, be a simple number. All that is necessary in the spreadsheet is to specify whether the number is an integer with no decimal places or a number with several decimal places. Be sure to specify the appropriate number of decimal places for the measuring instrument used. This is just a convenience for compact display, because most spreadsheets store any number as 32 or 64 bits of information.

To keep the spreadsheet as compact as possible, specify the width of a column of data to be the minimum needed to display all values and the variable name properly. Then many columns of data can be seen in one screen display and perhaps interesting patterns in the data will materialize. Printing will also be more efficient.

The situation for variables that contain letters or other symbols is different. Here it is wise to anticipate the kind of analyses and graphs that are needed. Whereas *variable names* need to be compact, *values* often should contain the full name for an object. This will be especially helpful when making a graph. Suppose the study involves several inbred strains of mice of both sexes treated with several doses of ethanol or saline. The preferred strain acronyms might be

C57, 129, or BALB, but it is far better to use the full, official strain name to show substrain. These not only serve well as labels in a graph, but they will avoid future uncertainty about exactly what genotype was used. For sex, type the full words Female and Male rather than using number codes such as 1 and 2. It is so easy to forget which was which when there is only a stark digit in the spreadsheet. If a number such as a dummy or effect code is required for a multiple regression analysis later in the process, they can be easily added to the data set. Dose of a drug can be incorporated into a value as well, such as Ethanol_3gKg for 3 g/kg. In addition, a variable named EthDose_gKg might have values such as 1.25 and 2.25 and then 0 for saline.

THE PLAGUE CALLED MISSING DATA

Everyone tries to obtain a good measurement for every variable, but in real life stuff happens: data are lost, power fails during a run, a hard disk crashes, a flash drive is lost or stolen, a flood carries away some samples, or your hamster makes a nest with some data sheets. If a measurement for one variable is missing for one object, leave the cell blank so that the columns of the same kind of data line up neatly down the sheet.

A spreadsheet can tolerate blank cells. A problem can arise, however, if one returns to the spreadsheet after a few days and is not sure whether a particular blank means genuinely missing data or just a value that has not yet been entered at the keyboard. One trick is to enter into each cell of missing data a peculiar symbol not used anywhere else in the spreadsheet. Suppose one decides to make ? the symbol for missing data. How clever. Then try to use a global Find and Replace command to replace the ? with the symbol that is compatible with a particular stat program. There is a problem with this because Excel looks at the ? symbol differently and will replace every ? as well as every digit in the spreadsheet with the new symbol. The user will need to back out of this mess quickly and try another solution. Hence, one clear lesson is to *test a clever idea on a spare copy* of the spreadsheet before changing the real thing.

One solution is to type "missing,""miss," or "ND" (for No Data) into any cell with missing data. Later use Find and Replace to replace "missing" with almost any symbol or even a blank in a spreadsheet saved with a different name. For several popular statistical programs, an empty cell will be recognized as missing data, and the program will then assign its own special symbol to the empty cell. Thus, an empty cell in the spreadsheet is an acceptable sign of a missing value.

It is highly recommended that information about missing data is recorded onto another sheet of the spreadsheet so that a future user will know *why* it was missing. This can sometimes influence the interpretation of results. It can also tell how to avoid missing values in future work.

There is one situation where an investigator might decide to adopt an extraordinary remedy for missing data. If several measures have been collected over a period of months on the same animals, they will most likely be analyzed with some kind of repeated measures or multivariate method. Unfortunately, the methods require that every individual has a measure at every time point. If only one out of ten or more is missing, the entire case will be ignored in the analysis. If the sample size is small because the study involves some rare, genetically engineered animals, this may make it much more difficult to detect real effects. Suppose one value, call it X_4, is missing out of 10 for one animal but the data are complete for seven other animals. Two options can be imagined, and there are others as well. (a) One could take the average of adjacent measures for that individual (mean of X_3 and X_5), which will place the estimated X_4 at an appropriate place in a steadily declining or increasing series without affecting the overall trend line for that animal. This will have no undue influence on any test of between-groups effects. (b) If the trend line is not generally smooth and distinct peaks are sometimes evident for the seven points with complete data, multiple regression can be used to find the best

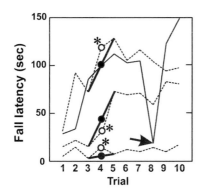

FIGURE 9.1

Latency to fall from the accelerating rotarod on 10 training trials given in one day for four mice of the NOD/LtJ strain tested in the author's lab. For data from the three animals shown as dashed lines, it is supposed that they are missing for trial 4 (open circles and asterisk) and the values shown as black dots are substituted, based on an average of values for adjacent trials 3 and 5. This method of linear interpolation will have little influence on the group average data except for one animal where trial 8 (arrow) was far below its otherwise excellent performance.

equation for predicting X_4 from a linear combination of the other nine measures from X_1 to X_{10}, omitting the one animal with the missing data point. Then plug in the nine X_1 to X_{10} values for that one animal into the equation and find the value of X_4 that is expected on the basis of its other nine. Enter this into the blank cell. This method will not bias the results, because all means will be just what they were before the substitution, but it will underestimate the variance within groups by a small amount. Any such substitution to salvage data where there is a missing value should be reported and discussed. If the substitution makes no difference to the conclusions about the study, it is probably best to drop the one animal from the data set, rather than resort to adroit fabrication. The more values that are missing for more animals, the less respectable any kind of data rescue will be. Some data are best laid to rest.

An example based on real data is shown in Figure 9.1 where the simple interpolation scheme is used to rescue one missing value for each of three mice learning to stay on the accelerating rotarod. In two instances with good learning, the interpolated value was a small amount off the actual line, whereas for a third that did not learn, it made no difference at all. Making all three substitutions would not alter group data in any meaningful way. The arrow shows the one data point where interpolation would make a major difference for one animal, a situation where a generally good runner fell quickly, then resumed excellent performance on the next trial.

IMPORTING THE SPREADSHEET INTO THE STATISTICAL PROGRAM

If every variable has a suitable name and every cell in the spreadsheet contains appropriate data, importing it into the stat package can be very easy. Suppose the spreadsheet was made in Excel and has file extension xls. Click on the File–Open tabs and choose files of type xls. Then double-click on the one to be imported. With luck, it will miraculously appear in the rectangular data array in the stat program and look like it did in Excel. The next thing to do is save it as the file type used by the statistics program.

Several things can be done to avoid pitfalls in this process. Make sure that Sheet 1 of the spreadsheet to be imported has nothing but the variable names and data. If the original Sheet 1 has more than one row of variable names at the top plus variable definitions — annotations about data collection, perhaps a few computations of means, and notes to pick up beer on the way home from work — open a new, empty spreadsheet and copy just the good stuff to it, then save it under a new name, close it, and import it. When copying from an Excel spreadsheet that used formulas to compute some values, be sure to check the Values Only option in the little menu at the lower right corner of the pasted data. Many statistics programs will refuse to import formulas.

147

Several popular stat programs can be very clever about imports. The program will look at all the values in a column for a variable and decide whether it is supposed to be a numeric or alpha (string) variable. There are several ways to find out the fate of the variables. Strings of alphanumeric information are usually left-justified, whereas numbers are right-justified. Right-click or double-click on the variable name to see its properties displayed.

If all is not well because one or a few values were not quite right, it may be easy to repair the problem in the stat program data matrix, but if there is any kind of outright error in the data, it is most emphatically recommended to return to the spreadsheet to fix the problem there, then repeat the import operation. Otherwise there will be versions of the "final" spreadsheet with data errors lurking in them. What if an esteemed colleague in a far off land requests a copy of the raw data after reading your brilliant article? You proudly fire off the xls file, forgetting that it still contains bloopers that were fixed in the stat program. The leading expert will find them. Just what a young scientist needs — to have his or her nose rubbed in a stupid data error by a famous person who knows everybody in the field.

CHECKING FOR ERRORS IN THE DATA

It is wise to do a little error checking in the spreadsheet before the import operation. The simplest way to check in Excel is to sort on a variable (working with a spare copy of the spreadsheet) and check the largest and smallest values. They should look reasonable, given the kind of measurement. Extreme values or impossible values or even letters in a field that should be all numbers will be located this way. Warning: always, always let Excel expand the selection when sorting, so that all data in a row remain together. Go through this process for *every* variable, even group and treatment variables where there might be typos.

If there are no glaring errors in the spreadsheet, the import operation should go smoothly and the array of data in the stat program should look good when scrolling up, down, and across. But that is no error-free guarantee. The worse kinds of mistakes are those that are difficult to see. Those insidious interlopers must somehow be detected and expunged before the real analysis and graph making commence.

Every variable in the data set must be carefully checked, from the first to the last variable, even the ID code. How could the ID be wrong? Let us count the ways. First, there might be two samples in the spreadsheet with the same ID. An object with its ID code might be missing from the data set entirely. Maybe someone typed the digit "1" when it should have been lower case letter "l"; see how similar they appear? What about the digit "0" versus the upper case "O"? What if a little ' or ' or . clung to the end of one value?

There is no fail-safe, foolproof method to find every possible error, but there are guidelines to help spot almost all of them. Keep in mind that it is rare to produce a flawless data file in the format of the stat program on the first attempt. If it seems too perfect, it probably is. Even if the data set is rather small and fits onto just one computer screen, be vigilant and do not trust a casual impression that all is well.

For *alphanumeric* variables that contain letters or other non-numeric symbols, the best method is to obtain a *frequency count*, the number of occurrences of each value of a variable. Alternatively, a *frequency histogram* chart can do this job well. One should already have a good idea of what values are permissible and roughly how many of each there should be. If, for example, the work was done with eight inbred strains, be suspicious if the table contains 10 different strain names, or there are three sexes. Most mistakes will be typos or misspellings, and once a wrong value has been identified, it is easy to find all other instances in the spreadsheet using Find and Replace.

One kind of error can be very difficult to spot. The frequency count may show an extra name that to even the best eyes looks identical to another name in that same group in the list. Why does the program see them as different values? In one study, we found that the person who entered the data into the spreadsheet typed a space at the end of one of the names, whereas most of the names had no trailing blank. Computers are dumb. They often see a blank as just another character, and in binary code it is. Sorting on the variable in Excel did not detect this problem. The only way to be sure is to place the cursor in the middle of a name and use the right arrow to go to its end. If it goes one space too far, then you have found that dastardly blank. In a large data set, this will be a tedious task. One can use Find and Replace, where the item in Find also has a blank at the end, but this could miss other similar errors. A safer alternative is to sort the data on that variable, then type what is certainly the correct name into the first cell in the group and copy it to every other cell in the contiguous part of the data set.

For *integer* numbers, a frequency count or histogram is useful, unless there are many values. If the number 2.1 found its way into the data set, it will appear in its own category between 2 and 3. What if ID codes were entered as integers and one is missing in what should be a continuous series of numbers? The number of categories in the table will be one less than the highest code number.

Just to be completely safe, after checking each variable by itself with a frequency count, examine pairs of variables with a *two-way frequency table*. This method will be most informative if one has some idea of how similar or different the frequencies in each pairwise category should be.

For *continuous numbers* with one or more decimal places, if there are many possible values, a frequency count is not going to be very helpful. A very good method to catch devious data is the *two-way scatterplot*, where one variable is on the X axis and the other is on the Y axis. This will find *extreme values* that are out of the range of possible or plausible values. It can also detect *outliers* whose values for the X and Y variables are not really extreme but are far from the cluster of points around a straight line that may typify the relationship for other data. The scatterplot is especially informative when something is already known about the relation between two measures. Then the educated eye will notice that something in the cloud of data points seems amiss.

Some stat programs use sophisticated algorithms to find outliers, but at this stage of the work, a graphical image viewed by a critical eye is probably the best tool. If a value looks strange and out of place, chances are good that there was something wrong in the data collection or entry into the spreadsheet. Stat programs have a tendency to identify at least one or two outliers in any large data set, even when there was nothing wrong at all with the data.

THE CRUCIAL DISTINCTION BETWEEN ERRORS AND EXCEPTIONS

When checking the data for errors, things such as typos need to be repaired forthwith, but just because a data point stands all by itself to one side of the neat cluster of other data in a graph does not mean it was a mistake. It could be a genuine exception, a real fact that refuses to obey the rules we set in our minds. Do not remove an observation merely because it does not fit with what one *wishes* the results to be. This could be construed as data fudging, faking, or even fraud.

Once an outlier has been identified, the next step is to find out as much as possible about the original data. Consult the raw, unprocessed data sheet with values entered by a person or the computer file produced by a machine. If there is a discrepancy between the original observation and what was typed into the spreadsheet, fix the spreadsheet and make a note of this.

What if all seems to be in good order on the original data sheet? Technicians do make mistakes when recording readings from instruments, and some of them make quite a few mistakes if their minds wander off. Recall that many ships have been launched or lost because of romance, for example, and research projects can be compromised if a technician thinks about sex too

149

much while on the job. The author once had an assistant who ran mice through a series of trials in a test of learning and recorded scores of 0 neatly in a column on every trial for three mice in succession. It seemed utterly impossible that they would do absolutely nothing trial after trial. A brief investigation uncovered two anomalies: first, the test apparatus had been turned off the whole time, so that nothing happened when a mouse was put into the box, and second, the young lady was trying to decide which of three boyfriends was the best for her. The data were worthless and needed to be tossed out, along with the assistant. (She eventually became a noted neurologist.)

Suppose no plausible reason for an outlier is found. It could be a mistake, but it could also be real. At this stage of the process, a real expert who knows those kinds of measurements very well needs to be consulted. An expert may know that a value is quite impossible because the object in question simply cannot have such a property. It may be outside the range of measures ever observed in any other lab over the course of recorded history. If this happens, there is just cause to designate that cell in the spreadsheet a missing value, especially if the expert happens to be the supervisor or boss and will not later complain loudly about deleting a datum. If someone else challenges the decision later, then blame the expert.

Now consider this perplexing situation. A two-way scatterplot uncovers an outlier far from the cluster of other data points around a straight line. Its X value is within the range for other X scores, and its Y value is within the range for other Y scores. Nothing in the data sheets appears to be out of the ordinary. The technician is happily married with two kids and a secure job. The supervisor, after dwelling on the matter for some time, thinks it just might be possible, although highly improbable. There is no cause to doubt the observation, except that it is an exception. If so, then it stays. *To remove or change a datum, there must be some definite, defensible reason to do so.*

Examples from a recent study in the author's lab of ethanol effects on behavior are presented in Figure 9.2. For the elevated plus maze (A), there was a rather peculiar relationship between time spent in the two open arms, often thought to indicate mouse anxiety, and number of entries into any arm for three of the strains. Closer inspection of the data revealed a common problem in behavioral tests: inactivity. Two A/J mice simply sat in one place for almost the entire 5 min trial, one in a closed arm and the other in an open arm. Several BALB/cByJ mice and one 129S1/SvImJ mouse also spent almost the entire trial sitting in an open arm. At first glance, this might suggest that BALB/cByJ and 129S1/SvImJ mice in particular experienced low levels of anxiety in the test, but the outliers point to an alternative interpretation: when some

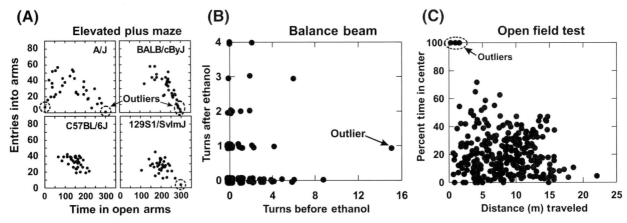

FIGURE 9.2

Examples of outliers in data from a recent study in the author's lab of ethanol effects on several behavioral tests. Data enclosed by dashed lines in panels (A) and (C) arose because of extreme inactivity by certain mice. The one mouse in panel (B) with an extreme level of turns on the balance beam was an exception that did not repeat its performance after receiving an ethanol injection. Data in panel (B) are jittered to reveal overlapping points.

of the mice enter an open arm, they tend to freeze. There were no errors in these data, but extremely low levels of activity make the test results impossible to interpret for this task. For the balance beam (B), one C3H/HeJ mouse had an exceptional number of turns on the narrow beam but only one slip of a hind foot off the beam, and nothing else in its data was unusual. It simply turned around frequently on the beam, a pattern that disappeared after receiving an ethanol injection. In an open field test (C), a problem of inactivity was also evident for three mice that moved very little but spent most of the time in the center of the apparatus. Time spent in the center is sometimes regarded as an indicator of low anxiety, but any such interpretation is hazardous when the animals move so little. All three examples entailed real outliers, not mistakes. For certain analyses of anxiety-related behavior, it may be best to exclude them because of inactivity, but their data are suitable for most other analyses. There was no good reason to exclude any of those animals from the entire study.

If the data set is large and there is only one outlier, results of the analysis will probably not be influenced very much, but that one datum could turn out to be a highly *influential* value, such that the results of the analysis are substantially different with and without it present. It is legitimate to analyze the data with and without the suspicious value and inform the readers of the consequences. If, happily, it makes little difference, then the results are *robust* with respect to that exceptional value. Just leave it in the data set and make no mention of this in an article, unless a reviewer notices it in a scatterplot and raises a question. If it does make a difference, it is fair and wise to explain this to the readers. They will think the author is an honest person. Some may also think you are naive and lacking ambition because you are unwilling to cheat just a little to make the results look better. If people say this, check their articles to see if their data look consistently too good. Perhaps they cheated their way to the top.

Finally, after all this error checking and soul searching, there should be complete confidence in the validity of the data set. Save it under a good name, and let the analysis begin.

To the uninitiated, this preliminary process may come as a surprise. It is not unusual for the entry of data into the spreadsheet, import into the stat program, and then thorough checking for and repairing of errors to take almost as much time and consternation as the formal data analysis. The prelude to data analysis must be performed well. Everything else depends on this.

LOOKING AT RESULTS

According to many texts on statistical data analysis, the next step is to examine tables of means and variances, then conduct some tests of significance of group differences. This part of the research process can be rather abstract and tedious. It is necessary before a manuscript can be submitted for publication, but it is good to look at the results before doing formal significance tests. To have a good look, we need to see pictures. Graphs and charts convey the scientific meaning of data far better than P values and F ratios. Making them can also be enjoyable for scientists with an artistic flair.

One excellent means of describing results is to show the score of every individual in a group with a frequency distribution. Four kinds of graphs are shown in Figure 9.3A for the same set of data on time spent near the wall of an open field. A *dot histogram* is effective unless there is a very large sample, in which case a *frequency histogram* is better. A compact distribution can be created by *jittering* a dot density plot so that overlapping points are moved by small, random amounts to make them visible. A dot display has three major strengths: it reveals the spread of scores, especially extreme values; it shows whether the distribution is symmetrical or skewed; and it presents sample size vividly.

Many publications describe results with bars to show group means and standard error brackets to indicate variability in the data. The standard error of the mean of n raw X scores is

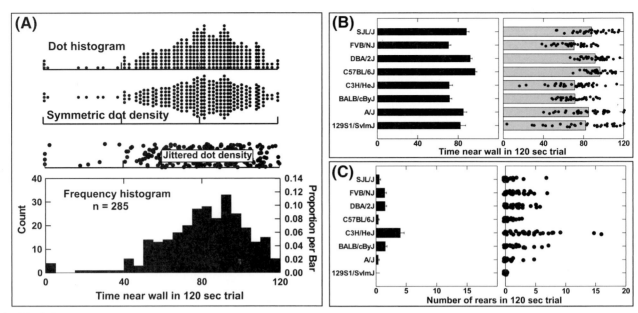

FIGURE 9.3

(A) Time spent near the wall of an open field during a 120 sec trial for 285 mice of eight inbred strains. Four kinds of frequency distributions of the same data are shown. (B) Bar chart on the left showing strain means and standard error brackets, versus bars to show means with individual scores superimposed and jittered to show overlapping points. (C) Number of episodes of rearing during the same 120 sec trial. The bar chart on the left provides a general impression of group differences, whereas the jittered frequency distribution shows individual scores that for certain strains were highly variable. (*Data from the author's lab.*)

$S_M = S_X/sqrt(n)$. Thus, standard error bars are smaller for larger samples. S_M is a good indicator of how far the sample mean M is likely to be from the true mean μ, because in most cases M will be no more than $2S_M$ from μ. The range $M \pm 2S_M$ is not a good indicator of the range of data in the sample, however. Consider the examples in Figure 9.3B and C for eight strains of mice. For many of the strains there is no overlap of their $M \pm 2S_M$ ranges, which suggests to the viewer that the strains are very different. By superimposing a jittered dot density on the bars, however, it is apparent that there is considerable overlap of scores in all eight strains. Another fact of potentially great importance can be perceived this way: certain strains, especially 129S1/SvImJ and C3H/HeJ, show vast individual differences that span the entire range of possible values. It is easy to misinterpret the bars showing $M \pm S_M$. The situation would be better if the graphing software displayed $M \pm 2S_M$, but the best option in many cases is the dot density.

Standard error bars do not describe data. They are a tool to assess statistical significance, the probability of wrongly concluding that the true group means are different, as discussed in Chapter 5 (Figure 5.9). The P value for a statistical test that compares two groups, perhaps a control and a treated group, is the probability that concluding a treatment had an effect is actually a mistake, a false positive "finding." If the P value is very small, then the treatment most likely did have a real effect.

There is an interesting relationship between the visual impression of the overlap of scores of two groups and the statistical significance of the group difference in means. As the investigator gains experience with many kinds of data sets, he or she will be able tell at a glance whether the experiment worked. A small utility program helps to visualize the relationship. In **Simulation delta**, the user simply enters the expected mean score of the control group and the standard deviation of scores within the two groups. It then draws random samples (normally distributed) of 100 scores for the control and treatment groups when there are four effect sizes: $\delta = 0.5$ (small effect), 1.0 (large effect), 1.5, and 2.0 (very large effects). After copying the results to an empty spreadsheet and choosing the Values Only option in the Paste menu, the data can be imported into a statistical program to make graphs and conduct tests. Because these data are randomly generated, a sample size less than 100 can be studied simply by selecting fewer cases prior to making the graphs and doing the tests.

Results for this exercise using samples of n = 100 and n = 10 are shown in Figure 9.4 for a control group mean of μ = 10 and σ = 2. Dot histograms are aligned with μ = 10 and bars with standard error brackets are shown in the same panel. Significance is tested with a t test that yields P values as indicated. For the larger sample size, it is perfectly obvious that the two groups are different when δ = 1.5 or 2.0; many scores in one group are outside the range of the other group. No formal t test is needed to prove the obvious. With δ = 1.0 the situation is not quite as clear because the ranges of scores in the two groups are almost the same. The mode of the treated mice appears to be higher than that of the controls, and the t test confirms this with P < 0.000001. For δ = 0.5, there is considerable overlap of scores and considerable doubt about the presence of a real treatment effect. The t test says there was an effect with P = 0.0001, something that most researchers would find significant in the statistical sense but disappointing in terms of the actual magnitude (d) of the effect. Generally speaking, when the sample size is fairly large, if one group appears to the viewer to have many more high scores than the other, the group difference will definitely be statistically significant. Usually the difference will be visually apparent to the educated eye when δ > 1.0. For δ = 0.5, there will be serious doubt, while δ = 0.25 will occasion much scratching of the head and muttering. The t test on large samples will sometimes indicate a clearly significant treatment effect when the researcher cannot perceive it in the graph.

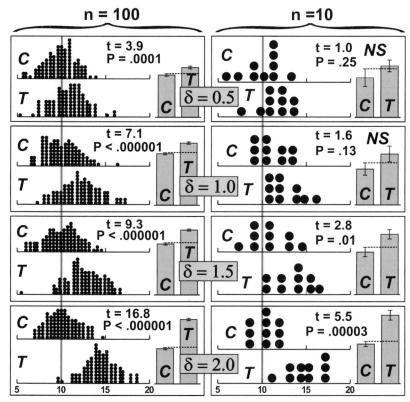

FIGURE 9.4

Dot histograms of randomly sampled scores in a control (C) and treatment (T) group where the true means of the C and T groups differ by 0.5 to 2.0 standard deviations. A t test of significance and the one-tailed P value are shown for each case. Data were generated with the Excel utility **Simulation delta**. To the right of the histograms are bars with standard error brackets for the same data. The dashed lines extend from one standard error above the mean for the control group. When sample size is small (n = 10), the standard error brackets overlap for effect sizes of 0.5 and 1.0, whereas they do not overlap when n = 100. For the larger samples, the group difference is clearly significant for all four effect sizes.

153

The right panel of Figure 9.4 gives results for samples with n = 10. When δ = 0.5 and 1.0, there is considerable overlap of scores and the group differences are not significant because P > 0.05. For δ = 2.0 there is almost no overlap and the effect is obviously real. For δ = 1.5 there is some overlap but there are several scores in each group that extend beyond the limit of the other group, so the effect is probably real. Generally speaking for small samples, if there is almost no overlap of scores of two groups, the difference in group means will be statistically significant. At the same time, there can be borderline cases with δ = 1.5 where significance is in doubt and a formal t test will be needed to aid one's decision. For the specific data in Figure 9.4 when n = 10, the value of P = 0.01 is not small enough to dispel all doubt; the result is marginal or borderline, something confirmed by inspection of the dot histogram.

When there is only a graph with bars and standard error brackets for guidance, a good guess about statistical significance can be based on the brackets. Figure 9.4 uses a dashed line to check for overlap. If there is overlap of the brackets for two groups when they display $M \pm S_M$, the group difference will definitely *not* be statistically significant with P < 0.05. If the group means differ by three standard errors, the difference probably is significant but the result is near the borderline. If they differ by four or more standard errors, the difference is definitely significant with P < 0.01, regardless of sample size. Thus, standard error brackets provide a good visual indicator of significance of a group difference. They do not, however, indicate effect size in any general sense because the standard error depends so strongly on sample size. They also do not show whether an assumption of the t test, such as normally distributed data, is well founded.

More complex relations in the data can sometimes be expressed effectively with a two-way scatterplot, as in Figure. 9.5, for eight strains tested in the open field. The strain FVB/NJ spent considerable time away from the wall and reared many times in 120 sec. An ellipse is drawn around the cloud of points for that strain and then copied to the other strains. Clearly, only BALB/cByJ resembled FVB in both respects. When many points for one strain are outside the range for another, the group difference for that measure is probably statistically significant. For time near the wall, it is apparent that 129, BALB, and C3H were quite similar to FVB, while C57 and DBA clearly had higher means than FVB. For A/J and SJL there was considerable overlap with FVB but both strains had many more high values. They were probably significantly higher, but a formal t test might be needed to confirm the impression. Rearing was obviously less than FVB for 129, A/J, and C3H, and probably for DBA too.

This lesson on perceiving effects in data is not meant to replace formal significance tests. The goal of a preliminary data analysis is to find ways to show the results graphically in order to highlight the major findings of a study. If there are major findings, this should be evident from

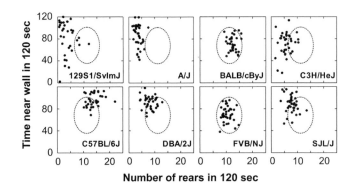

FIGURE 9.5

Two-way scatterplots of number of rears versus time spent near the wall of the open field in a 120 sec trial for eight inbred strains. For certain strains there is little or no overlap between the clouds of points. The dashed ellipse encompasses almost all scores in the FVB.NJ strain. It is superimposed on the other strains at the same location with respect to X and Y axes. Data are from a separate study; mice were not the same ones used for Figure 9.3. (*Data from the author's lab.*)

a few key graphs. Discoveries are almost always made by looking at the data, whereas significance tests are needed to confirm things that leave the educated perceiver in doubt. If the principal results of a study cannot be seen in the graphs, either a better kind of graphical display is needed or the effects are too small to be of interest to most colleagues.

There are situations in behavior genetics where researchers are genuinely interested in rather small effects, such as quantitative trait loci (QTLs) that increment a phenotype by only a small amount. In that quest, they typically graph a series of probabilities using specialized software rather than the phenotypes.

STATISTICAL DATA ANALYSIS AS A FINAL STEP IN THE PROCESS

Having achieved a good understanding of the major results of a behavioral study, the investigator then needs to do a thorough statistical analysis of the data. As suggested in Figure 1.1, this phase of the work is not explored in depth here. There are many fine texts on data analysis, and every university offers courses on this topic as well as expert consultants who are eager to work with new data. Choosing the correct statistical test for specific kinds of data depends on many factors, including the research design, shapes of the distributions of scores, the numbers of values that may be jammed against upper or lower limits on the measurement scale, and confounded or correlated measures on the same individuals. There is nothing about behavioral data per se that requires special treatment from a statistical standpoint. Almost all real data in every field are messy and violate assumptions of finely crafted and elegant tests of significance. Tests based on lovely bell shaped curves are convenient tools to aid the researcher in reaching correct conclusions, but they may not be the best option when working with behavioral or neural data. There are no simple rules for deciding how best to analyze data. This is why texts on the topic are usually very thick and dense, while software to digest the numbers is expensive and prone to hiccups.

Domains and Test Batteries

The laboratory mouse can express a remarkably diverse repertoire of behaviors. Some of these are evoked and measured by a specific kind of test, whereas others are expressed spontaneously. Some tests are carefully engineered to assess a particular kind of construct such as anxiety, whereas some researchers simply watch the stream of behaviors and strive to minimize interference with their furry subjects (Michel, 2010). The field of behavioral and neural genetics with its emphasis on mouse models of human functions is dominated by the former approach. Increasingly, elaborate batteries of automated tests are being used to screen large numbers of mice for behavioral changes caused by genetic defects (Paylor, 2008).

A comprehensive inventory of everything a mouse can do or every kind of test that has been used in research is beyond the scope of this book. Instead, the focus is on the most common kinds of behaviors and tests presented in Chapter 3. The tests and behaviors can be classified with the aid of several dichotomies or dimensions that the researcher uses when choosing a test.

TYPES OF OBSERVATIONS
Whole behavior versus criterion response

Many behaviors can be identified from a pattern of actions performed by the whole mouse, and these are typically scored by a human observer trained in the ways of mice. Others can be detected by photocells, microswitches, strain gauges, or video tracking programs that provide automated assessment. Any behavior that meets a specific criterion such as photocell beam break is then a *criterion response*. Automation comes with a cost, because there are often several different patterns of behavior that can cause the same tally of an event by an electronic device. For example, in an open field test the photobeams can be interrupted by an animal that walks from place to place, but they can also be broken by a mouse that is grooming or having a seizure. Absence of photobeam breaks in a short period of time might indicate immobility or freezing, but the mouse could just as well be grooming itself while sitting in front of one photocell or perhaps sniffing intently at a speck of dirt on the floor. If an apparatus has a lever connected to a microswitch, any action that depresses the lever — a deliberate, well-coordinated press with one paw or an awkward landing with the rump after a wild jump — will register in the computer as a lever press. It is therefore wise to record the whole stream of behavior on videotape, even when an automated device records criterion responses.

Free expression versus highly constrained actions

When a mouse is allowed to explore a complex environment, either alone or with other mice, many kinds of actions will be performed freely, voluntarily, or spontaneously in elaborate and highly variable sequences. The mouse may defeat the purpose of a test designed to assess some specific construct by remaining in just one part of the apparatus or even escaping altogether. At the other extreme of a continuum are tests that severely limit what the mouse can do. The tail flick test used to assess response to pain entails an experimenter holding the animal firmly over a hot plate so that its tail touches the plate (Mogil, 2009; Mogil et al., 1999a). Just about the only possible action for the mouse is to withdraw its tail from the plate. Because of the intimate relation with a human handler, this kind of test seems to be prone to large experimenter effects (Chapter 15). Pavlovian eyeblink conditioning, where a puff of air onto the cornea follows a signal, requires that the mouse be rigidly restrained so that its body and head can move very little (Vogel, Ewers, Ross, Gould, & Woodruff-Pak, 2002). The measurement of optokinetic eye movements that compensate for image motion on the retina also requires tight restraint (van Alphen, Winkelman, & Frens, 2010). Other kinds of apparatus, such as the elevated plus maze, constrain the places where a mouse can walk but do not restrain the animal directly. Consequently, the details of apparatus design and test parameters can be very important for that kind of test.

Microscopic analysis versus the whole picture

Every kind of behavior pattern is a sequence of smaller motor movements, and these can be measured or analyzed at several levels in relation to the fine details. For example, a mouse may pause during exploration of a new environment to groom itself. Grooming involves several different patterns for cleaning the flanks, rubbing the face or ears, licking the anogenital region, and nibbling the nails (Kalueff, Aldridge, LaPorte, Murphy, & Tuohimaa, 2007). Rubbing the face can in turn be assessed as a rapid sequence of coordinated movements of muscle groups (Fentress, 1992; Golani & Fentress, 1985; see Figure 10.1).

Short duration versus full day observation

The fine details of a single action such as a reflex are best assessed over a short period of time when the behavior is expressed with some consistency. Over a longer period of time, the behavior, such as grooming, may occur intermittently and the animal may shift positions, making it very difficult to measure fine movements. For compilation of behavior over a long period, fine details must be ignored while macroscopic indicators such as revolutions of a running wheel or grams of food eaten provide convenient measures that can be automated. A noteworthy exception is the IntelliCage, which collects detailed information continuously for

(A)

(B)

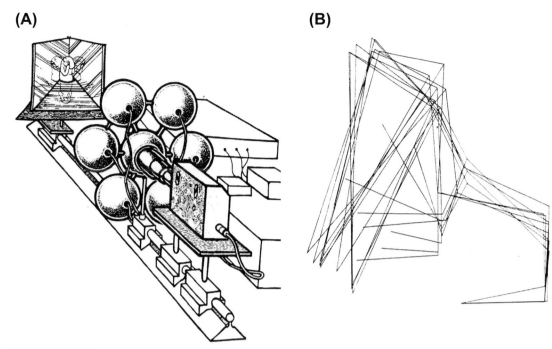

FIGURE 10.1
Microanalysis of motor movements involved in grooming the face of a mouse. (A) Arrangement of camera, lights, and mouse to record behavior. (*Reprinted with permission from Golani & Fentress, 1985*). (B) Computer representation of forearm and head of weaver (*wv/wv*) mouse grooming its face. (*Reprinted with permission from Fentress, 1992*).

24 h or more on operant learning and spatial memory from identified individuals in a social group of mice living in a large cage (Barlind, Karlsson, Björk-Eriksson, Isgaard, & Blomgren, 2010; Galsworthy et al., 2005; Rudenko, Tkach, Berezin, & Bock, 2009). Automated observation around the clock in the animal's home cage (de Visser et al., 2006; Goulding et al., 2008) can be informative. It will be important to assess inbred strain correlations of home cage behaviors with scores on many of the discrete trial tests (Tables 3.1 and 10.2) that are administered in unfamiliar surroundings.

THE ETHOLOGICAL METHOD

Classical ethology, as exemplified by the work of Lorenz (1965, 1981), maintained that most behaviors important for adaptation to an environment are innately determined fixed action patterns, each performed in a stereotyped way whenever it is released by a specific configuration of stimuli. The theoretical content of Lorenzian ethology has been criticized by Lehrman (1970) and many others. Nevertheless, the emphasis on patterns of behavior expressed by freely moving animals has much to recommend it. The modern ethological approach (Lehner, 1996) emphasizes methods for observing behavior, while eschewing the genetic determinist ideology of Lorenz (Gerlai, 1999; Martin & Bateson, 2007). Ethologists conduct experiments that alter stimulus conditions to analyze causes of behavior patterns (ten Cate, 2009), but they strive to make the stimuli very similar to what the species would encounter in the wild.

The ethological method stresses the importance of watching the animal and identifying distinct postures or patterns of movements exhibited in certain situations. The *ethogram* is the collection of all distinct behaviors expressed by a mouse (Schellinck, Cyr, & Brown, 2010). Eisenberg (1968) identified 78 behaviors. Each is given a unique name, as shown in Table 10.1, for an abbreviated list of more than 60 distinct behaviors. The typical duration of a distinct behavior is short, ranging from about one second for a tail rattle, to several

TABLE 10.1 Categories of Behaviors in the Ethogram of the Mouse

Class	Instances
	Activities of an individual
Exploration	Walk, run, dart, jump, lean, rear, climb, reach (1 paw), object sniff, stretch-attend, head dip, swim
Feeding	Carry food, hold food, nibble/chew, lick, teat attach (pup)
Elimination	Defecate, urinate, sneeze
Self-care	Face rub, back/flank rub, tail rub, anogenital lick, wound lick, sleep
Shelter	Carry material, shred material, weave material, kick dig, push dig
Distress	Freeze, repeated circle, piloerection, hunch, shiver, squeal
Seizure	Wild running, rapid jumping, clonic convulse, tonic extend, wet dog shake
	Social activities of two or more individuals
Non-agonistic	Anogenital sniff, huddle, allogroom, barber, vocalize, wink
Agonistic	Aggressive groom, chase, wrestle, flank bite, tail bite, tail rattle, defensive kick, standing submit
Mating	Chase, anogenital lick, mount/grasp, intromit, pelvic thrust, lordosis, ejaculate
Parental	Crouch over pups, anogenital lick, pup carry, pup in mouth
Other	Bite experimenter, wrap on rotarod

seconds for chewing while holding a piece of food. For almost all of these, it is essential that a well-trained human observer score the stream of behavior. Mice can move quickly from one behavior to another, and the scoring may be facilitated by a video recording that can repeat the action at a slower speed (Branchi, Santucci, Puopolo, & Alleva, 2004). To a human, a mouse fight may appear to be a blur of paws, tails, and fur with bedding material flying around the cage to complicate things even more. In slow motion, however, the bout appears to be an elaborate sequence of thrusts, kicks, and sometimes bites. Usually one mouse is the initiator of a bout of agonistic behavior and the other defends itself, but these roles can change dynamically during the course of a fight. A good ethological record gives the sequence of each kind of behavior, rather than just stating that there was a mouse fight or that two mice were mating. Later, during the analysis phase of the study, it may be convenient and helpful to aggregate fine-scale behaviors into molar categories.

An epoch of behavior can be scored by tallying the number of occurrences of each distinct pattern. In principle it is also possible to determine the duration of each behavior, but in practice this is very difficult and almost certainly requires slow motion video playback. When examined in microscopic detail, the transition between one behavior and another may not always provide a convenient cue that one has stopped and the next has begun.

If the test situation is relatively simple and there is only one mouse to be scored for a short time, *continuous sampling* can be employed; a key on a computer or handheld device is depressed every time a different behavior begins. For more complex situations with more than one animal, the demands of continuous sampling on the observer become onerous. *Time sampling* in which behavior is scored on cue every few seconds is better suited for longer and more complex streams of behavior. It yields a relative frequency count of different behaviors, estimates the duration of particular kinds of activities, and documents transitions from one kind of behavior to another. Software such as Observer XT from Noldus Information Technology is useful for recording distinct behaviors with a time stamp for later analysis, and it functions well for both continuous and time sampling methods.

Several computerized video-tracking systems purport to distinguish distinct behaviors in a simple environment such as an open field (see Chapter 14), while careful placement of photocells may detect specific actions in some kinds of apparatus, such as a head dip on an

TABLE 10.2 One Hundred Behavioral Tests Categorized by Domains and Subdomains

Domain	Subdomain	Abbrev.	Test Name	Domain	Subdomain	Abbrev.	Test Name
Anxiety	Bright	LD	Light–dark box	Learning	Shock	SDA	Step-down avoidance
Anxiety	Height	EPM	Elevated plus maze	Learning	Shock	SHA	Shuttle avoidance
Anxiety	Height	ESQM	Elevated square maze	Learning	Shock	TMAA	T maze active avoidance
Anxiety	Height	ZM	Elevated zero maze	Learning	Shock	YMA	Y maze avoidance
Anxiety	Object	DBT	Defensive burying test	Learning	Spatial	BM	Barnes maze
Anxiety	Object	NEOO	Neophobia to object	Learning	Spatial	MWM	Morris water maze
Anxiety	Threat	SUPE	Suppression of eating	Learning	Spatial	4WM	4-arm water maze
Anxiety	Shock	VCT	Vogel conflict test	Learning	Spatial	RAD	Radial maze — 8 or 6 arms
Anxiety	Shock	GCT	Geller conflict test	Motor	Coord.	CLIMB	Climbing limb analysis
Arousal	Sound	FPS	Fear potentiated startle	Motor	Coord.	GAIT	Gait or footprint analysis
Arousal	Sound	PPI	Prepulse inhibition	Motor	Coord.	SWIM	Swimming limb analysis
Depression	Shock	LH	Learned helplessness	Motor	Dynamic	ARR	Accelerating rotarod
Depression	Tail	TS	Tail suspension	Motor	Dynamic	FSRR	Fixed speed rotarod
Depression	Water	PFS	Porsolt forced swim	Motor	Dynamic	WHR	Wheel running
Drug	Method	SAIC	Self-administration, intracranial	Motor	Fixed	BB	Balance beam
Drug	Method	SAIV	Self-administration, intravenous	Motor	Fixed	DT	Dowel test
Drug	Method	SAOR	Self-administration, oral	Motor	Fixed	GT	Grid test
Drug	Withdrawal	HIC	Handling-induced convulsions	Motor	Fixed	HB	Hole board
Exploration	24 h	CRA	Circadian rhythm of activity	Motor	Fixed	HW	Hanging wire
Exploration	Home	HCA	Home cage activity	Motor	Fixed	ST	Staircase test
Exploration	Object	BT	Barrier test	Motor	Fixed	VPT	Vertical pole test
Exploration	Object	NP	Nose poke	Motor	Forepaws	GRIP	Grip strength
Exploration	Object	OEX	Object exploration (memory)	Motor	Forepaws	PAWPR	Paw preference in reaching

Continued

TABLE 10.2 continued

Domain	Subdomain	Abbrev.	Test Name	Domain	Subdomain	Abbrev.	Test Name
Exploration	Empty	OFA	Open field activity (anxiety)	Motor	Observ.	ATAX	Ataxia observation
Exploration	Empty	YMSA	Y maze spontaneous alternation	Motor	Reflexes	REFL	Reflexes
Ingestion	Liquid	DRNK	Daily liquid intake	Motor	Reflexes	SHIRPA	SHIRPA battery of reflexes
Ingestion	Solid	FOOD	24 h food consumption	Pain	Paw	FORM	Formalin test
Ingestion	Solid	FPCD	Food preference — cafeteria diet	Pain	Paw	PWHP	Paw withdrawal — hot plate
Learning	Classical	CPP	Conditioned place preference	Pain	Paw	VIS	Vocalization induced by shock
Learning	Classical	CTA	Conditioned taste aversion	Pain	Tail	TFCW	Tail flick — cold water
Learning	Classical	EBC	Eyeblink conditioning	Pain	Tail	TFHP	Tail flick — hot plate
Learning	Classical	FCA	Fear conditioning — auditory cue	Pain	Tail	TFHW	Tail flick — hot water
Learning	Classical	FCC	Fear conditioning to context	Pain	Tail	TW	Tail withdrawal
Learning	Habituation	OEX	Object exploration >1 trial	Pain	Somatic	VP	Vocalization from somatic pain
Learning	Maze	HWM	Hebb-Williams mazes	Parental	Maternal	PUPR	Pup retrieval test
Learning	Maze	L3M	Lashley 3 maze	Sensory	Hearing	AS	Auditory startle
Learning	Maze	LABM	Labyrinth maze	Sensory	Hearing	AUD	Auditory acuity
Learning	Maze	TMF	T maze for food reward	Sensory	Smell	FF	Food finding
Learning	Maze	TUM	Tunnel maze	Sensory	Smell	OLF	Olfactory discrimination
Learning	Maze	WMD	Water maze discrimination box	Sensory	Taste	TBT	Two-bottle choice test
Learning	Maze	YMF	Y maze for food reward	Sensory	Touch	VFT	Von Frey touch test
Learning	Operant	IC	IntelliCage	Sensory	Vision	VC	Visual cliff
Learning	Operant	OPC	Operant conditioning	Social	Agonistic	AGRI	Agonistic — resident intruder
Learning	Operant	PB	Puzzle box	Social	Agonistic	AGRR	Agonistic — round robin
Learning	Operant	SILT	Serial implicit learning task	Social	Agonistic	AGSO	Agonistic — standard opponent
Learning	Shock	AA	Active avoidance	Social	Fixed	TDT	Tube dominance test
Learning	Shock	GNG	Go–no go avoidance	Social	Fixed	SRP	Social recognition or preference
Learning	Shock	IA	Inhibitory avoidance	Social	Free	SDT	Dominance /interaction >2 mice

elevated plus maze. Any attempt at automating the ethogram must be rigorously validated against the opinion of skilled observers.

It is best practice to label behaviors according to the topography of the action, rather than imputing a purpose or function to a sequence of actions. Purpose is an inference from facts, whereas data should consist of facts. If a mouse urinates in a novel environment, the act of urination is observable. Whether this may also qualify as territorial marking cannot be determined from the mere fact. Several other kinds of behaviors impute motive or goal: foraging, hoarding, burrowing, aggression, and reproduction. Each of these can be better recorded in terms of distinct behavior patterns that may or may not culminate in some end result. For example, digging may result in a burrow, but the mere act of digging is not proof that the animal is trying to construct a burrow. Likewise, reproduction is sometimes the consequence of mating behavior, but it is unlikely that the cavorting mice have a well-formed idea of what is going to happen 19 days later. On a descriptive level, mice engage in sexual behavior, not reproduction.

Anyone who watches the antics of a mouse or a pair of mice for only a few minutes in different situations will note how the range of behaviors expressed depends strongly on the situation. This is obvious for things that occur only in a social context, such as mating and agonistic interactions; isolated individuals do not perform social behaviors. Likewise, digging behavior will be seen only when there is a medium in which to dig. When several mice are released into a large and complex environment, as was done in McClearn's hypothesis generation apparatus at the Institute for Behavioral Genetics (Chapter 1), almost the complete range of all mouse behaviors may be apparent over a period of several hours or days. This is an excellent means to become familiar with mouse behaviors.

For many kinds of behavioral tests, automated measurement of the stream of behavior should be supplemented with ethological observation of postures and patterns that may have special importance in a test situation. In the simple open field, distance traveled, rearing while far from a wall, or leaning against a wall can be transduced effectively with photocell beams. Patterns such as grooming or the "stretch-attend" will not be revealed in the sequence of beam breaks, however. It also is impossible to tell from photocell beam breaks whether a mouse is truly "freezing" or just sniffing intensively at a drop of urine or other interesting object.

When using the ethological method to score sequences of behaviors, the test usually is not standardized and given some special name. If the researcher is interested in courtship and reproduction, it is usually sufficient to place an adult male and adult female into a cage and then watch the action. Placing two males in the same kind of cage can lead to agonistic interactions, whereas housing several females together may result in barbering. It is very important to use the same situation on every test occasion: the same size of box, the same kind of bedding and food, and the same trial duration. The test parameters should then be described in detail in the methods section of a report.

DOMAINS AND SUBDOMAINS

Whereas the ethological approach is best applied unobtrusively with mice living in a large and complex cage in the colony room that may also serve as the home cage, tests that seek to assess specific behavioral processes often involve special apparatus that limits what can be done by the mouse and allows convenient, automated measurements. Such tests are usually given in a room separate from the colony room. Often the test has a special name that is widely recognized among mouse researchers (Table 10.2). Most of those tests try to study a particular domain or class of behavior.

For the animal, behavior is a continuous stream of actions that help it adapt to a situation. As the situation changes, the behaviors that are expressed also change. To a considerable extent,

163

a classification of behavioral domains is a classification of situations. Some tests are designed to assess learning and memory, for example, but the mouse will learn things during the course of virtually any test and recall aspects of the experience the next day. In any test administered by a human experimenter, there will be some degree of anxiety evoked by the situation, even if the test is not designed to assess anxiety. Nevertheless, many tests pertain mainly to a specific domain of behavior.

Categories of behavioral domains have been established in the ontology of behavioral and neural function of the Mammalian Phenotype Browser. The categories shown in Box 10.1 are designed primarily to classify abnormalities arising from genetic defects. They are less useful for classifying tests that pertain to the normal range of mouse behavior, and they do not always reflect the categories of tests in common use. For example, grooming of self is a motor behavior requiring coordination and is not specifically measured by any test. Nociception or pain is an important research topic that warrants its own category, even though it is a subcategory of the domain of sensory function.

Crawley and Paylor (1997), in their Table 1, presented a categorization of behavioral tests that emphasized the role of mice as models for human neuropsychiatric disorders (Box 10.2). For each broad category, several tests in common use were cited.

10.1 BEHAVIORAL CATEGORIES IN THE MAMMALIAN PHENOTYPE BROWSER

Circadian rhythms
Eating/drinking
Emotion/affect
Grooming
Learning/memory
Motor capability/coordination
Sensory capability/reflexes/nociception
Sleep patterns
Social/conspecific interaction
Vocalization
Seizures

Source: www.informatics.jax.org/searches/MP_form.shtml

10.2 CLASSES OF BEHAVIORS

Learning and memory
Feeding
Pain/analgesia
Aggression
Reproduction
Anxiety models
Depression models
Schizophrenia models
Drug abuse models

Source: Crawley & Paylor (1997).

An alternative classification of domains in Table 10.2 shows the primary domain and subdomain for 110 behavioral tests used with mice. It was devised by first listing the most common tests, then assigning them to categories largely as they are described in published articles. The classification divides some domains that are a single entry in Box 10.1, such as emotion/affect that is separated into tests of anxiety- and depression-related behaviors. Learning and memory is simply designated as learning because there can be no memory without learning and vice versa.

COMPLEXITY OF DOMAINS

It is apparent from many inbred strain surveys and other studies of genetic correlation that all behavioral domains are complex and heterogeneous (Crabbe et al., 2005; Finn et al., 2003; Mogil et al., 1999b). No domain can be adequately assessed with a single behavioral test. Different tests that pertain to the same domain almost always show less than perfect correlation with each other. Consequently, an investigation that seeks to understand how a new mutation or a substance such as ethanol affects a particular domain must employ more than one test of behavioral functions in that domain.

Domain complexity is apparent in studies of anxiety-related behavior in mice. Several tests are thought to reflect levels of mouse anxiety to some extent (e.g., elevated plus maze, EPM; light–dark maze, LD; open field activity, OFA; and elevated zero maze, ZM). Each involves slightly different motives as well as sensory and/or motor capabilities, and no one test can serve as the gold standard for anxiety against which all others may be judged (Crawley, 2008). If the investigator believes that a genetic mutation or experimental treatment alters anxiety, this is best evaluated by using more than one test. If the treatment does indeed strongly affect anxiety levels, it should change scores on all the tests, although probably not to the same degree.

A large study of 1671 genetically heterogenous mice by Henderson, Turri, DeFries, and Flint, (2004) administered five tests to each animal (OFA, LD, EPM, mirrored chamber, and ESQM) with one test per day. After all five tests had been given, OFA and LD tests were given again on separate days. A factor analysis of phenotypic correlations found that almost all factors were loaded only on measures within a single test session, and there were few commonalities across different tests. Genetic analysis detected many commonalities across tests, but the pattern was complex and could not be accounted for by a single genetic factor or set of factors. Quantitative trait loci (QTLs) linkage analysis detected many highly significant linkages on seven chromosomes. Logarithm of odds (LOD) scores are shown in Table 10.3 for total activity

TABLE 10.3 QTL Analysis (LOD Scores) of Anxiety-Related Behaviors[a]

	Total Activity Counts					Percent Activity in Threatening Areas				
Chromosome:	1	4	7	15	18	1	4	7	15	18
cM range:	74–80	34–47	29–59	20–22	23–32	74–83	35–51	50–65	22–28	27–34
OFA 1	27	3	13	12	5	3		8	3	5
LD 1	30	6		14	5	10	3		11	4
EPM	17	7	11	6	11				3	
ESQM	22	4		11	10		3	17	10	
OFA 2	24	10	9	5	9		7	22	6	4
LD 2	23	12	8		5	4	4			3

[a]The cM interval is the 15th to 85th percentile range. LOD scores are rounded down to the nearest whole number. More detailed information is provided in the original source (Henderson, Turri, DeFries, & Flint, 2004). Abbreviations for tests are given in Table 10.2.

counts and percent time in threatening areas for all tests. Many LOD scores were very high, leaving no room for doubt about the significance of linkage with genetic markers. For total activity in an apparatus over an entire trial, very consistent linkage across types of apparatus was seen for QTLs on four chromosomes, indicating common processes that contribute to overall activity levels. For the indicators of anxiety, on the other hand, there was far less consistency across tests; only the QTL on chromosome 15 influenced anxiety-related behavior in the same way on all tests. Open-arm exploration of the elevated plus maze, a popular indicator of anxiety adopted in many other studies, showed little evidence of genetic linkage.

Heterogeneity of the domain of anxiety is also evident in studies that manipulate anxiety-related behavior experimentally and assess effects with multiple tests. A knockout of the serotonin transporter gene (*htt*) increased anxiety-related behavior and reduced exploratory activity by large amounts on all tests in a battery (EPM, LD, emergence test, OFA; Holmes, Yang, Lesch, Crawley, & Murphy, 2003). On the other hand, many other studies have reported treatment effects on one test in a battery but not others. For example, isolation housing increased anxiety-related behavior on two tests (LD, neophagia), while it decreased on the EPM (Voikar, Polus, Vasar, & Rauvala, 2005). Effects interacted strongly with genotype and were generally greater for DBA/2 than C57BL/6 mice. In a study of chronic mild stress administered to three inbred strains, increases in anxiety-related behavior were substantial in the LD box but not significant at $\alpha = 0.05$ on the EPM (Mineur, Belzung, & Crusio, 2006).

The domain of motor coordination is also complex, especially when effects of ethanol are considered. When seven motor tests were given to eight inbred mouse strains, the pattern of strain correlations revealed the complexity of the domain (Crabbe et al., 2005). An abbreviated battery of five tests was later given to 20 inbred strains, and strain correlations among tests were low to moderate (Figure 10.2). After a saline injection, slips on the balance beam were almost

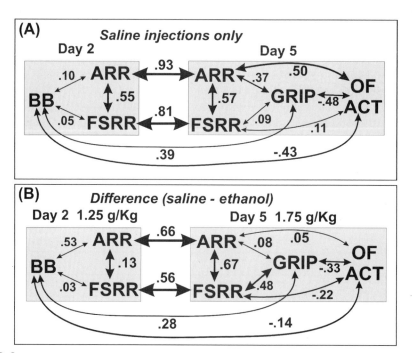

FIGURE 10.2
Correlations among strain means on five motor tests for 20 inbred mouse strains (eight mice per strain) studied in the author's laboratory at the University of Windsor. (A) After a saline injection, correlations between the accelerating rotarod (ARR) and fixed speed rotarod (FSRR) tests were much lower than between the same kind of test given on different days. Correlations of balance beam (BB) and grip strength (GRIP) with other tests were low to moderate. Open field activity was relatively high in mice that stayed on the ARR longer and lower in mice that made more slips on the BB. (B) Strain correlations of ethanol effects were positive for most tests, except for open field where ethanol caused an increase in activity for several strains.

unrelated to performance on two versions of the rotating rod test, while grip strength and open field activity were uncorrelated with the fixed speed version of the rotarod. When ethanol was administered, the degree of impairment of motor coordination was positively correlated for the balance beam and accelerating rotarod (ARR) at a lower ethanol dose, whereas at a higher dose effects on the ARR were not correlated strongly with those on open field activity or grip strength.

TEST BATTERIES

Because of the complexity of any domain, it is essential that more than one kind of test be given in an experiment. Should the tests be given to separate groups of mice, or should one mouse see all tests? Economy of animals is best served by repeated testing of one mouse, but there is concern that the experience of one kind of test will influence scores on subsequent tests of a different kind. Repeated testing could affect the validity of the tests. It could also make it more difficult to interpret patterns of scores across genotypes. There are good reasons for these concerns.

Every test changes the mouse

Every test of a live animal is crafted to measure some process or construct. The very act of taking a test, however, will change the mouse. Weighing a mouse on a balance for a few seconds does not change its weight very much, but the telltale boli remaining on the pan remind us that there has been some change. Behavior presents special challenges because the change wrought by the test can be substantial, yet invisible. The test measures the state of an animal that exists at the time of the test, but that state reflects the animal's history to some extent. If the test seeks to assess anxiety-related behavior, for example, events before the test that alter anxiety will alter test scores. If the test assesses learning and memory, performance will depend greatly on prior experience with the same task. The exquisite sensitivity of mouse behavior to prior experience poses special challenges when the investigator wants to assess several phenotypes in the same study.

This reality is apparent from the QTL analysis of anxiety-related behavior (Henderson et al., 2004) that repeated open field and light—dark box tests after five tests in a battery had been given (Table 10.3). The patterns of linkage for general activity level in the open field were very similar on both occasions. Activity in the LD box showed a clear QTL effect on chromosome 15 on the first trial but none on the second trial, versus no QTL on chromosome 7 on the first trial and a high LOD score for that chromosomal region on the second trial. Similarly for anxiety-related behavior, both OFA and LD tests showed different patterns of linkage on the first and second tests.

Any test is done as part of a daily routine in which the mouse is removed from the home cage, placed into a holding cage, transported to the test room, and placed near the apparatus for several minutes. To the experimenter, this routine is performed so often that its effects seem negligible. For the mouse that experiences it for the first time, there is a radical change from the daily routine in the home cage. After several days of testing, the effects of routine handling will change. Some investigators think that the animal habituates to repetitive events and the routine no longer changes behavior. This might be true if only the handling occurred each day. Habituation happens when some salient stimulus is followed by nothing of significance for the mouse. In the mouse lab, on the other hand, routine handling is inevitably followed by some kind of test, and some of the tests can be beneficial or stressful and most certainly memorable. The mouse learns that the routine journey to the test room will be followed by something else, so that routine handling becomes a conditioned stimulus that signals it is time to get ready for school. The net result of these interactions is that scores on any test always reflect things that happen to the mouse before the test begins. There is simply no way to assess any behavioral process in isolation from the history of the individual mouse.

Advantages of using different mice for different tests

Suppose the researcher wants to compare the scores of several inbred strains on two kinds of tests, perhaps a conditioned place preference and a water escape test. The longer term goal is to discover polymorphic genes that have similar influences on the two tests. One of the processes is likely to be memory for prior experiences. Both tests require good memory to achieve good performance, but they rely on different kinds of motives: fear of electric shock in a dangerous place and aversion to water. Researchers would like to know whether the kinds of memory processes utilized by the two tasks are essentially the same or instead differ substantially when there are different motives at work. If the same kind of memory is crucial for both, then the strain performing well on one task should perform well on the other. There is no doubt that the data most pertinent to this question will come from testing different animals on the two tests. The optimal design would have many genotypes, with each individual measured on only one task. The strain correlation between tasks would then reflect mainly genetic sources of similarity.

There is one other major advantage to giving just one kind of test to one mouse. The logistics of the study can be simplified, and several kinds of tests can be given in one day, such that throughput of the entire study can be accelerated, especially if some tests can be administered simultaneously to more than one mouse by one technician using automated apparatus. There would be no need to space the different tests for one mouse across several days or weeks. An intensive test schedule might get the entire job done in just a week.

Efficiency and economy of numbers comes at a price

In the interests of efficiency, the experimenter might wish to test one mouse on both tasks. If the study involves inbred strains, there is no relevant information about genetic correlation added by this option because all animals in a strain have the same genotype. The same consideration applies to a study of a targeted mutation where the mutation has been placed on an inbred background. If separate mice with the same mutation are to be tested on two tasks, twice as many animals will be needed.

Efficiency is purchased at the cost of clarity and ease of interpretation of data. We must presume that giving one mouse two different tests will yield different data than separate tests on different mice. Everything known about the psychology of learning and memory supports this presumption. If a test changes the mouse, as it always does in some way, behavior on the test that is given second will be changed by experience with the first test. The only question is whether the change will be large enough to influence the conclusions from the research.

If the study were confined to only two kinds of tasks, doubling of the sample size might be contemplated to achieve unambiguous data. Given the current stage of investigation in the field of mouse neurobehavioral genetics, however, it is rare that only two tasks are given. Usually there are many tests given to one animal over a period of weeks (Table 10.4). The option of using separate groups of mice for each test is generally not realistic because of the vastly larger samples that would be needed.

Current wisdom about test batteries

Current wisdom in the field is that one mouse should take many tests. The order of tests should be the same for every mouse and should be ordered from least to most stressful (Crawley, 2008). The spacing of tests should be as wide as possible to minimize carryover effects (Paylor, Spencer, Yuva-Paylor, & Pieke-Dahl, 2006). If this is done, objections from reviewers and referees are not likely to be vehement. Nevertheless, it would be comforting to have some objective evidence of the consequences of testing mice in batteries.

TABLE 10.4 Recent Studies Utilizing Batteries of Mouse Behavioral Tests[a]

Study	Test Spacing	Genotypes	n per Group	Test Order
McIlwain et al. (2001)	1 week	1	8	REFL OFA LD ARR PPI AS FCA MWM TF HP
Goddyn et al. (2006)	?	1	12	GRIP ARR HCA OFA EPM IA MWM FCC FCA
Moy et al. (2007)	1 week	10	20	HCA REFL OFA ARR FF EPM TM:F MWM SRP
Mitsui et al. (2009)	5 days	2	12	4 cohorts (OFA LD EPM BB GAIT FF) (OFA HCA YM SDT SRP EPM ARI) (IA MWM) (SRP)
Lad et al. (2010)	2 days	8	11	BT HCA OFA OE XL EPM LD SHIRPA PB MWM TS
Foldi et al. (2010)	1 day	1	40	EPM HB SDT SRP PFS PPI
Philip et al. (2010)	1, 2/day	70	10	Several batteries[b](CPP OFA COC) (OFA EPM ETH) (TR HP LD ZM OFA PPI AS) (HIC ETH) (OFA MOR) (HP TW VFT) (DT PFS VP)
Solberg et al. (2006)	3, 4/day	8	12	OFA EPM FF HCA AS FCC FCA

[a]Value for test spacing applied to most but not all tests in battery. Sample size (n) per group is an approximation based on average for several tests. Abbreviations for tests are given in Table 10.2.
[b]Tests in parentheses were given in specified order to one group of mice. Separate samples were tested in four different sites. Abbreviations are mortly given in Table 10.2. COC, ETH, MOR denote respones to cocaine, ethanol, and morphine, respectively.

There are two kinds of experiments that could provide such evidence. (a) Within the scope and time frame of a single study, it is possible to have separate groups of mice that receive only one test evaluated alongside those that are run through an entire battery of the same tests. As indicated in Figure. 10.3A, it is important to test mice in the different groups at the same time of the day and week, and to make them comparable in every other way that can be controlled. (b) All mice could receive all tests but in different orders, such that each test is given first to some of the mice and could not be influenced by previous tests (Figure 10.3 B—E).

A formal test of both approaches was done by McIlwain, Merriweather, Yuva-Paylor, & Paylor (2001) in a widely cited experiment. The first phase compared mice given eight tests in a fixed order with independent groups of mice that received just one of the eight tests. The first test in the battery was a series of tests of simple reflexes that involved considerable handling by an experimenter (Table 10.4). That experience reduced activity in the open field

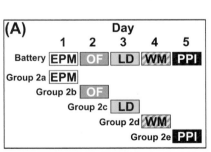
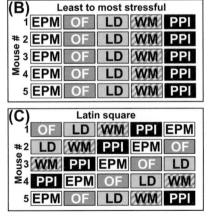
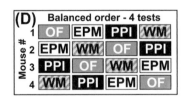
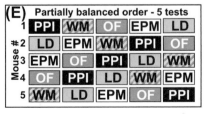

FIGURE 10.3
Designs of studies to assess the importance of test order on behavior. (A) Battery testing is compared with the same tests given on the same days to mice that are tested only once. (B) The conventional arrangement of tests from least to most stressful for all mice. (C) Latin square places each test in a different ordinal position for different mice, but any one test always follows the same test. (D) Balanced order of four tests has each test in each ordinal position once for every set of four mice, and each test follows every other test just once. (E) For five tests, the order cannot be perfectly balanced regarding order for a set of five mice. Abbreviations: EPM, elevated plus maze; OF, open field; LD, light—dark box; WM, water maze; PPI, prepulse inhibition.

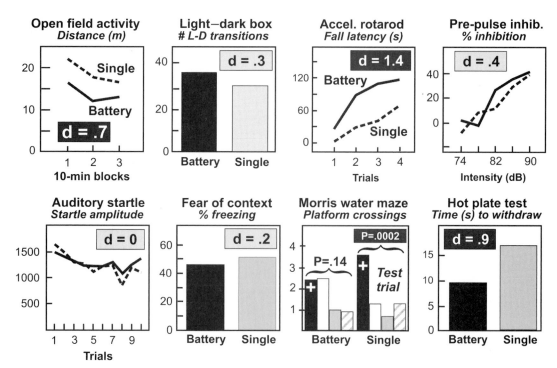

FIGURE 10.4

Results of a study of eight behavioral tests given to C57BL/6J mice either in a battery in the order shown from open field to hot plate or singly. For the battery condition, the open field test was always preceded by assessment of reflexes using a technique involving human handling, whereas in the single test condition mice received only the open field test. Effect size d of the battery testing effect is shown for seven tests, while the water maze is summarized by significance of the search in the correct quadrant (+) on a test trial. *(Redrawn with permission from McIlwain, Merriweather, Yuva-Paylor, & Paylor, 2001).*

by a substantial d = 0.7 standard deviation (Figure 10.4). Open field testing had little influence on behavior in the light–dark box one week later, but previous tests dramatically improved performance on the accelerating rotarod by d = 1.4. The next three tests in the battery showed little influence of prior testing, whereas battery training greatly impaired the search for the platform in the Morris water maze while reducing time to withdraw the tail from a hot plate by d = 0.9. Thus, for four of the eight tests, use of a battery altered scores significantly and substantially.

The first phase of the study was somewhat difficult to interpret, because each test given singly was compared with the same test in the battery for the same sample of mice. Thus, comparisons for the different kinds of tests were not independent from each other. Strictly speaking, from a purely statistical point of view, it would have been better to have a separate battery group receive the battery up to and including the test to be compared with the same test given only once. A larger sample would also have been informative, because the study done with n = 10 was generally capable of detecting only large effects. Furthermore, the data applied only to the inbred strain C57BL/6J. Nevertheless, results showed that test order effects can be noteworthy and should never be ignored when planning an experiment on behavior.

The second phase of the study compared four balanced orders of the same four tests given to different groups of mice (Figure 10.3D). Differences between scores on a single test in relation to its order in the battery were not large and in most instances were not statistically significant, but the sample size for any one order was small (n = 9 or 10).

A recent study by Lad et al. (2010) also compared battery testing to single tests for C57BL/6J mice with n = 7 to 10 mice per group. Effects of battery testing were generally small on six tests.

All mice in the comparison groups were housed singly, a procedure that is not used in most labs because it tends to alter a wide range of behaviors (Brain, 1975; Koike et al., 2009; Lagerspetz & Lagerspetz, 1971; Martin & Brown, 2010).

It thus appears that there is very little experimental evidence to substantiate common wisdom about battery testing. The most thorough study reported in the literature found that half of the tests were influenced to a major extent by inclusion in a battery. It seems likely that certain tests, especially those involving reflexes elicited by strong stimuli, are not strongly influenced by prior experience with other tests. This notion is intuitively appealing, and is supported by data on auditory startle and prepulse inhibition. Nevertheless, it is contradicted by data from the same study on hot plate tail withdrawal that is reflexive (Figure 10.4). It appears that the issue of battery testing needs further investigation. It is important to do this, because almost every study of inbred strains and targeted mutations now employs several behavioral tests given to the same mice.

Testing effects of batteries

The most persuasive evidence about battery testing has come from comparisons of mice given only one test with those receiving several tests in a battery (Figure 10.3A). The elegant simplicity of the design is satisfying. Unfortunately, the evidence is moot. The field is never going to adopt a universal policy of one test per mouse. When an investigator creates a novel mutation, he or she wants to know in a very broad sense "what's is wrong with my mouse?" (Crawley, 2000). This can be answered only by multiple tests of behavior (Crawley, 2008), and with rare knockout animals this inevitably means multiple tests on one animal. Thus, the two critical questions for practical research with a test battery are (a) the ordering of tests and (b) the spacing of tests. A third question of great importance is whether the effects are closely similar for all genotypes or if some strains are much more sensitive to test order effects. If the battery entails a very large number of tests, age differences at testing may also be a concern.

Figure 10.3B—D shows three methods of ordering several tests on the same mouse. The customary method is to give all animals the same tests in the same order, which confounds test with order of testing. The Latin square design eliminates this confound but each test is always preceded by the same test(s). A perfectly balanced order can be arranged when there are four tests; each test occurs equally often at each position and is preceded once by every other test. This design was used by McIlwain et al. (2001). For five or more tests, the second order contingency cannot be perfectly balanced; certain sequences of two tests occur twice while others occur only once in a set of five mice.

No matter how elegant the balancing act, studies that compare many test orders run afoul of logistics. In the simple design of one battery versus single tests (Figure 10.3A), any order could be used in the battery because any one test would be given during the same day or week to mice in both groups. For example, the lab could be arranged for training in the Morris water maze for all mice in a one-week period. If different mice are to be given different test orders, on the other hand, it will be necessary to have all tests in operation at the same time. Some mice might receive their three brief trials of grip strength just a few minutes before others receive four trials in the water escape tank or vice versa. A study of balanced orders of testing could only be done with very wide test spacing of at least one week.

An alternative approach could be adopted for assessing the sensitivity of any test to battery effects. This approach derives from the fact that most batteries involve certain common tests. Table 10.4 shows that among eight studies, seven used open field activity, six the elevated plus maze, five the Morris water maze, four the light—dark box, and three the accelerating rotarod and prepulse inhibition. Of course the mix varies for research on learning and memory as opposed to pain sensitivity, but here the concern is with batteries that span a broad phenotypic range. Suppose there will almost always be a core of five tests such as

OFA, ARR, PPI, MWM, and HP in that order. On day 1 the mice might receive an intraperitoneal injection of saline, then a 5 min OFA test, then 10 trials of ARR. On days 2 to 4, just one of the other tests could be given. Finally, on Friday a new test might be administered with half of the mice being naive and half having four days of experience with the battery (Figure 10.5A). If 20 mice were studied in each of the two conditions, power to detect an effect size of $\delta = 0.6$ would be 80% when $\alpha = 0.05$ one-tailed. By using this design to assess several kinds of tests, it would be possible to determine which tests are markedly sensitive or insensitive to inclusion in a battery.

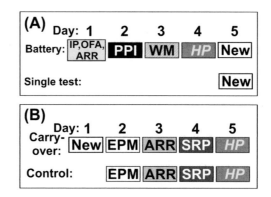

FIGURE 10.5

Designs to study test battery effects. (A) A new test to be evaluated is given either singly or on the same day after mice had received an intraperitoneal injection (IP) and then several tests on previous days. (B) The new test precedes four tests known to be prone to battery testing effects, and the scores on those four tests are then compared to those of mice that do not receive the new test first. Abbreviations: ARR, accelerating rotarod; EPM, elevated plus maze; HP, hot plate tail withdrawal; OFA, open field activity; PPI, prepulse inhibition; SRP, social recognition and preference; WM, water maze.

A variation on this theme could assess which tests have major or perhaps very little effects on tests that follow (Figure 10.5B). The fixed battery of tests should be comprised of those known from the first phase (Figure 10.5A) to be sensitive to test order. After a test is shown to be sensitive to prior tests in the battery or influence other tests that come after it, the importance of spacing of tests could be evaluated with the designs shown in Figure 10.6.

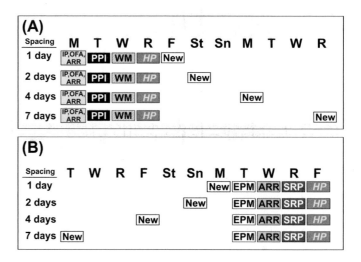

FIGURE 10.6

Designs to evaluate the importance of time interval between tests. (A) Groups all receive the treatments described in Figure 10.5A, but time between the battery and the new test varies. (B) Groups all receive the tests shown in Figure 10.5B, but time between the new test and the battery varies. Abbreviations are the same as for Figure 10.5.

The spacing of individual tests in various published studies ranges widely (Table 10.4): one week, two days, one day, and even several tests in one day. Some reports do not even mention the spacing of tests, which implies that some researchers do not believe this is an important variable. Insufficient research on the impact of test spacing has been done. One study found little difference between 1- or 2-day intervals and 1 week between tests (Paylor et al., 2006). This kind of evaluation needs to be extended to a wider range of tests, because the issue is important for test validity as well as throughput of the battery.

Effect size and importance of test order effects

When investigators employ samples of 10 mice in each of two independent groups, this shows that they are interested only in detecting large effects of any treatment, including test order effects. Power to detect smaller effects is generally quite low and may pass undetected. Everyone knows from introductory statistics courses that failing to reject the null hypothesis of no effect does not prove that the null is true (Chapter 5). Nevertheless, there seems to be a tendency to do just this when considering test order effects. Both studies of battery testing summarized their findings in tables that utilized the "=" sign to denote no significant difference between battery and single testing. The clear implication is that order had no effect at all and can safely be ignored. In fact, the studies demonstrated no such thing. Power was too low to proclaim order effects nil.

What magnitude of test order effect is worthy of note? If order effects are regarded as just another treatment effect, then the same criterion for other kinds of treatments should be used. If investigators would like to know that order effects are not worth a worry, they need to show that even small to moderate effects are generally absent. A sample size of 20 mice per group would be a reasonable choice.

That sample size pertains to a single behavioral test. When the same mouse is given multiple behavioral tests, many statistical tests of significance are done. If the Type I error criterion is set at $\alpha = 0.05$ for each test, the probability of making one or more false positive discoveries in a study will exceed 5% by a substantial amount (Chapter 5). Good arguments can be made for using a more stringent criterion where $\alpha < 0.05$. This would require a larger order effect to reject the hypothesis of no effect. For many investigators, the consequence would be pleasing. Test order effects are a complication that is not desired in work with mice. It does not seem wise to use a stringent Type I error criterion, when the researcher hopes to find no significant order effect, but a more lenient $\alpha = 0.05$ for tests of an experimental treatment that is the main focus of the study. That would amount to cheating with numbers.

Interaction of genotype or experimental treatment with test order effects is perhaps the most important question. If all test order effects are additive with respect to genotype, they will generally raise or lower scores on a test by about the same amount for all genotypes, and the principal results of the experiment will not be altered. As shown in Chapter 5, larger samples are needed to evaluate genotype by treatment or test order interactions than to show only average or main effects. A rigorous evaluation of this kind of interaction would require a study of very large scope and sample size. No pertinent data have been reported to date.

One solution: Standard test orders

The most common practice when using a test battery is to keep the order fixed for studies done in the same laboratory and describe the order in detail in a methods section of the report. Order effects may be present, but they will be more or less the same in every study. Other investigators will be able to copy the methods and use the same order.

Unfortunately, ordering of tests is almost always unique to a specific lab. As shown in Table 10.4, there is no consensus on the precise order across labs. A general pattern of placing less stressful tests earlier in the order is evident, but there are exceptions. If good data on the importance of test order were available, perhaps using designs like those in Figures 10.5 and 10.6, different labs might be more willing to adopt a common test order.

How stressful are tests?

Placing less stressful tests earlier in the battery makes good sense because effects on subsequent tests are likely to be smaller. Many of our notions about stress, however, are based on human intuition and informal observations of mice during testing. If mice are markedly difficult to handle during a particular kind of test (Wahlsten et al., 2003a), it is reasonable to suppose they find it quite stressful. It may be that the mice are telling the experimenter something important when they try to run away from a test situation. This phenomenon could be utilized to obtain an inventory of more and less stressful tests. Several general methods can be envisioned.

Apparatus is usually constructed so that the animal cannot escape during testing. Suppose instead that it is altered so that the mouse can leave the situation altogether and return to its holding or home cage. This would be done only during formal tests of stressfulness. After the mouse discovers the escape route, latency to depart on successive trials or days might reveal how the animal perceives the test. It is possible that some tests are actually preferred over a barren and brightly lit holding cage. The day's trials could be halted when the mouse departs, so that escape from the test serves as a reward.

A variation on this theme could be applied after the mouse has experienced several trials on an apparatus. A holding cage with an exit hole could be placed next to the apparatus so that the mouse could walk onto or into the apparatus. Emergence time from the cage might then reveal the mouse's view of the test. There would be an upper limit where the animal would never leave its cage.

Relative stress levels induced by two tests might be compared using a two-choice paradigm. The animal could be placed on an open runway or platform with different test apparatus at

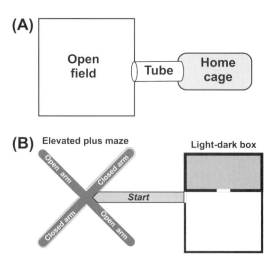

FIGURE 10.7
Arrangements of apparatus to evaluate preference of mice for a test apparatus. (A) Time to emerge from the home cage and enter a test apparatus or exit the apparatus and retreat to the home cage can indicate desirable or aversive features of a test from the perspective of a mouse. (B) Relative features of two tests can be assessed by joining them with an open walk way that is brightly lit. The mouse should tend to prefer the less aversive test.

each end or a clean holding cage at one end (Figure 10.7). The mouse could then choose which kind of test to take that day. If the animal balks at entering either apparatus, it might be informative to measure how much time is spent near each one. This approach would be meaningful only after the mouse has had considerable experience with both kinds of apparatus.

Motivating Mice

Among the tests listed in Table 10.2, several require no specific method to motivate action by the mouse. Mice do many things spontaneously, such as explore a novel environment. Whether the tendency to explore should be called a "drive" or motive can be debated at length, but the tendency is inherent in the test situation. The experimenter does not need to do anything to get the mouse ready for this test. For some of the other tests, special procedures are needed to motivate the mice, especially when learning and memory are to be evaluated. Mice will explore a maze even when they are not hungry and no food reward is provided, but they will traverse the maze more quickly and consistently to locate places where there is food if they are hungry. Motivating procedures can be applied with many kinds of tests and

Mouse Behavioral Testing. DOI: 10.1016/B978-0-12-375674-9.10011-4

therefore warrant their own discussion. Methods to manipulate four kinds of motives are presented here: hunger, pain from electric shock, immersion in water, and aversive puffs of air.

Three issues are especially important when manipulating motives. (1) There are several ways to alter each kind of motive, and it would be helpful to identify one method that is best for most studies. (2) The method must work well with a wide range of genotypes. Fine-tuning the parameters of a protocol with just one inbred strain may not yield a method that is generally useful. (3) It would be valuable to have a way to measure the degree of motivation, not just manipulate it. It would aid the interpretation of studies of learning and memory in particular if it were possible to assert with confidence that two mice or two strains are equally motivated. For example, it is well known that inbred strains differ widely in the rate of learning most tasks (Wahlsten, 1972a). Do these differences arise entirely from neural processes that constitute learning ability, or are they to some extent caused by different degrees of motivation? More highly motivated animals tend to learn a task faster and perform more consistently. It is unlikely that all differences in learning arise solely from variations in motivation, but some portion of them might. If there is a valid means to equate levels of motivation, this question can be answered.

HUNGER AND FOOD DEPRIVATION

Finding and storing food occupy much of the waking period of most species of animals. Many animals will gather food even when they are not hungry. Food consumption tends to occur in a 24 h cycle and is regulated by a complex system of molecules and receptors (Abizaid et al., 2006; Funahashi et al., 2003). At the same time, restricting the times when feeding can occur alters the circadian cycle of many behaviors (Froy & Miskin, 2010; Minana-Solis et al., 2009). Thus, there is a feedback relationship among the sleep—wake cycle, exploratory activity, and feeding.

In the laboratory, mouse life is somewhat different from the feral state. An abundance of nutritious food is almost always available. Mice of different genotypes are usually given the same kind of food, rather than allowing strains to self-select different proportions of protein, fat, and carbohydrates (Smith, Andrews, & West, 2000). Leisurely feeding occurs throughout the active phase of the light—dark cycle in many short bouts. Mice often eat before they are really hungry in the physiological sense. When a well-fed mouse is removed from its home cage and placed into a maze that contains pellets of food, it usually does not eat at all and spends the time exploring every nook of the new surroundings in a highly variable journey. If the mouse is supposed to learn a specific route to find a food reward, it will do this much better if it is hungry. There are several ways to make a mouse hungry, and different degrees of hunger can be generated.

Appetite without deprivation

Food deprivation procedures are time-consuming for the experimenter, and it would make life for human and mouse much easier if learning would occur rapidly in non-deprived mice when tasty food is its own reward. This option was evaluated in the author's lab with four strains of mice. In a simple test, an animal was placed in a clean cage, presented with a piece of food in a glass dish, and allowed to eat for two minutes. The experimenter observed the animal and recorded the time it spent eating. Each animal saw eight kinds of food in one session in a balanced order. The test was given to mice before food deprivation and then after 24 and 36 h of deprivation. As shown in Figure 11.1, there were marked strain differences in almost every aspect of eating. Live mealworms were highly favored by two strains, but the RF/J strain often dropped the worm as soon as it began to squirm. Walnuts were consumed more consistently by all four strains, while all other foods showed marked strain-dependent effects on the degree of hunger. Strain differences in the consumatory response were marked when they were not deprived and much less varied when they were hungry. Thus, the palatability factor had a lesser

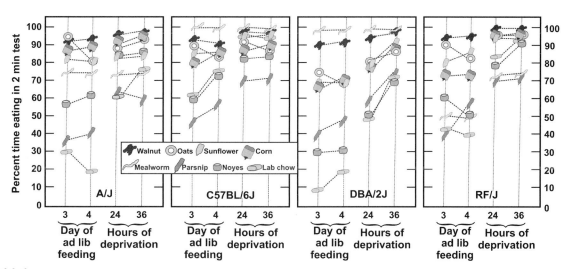

FIGURE 11.1

Time spent eating eight kinds of food by four inbred strains from the Jackson Labs. Each mouse received one food per 2 min trial and all eight foods in a balanced order in one day. The same foods were presented on four consecutive days, two before deprivation, one after 24 h deprivation, and the next after 36 h deprivation. *(Based on data from the author's lab, collected with the assistance of David Sternthal.)*

role when mice were hungry. The food that showed the greatest difference between deprived and non-deprived states was the familiar lab chow. Other tests were done to train mice in a maze, and performance was highly varied when they were not deprived, even with walnuts as a reward. After the preliminary study, the option of studying learning without food deprivation was abandoned.

The 2 min eating test showed considerable promise as a means to assess hunger, and it was used in a series of studies to find the best means to manipulate and assess hunger. An early study of learning in rats had also used eating time as the indicator of hunger (Washburn, 1926). Several eating times were tried in preliminary work, and it was found that hungry mice usually ate during most of a 2 min period before taking a break to explore or drink. A further complication was found, however, when a mouse saw several kinds of food in one day. Except for the DBA/2J strain, mice actually ate lab chow with some enthusiasm in a session where they also were served other tasty morsels (Figure 11.1). The incentive value of the palatable foods appeared to increase the appetite for boring lab chow that filled the food hopper in the home cage. It appeared that results would be most clear if just one kind of food was available in a study of learning — the familiar lab chow.

A recent study of retinal function assessed the role of cones in mouse vision by rewarding the animal with a drink of highly palatable soy milk when it pushed the correct panel (Williams, Daigle, & Jacobs, 2005). The mice liked the milk so much that they pressed more than 300 times in one session. Even under the best stimulus conditions, however, they never achieved better than 80% correct, and levels of 60 to 70% correct were common for lower light intensities that were still visible. As a result, the entire series of tests to determine sensitivity curves took several months for one mouse. The study was done only with C57BL/6 mice that are particularly good at this kind of operant task.

IntelliCage

The IntelliCage (Figure 11.2A) from New Behavior (www.newbehavior.com) offers an entirely different approach to the study of mouse behavior that does not require any specific manipulation of motivation or handling by a human during the test (Galsworthy et al., 2005). The animals are housed in a social group but individually identified by a radio-frequency chip under the skin. Water is available from one of two drinking spouts in each of four corners.

(A)

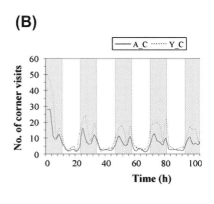

(B)

(C)

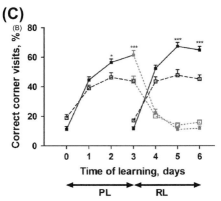

FIGURE 11.2

(A) The IntelliCage. (*Reproduced with permission of New Behavior, www.newbehavior.com.*) (B) Visits by 10-week-old mice (Y) and 14-month-old mice (A) to all corners over six days when all water bottles were available for drinking, showing a crepuscular pattern of daily activity. Young and adult mice were housed in the same IntelliCage. (*Reproduced with permission from Mechan, Wyss, Rieger, & Mohajeri, 2009.*) (C) Correct corner visits during place learning (PL) and reversal learning (RL), showing visits to formerly rewarded corners during extinction as fine dashed lines. Open squares are wild-type controls and solid squares are mice of the R6/2 model of Huntington's disease. (*Reproduced with permission from Rudenko, Tkach, Berezin, & Bock, 2009*).

The experimenter can control which corner and bottle is available to the animals, and the contingency can be different for each animal. Incorrect choices can also be punished with a puff of compressed air. The mice are able to pace themselves and visit corners any time of day, whenever they want to drink. Visits to the drinking stations show strong variations with time of day (Mechan, Wyss, Rieger, & Mohajeri, 2009; Rudenko et al., 2009; see Figure 11.2B). When only one corner has water available (place learning; PL), mice gradually increase visits to that corner, while those visits end when a new corner is made positive (reversal learning, RL; Figure 11.2C). Discrimination between rewarded and non-rewarded corner visits and bottle choice can be sharpened with mild punishment for wrong choices (Mechan et al., 2009; Rudenko et al., 2009). While it is not entirely clear what cues are being utilized by mice when making choices between corners (Barlind et al., 2010), the IntelliCage offers an alternative to discrete trial tests involving human handling and food deprivation.

Methods of food deprivation

Three general approaches to depriving an animal of food are prevalent (Table 11.1). Each has certain positive features and disadvantages, especially when strains are compared. Inbred strains and the sexes differ in their basal metabolic rates (Overton & Williams, 2004) and activity levels in the home cage (Goulding et al., 2008), the amount of food they consume each day (Bachmanov, Reed, Beauchamp, & Tordoff, 2002), and the rate at which they gain and lose weight (Gelegen, Collier, Campbell, Oppelaar, & Kas, 2006; Rikke, Battaglia, Allison, & Johnson, 2006).

TABLE 11.1 Three Methods of Food Deprivation (All Require Mouse to be Weighed Daily)

Method	% Initial Body Weight	% Average Food Consumption	Fixed Number of Hours
Requires baseline data (1 week)	Yes	Yes	No
Weigh food daily and adjust g	Yes	Yes	No
Bias from developmental stage	Yes	Yes	No
Applicable with group housing	No	No	Yes
Periodic recovery period	No	No	Yes
Tailored to individual metabolism	Yes	Yes	No

PERCENTAGE OF FREE FEEDING BODY WEIGHT

A method popular in psychological studies of learning in rats is to weigh the animal at the start of a study and then give it just enough food each day to reduce it to some set percentage of that weight and keep it there for several days or weeks. That level might be set at 85 or 90% of original body weight. This method is sometimes used with mice as well (Johnson, Bannerman, Rawlins, Sprengel, & Good, 2005; Stanford & Brown, 2003). The advantage of this approach is that it takes into account individual differences, even those within a strain, in many aspects of appetite and metabolism. A marked disadvantage is the considerable labor required by the experimenter to weigh the animal and calculate what amount of food to give it each day. It may require several days of titration to reach that stable level of 85% weight before the study of learning can even begin, and asymptotic body weight will never be a stable quantity; small adjustments must be made regularly. This method can only be employed with animals housed singly, and isolation rearing can alter many behaviors (Brain, 1975; Voikar et al., 2005). The free feeding weight may be biased by physiological and developmental factors. If deprivation is started when mice are only four or five weeks beyond weaning, 85% of their free feeding weight will eventually amount to a much lower fraction of how large they would be if allowed to grow freely (Figure 11.3), so that appetite is increased at the expense of their overall growth. If deprivation is begun in older mice, initial body weight will tend to be quite high in strains such as C57BL/6 and NOD/LtJ, which develop obesity and Type II diabetes when maintained with free access to lab chow. Some mice would be markedly overweight before deprivation (Martin et al., 2010), and 85% of that weight might restore them to better health. A further disadvantage is that mice on this regimen will always be mildly hungry for days or even weeks at a time; there is usually no recovery period during a study.

PERCENTAGE OF FREE FEEDING FOOD CONSUMPTION

It is also possible to monitor the amount of lab chow eaten by weighing the amount of food instead of the animal each day for at least a week. Then the animal can be given a fixed ration that is some percentage of the free feeding consumption. This method works best with mice housed singly. It also shares the same biases affecting the method based on body weight: young mice eat less than they would after several more weeks of growth, so the 60% of food at six weeks of age would amount to a lower percentage of what they would normally eat at

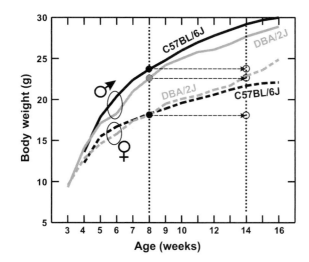

FIGURE 11.3

Growth curves for male (solid lines) and female (dashed lines) mice of two inbred strains, based on information in the Mouse Phenome Database (www.jax.org/phenome). Note that body weights at 14 weeks are about 5 g higher than at 8 weeks. If mice are placed on a food deprivation regimen that holds them at 85% of pre-deprivation body weight when they are 8 weeks of age, that value will be less than 75% of what they would have been at 14 weeks.

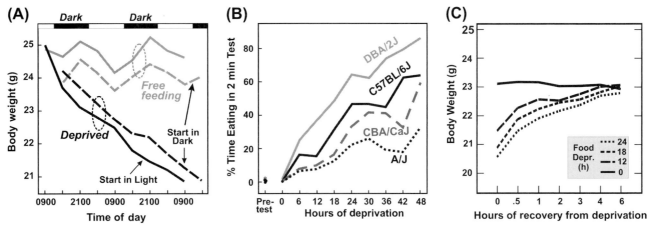

FIGURE 11.4

(A) Average body weights taken every 6 h for 4 inbred strains either on free feeding or total food deprivation beginning at either 0900 or 1500 for separate groups of mice. Note the considerable fluctuation in body weight of non-deprived mice. (B) Percent of a 2 min eating test during which mice that consumed food showed a steady increase with deprivation time for all four strains, but strains had different slopes. (C) Average body growth of the same four strains observed during recovery from different deprivation times. Recovery was nearly complete in all groups after 4 h. *(Based on data from the author's lab collected with the assistance of Diane Cygan and Clara Klieb.)*

10 weeks. This bias could be overcome to some extent by compiling growth profiles for each strain and sex, then assigning an amount of food equal to 60% of what they most likely would have eaten at that age if fed freely.

FIXED TIME PERIOD OF DEPRIVATION

All mice in the study are deprived of food for the same number of hours before the day's test (Dawson, Steane, & Markovich, 2005; Wendt et al., 2005). Figure 11.4A shows the rate of loss for four strains deprived of food for 48 h and weighed every 6 h around the clock. Non-deprived mice typically lost a gram or more of body weight during the light phase of the cycle when they did not eat much. Thus, 12 h of food deprivation is a normal, self-imposed event in the daily life of a lab mouse. The deprived animals lost about 10% of their pre-deprivation body weight in the first 24 h and another 10% on the second day. After 48 h, all mice were active and still in good health, but they sure were hungry. Recovery from 24 h or less of deprivation is very rapid (Figure 11.4C) and almost complete within 4 h after hungry mice are given free access to food.

If 24 h deprivation is used, it occurs on alternate days with a day or a weekend of recovery between days of deprivation (Figure 11.5). Weight loss is greater on the second cycle of deprivation (Fig. 11.5B), but mice recover quickly to their pre-deprivation weight (Fig. 11.5A) because they eat much more after being deprived (Fig. 11.5C, D). Usually there is a more complete recovery period over the weekend. This method is by far the easiest to apply, and it avoids several complications and biases of the other two methods. Because there is a recovery period between each episode of deprivation, overall growth is not impaired. Mice can be housed in groups, avoiding problems of isolation rearing and making more efficient use of cages. It is likely that a fixed period of deprivation will cause greater body weight loss in some strains, and they may not be equally hungry. This bias does not appear to be very large, but it clouds interpretation of many studies, as discussed next.

Experience with these and other studies of hunger and feeding in inbred mice supports a simple method of food deprivation (Box 11.1) that can be applied with a wide range of studies employing food reward and can be continued for several weeks. The duration of each cycle of deprivation can be as much as 48 h without noticeable harm to a mouse, but these data indicate that there is no real advantage to such a long period of deprivation. After 24 h, all mice

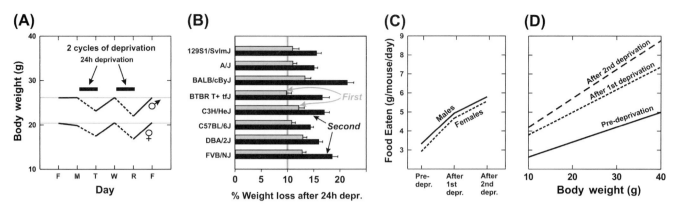

FIGURE 11.5
(A) Average body weight of male and female mice of 8 inbred strains before and after two cycles of 24 h deprivation. Mice lost more weight during the second cycle but also gained more the next day. (B) All strains showed greater weight loss during the second cycle. (C) More food was eaten on average by males (solid line) and females (dashed line) after the second cycle of deprivation than the first, which compensated for the greater weight loss. (D) Slope of relation between food consumption and body weight was steeper for deprived mice than before deprivation. (Based on data from the author's lab.)

11.1 FIXED DURATION OF FOOD DEPRIVATION

Duration: 24 h with no food
Timing: Remove food from cage at about the same time of day when test will be given
Recovery: 24 h free feeding before next cycle
Weighing: (1) At start of deprivation, (2) just before test, and (3) 24 h after recovery

---

will be hungry enough to motivate consistent eating and performance in a maze. It is best to remove the food at different times during the day for different cages of mice. Logistics will be a bit more complicated, but control will be enhanced. By weighing the mice at the start of deprivation, just before the behavioral test session, and then after 24 h recovery, it will be easy to detect any animal that may be suffering from the deprivation because it will fail to recover its body weight quickly.

A period of 18 h deprivation can also be used, but it yields a slightly lower level of hunger on average. Animals can be given free access to food for 6 h after the day's testing but before the next deprivation period begins. This might allow testing on a maze with food reward each weekday. Recovery of body weight is not always complete under this regimen, and mice may gradually lose weight over a period of several days. For mice that complete their behavioral testing late in the work day, an experimenter must come back to the lab late in the evening to remove their food. Furthermore, mice tested early and late in the day experience the deprivation during different phases of the light–dark cycle. As shown in Figure 11.4, if deprivation begins half way through the 12 h light phase, there is usually little eating anyway. Because a difference of 6 h deprivation time has relatively little impact on hunger at 18 and 24 h, one might want to take food away from all mice at the same time at the end of the work day. Then the mice tested early in the next work day would be 18 h deprived, while those tested later the same day would be 24 h deprived. This is not recommended. The 6 h difference does influence hunger and could increase variability in the results of a study of learning.

A refined method for daily deprivation

The greatest drawback to a fixed 24 h deprivation time is that mice cannot be tested daily, and someone must come to the lab on Sunday if a test is to be given Monday, Wednesday, and Friday. Accordingly, a method was devised to overcome this limitation. The method is outlined

TABLE 11.2 Refined Method for One or Two Weeks of Food Deprivation

Step	Procedure
1	House mice two per cage of same strain and sex; one week before tests
2	Feed excess of Bio-Serv grain-based pellets; 1 g and 200 mg sizes; divide pellets between two glass dishes on floor of cage
3	Friday afternoon — weigh mice; give counted fresh pellets for weekend
4	Sunday afternoon — weigh mice; remove and weigh food; find mean mouse weight
5	Monday — test mice after 18 to 24 h deprivation
6	Monday — weigh after both mice are tested; give counted pellets according to graph
7	Tuesday — test mice; weigh them; give counted pellets (adjust if below 85% weight)
8	Wednesday, Thursday, Friday — repeat the process
9	Friday afternoon after all tests complete — give excess of food for the weekend
10	Next week — repeat steps 5–10 beginning Sunday afternoon

in Table 11.2. To avoid isolation-induced behavioral abnormalities, mice are housed two per cage and food is placed in two glass dishes located in different parts of the cage, which effectively prevents one animal from monopolizing the food. There will be no way to determine how much food was eaten by which mouse, although divergence of body weights over days will indicate a need to intervene. In practice, this usually does not pose a major problem. By using grain-based pellets of fixed size (Bio-Serv, 1 g, 200 mg), it is easy to give a specified amount of food without weighing food cubes; simply counting out the daily ration is sufficient and quick. Because the food will be novel, it is important to put pellets in the home cage the week prior to the start of deprivation so that neophobia is overcome.

The study begins when mice are weighed on Friday and then given an excess of pellets in the two dishes. They are weighed again on Sunday and all remaining food is collected and weighed. These data are then used to derive a linear equation for predicting food consumed from body weight. Behavioral testing begins Monday after 18 to 24 h deprivation, depending on when the specific mouse is tested. Thus, the first day entails a fixed time of deprivation. After both mice in a cage have received their daily test, they are given a number of pellets determined by an equation based on their average body weights for Friday and Sunday. They receive this same ration for four consecutive days, unless the body weight of either animal falls below 85% of its pre-deprivation value. In that case, additional food should be given. On Friday afternoon, the mice are returned to free feeding for recovery over the weekend. They will eat more than on the previous weekend, but any deprivation the next week should be based on consumption data for the first weekend.

There are two important aspects of the equation used to determine the daily ration. First, it has been observed that food consumption varies with a wide range of factors, including the genotypes of the mice in the study, the kind of mouse chow, and factors unique to each lab. Figure 11.6 shows regression lines for four separate studies in the author's lab, one of which was replicated at the same time in John Crabbe's lab in Portland, Oregon. Mice of 21 inbred strains ate almost 0.5 g more on average in Edmonton than Portland. Slopes of the regression lines were substantially different in some of the other studies. Consequently, for the refined deprivation method, it is essential to gather baseline data in the same lab with the same strains and methods that are to be used in the deprivation study. It might be ideal to estimate curves for each genotype, but this will require very large samples because the range of weights within a strain is often quite narrow. It is more practical to use the same equation for every mouse and then make small adjustments as needed. The equation will account very well for any sex difference; the same equation is used for males and females.

The second aspect is the manner in which mice can adapt to a reduced ration. Under free feeding, it is apparent that mice consume more food than is really needed for good health and

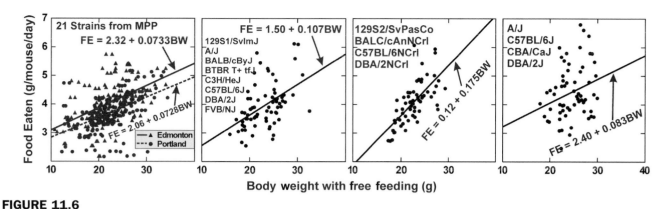

FIGURE 11.6
Food eaten (FE) versus body weight (BW) of adult mice in four studies with different sets of inbred strains having somewhat different ages. Slopes are markedly different across those studies. *(Based on data from the author's lab.)*

growth (Martin et al., 2010). If they are given a fixed ration each day that is a little less than the free feeding consumption, they can adapt to the new regimen quite well. Indeed, with a ration of only 70% of their free feeding consumption, mice do not usually fall below 90% of their former body weight (Figure 11.7). Only when the daily ration is limited to 50% of free feeding consumption does body weight descend below 90% of the initial weight. Thus, for the method using a fixed ration determined from free feeding body weight (Table 11.2), the equation should be derived in the local lab (Figure 11.5), and then each mouse should be given half of its expected free feeding amount. Because there are actually two mice in the cage, in effect the two mice are given an amount of food that was consumed daily on the weekend by just one mouse.

Whether to use the more elaborate fixed ration method (Table 11.2) or the much simpler 24 h method (Box 11.1) is a decision made in each lab based on several considerations. Both methods produce hungry mice that are motivated to learn a task to obtain food rewards. Whether they produce equally hungry mice is a very difficult question.

Degree of hunger and eating

For all strains of mice, longer periods of food deprivation increase hunger. Is it possible to be sure, however, that mice of different strains deprived of food in the same way are in fact equally hungry? In the test of 48 h deprivation (Figure 11.3), the 2 min eating test worked very well. Mice that had not been deprived of food ate almost nothing during the test, so the scale of hunger had a lower limit of zero. In each strain, there was an almost linear increase of time

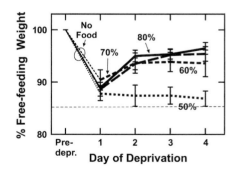

FIGURE 11.7
Body weights as a percentage of pre-deprivation weight for mice averaged over 4 inbred strains (129/SvPaslco, BALB/cAnNCrl, C57BL/6NCrl, DBA/2NCrl) totally deprived for 24 h and then fed daily either 80, 70, 60, or 50% of free feeding consumption. Only those maintained on the 50% ration remained near 85% of free feeding weight (thin dashed line). *(Based on data from the author's lab.)*

185

eating in relation to hours of deprivation, and this was also promising for a scale of hunger. Unfortunately, the slopes of the lines for the four strains were so different that the goal of equating levels of hunger was not realized. No researcher could believe that the A/J strain was as hungry after 48 h without food as DBA/2J after only 12 h of deprivation. It appears that the A/J strain does almost everything with less intensity than most other strains, including eating.

Eating involves a balance of competing tendencies. The 2 min test is always given in a clean holding cage, the same kind used as a prelude to any behavioral test. The mouse explores the cage actively for several minutes and then the experimenter places a glass dish with some food into the cage. Hunger is apparent when the mouse prefers to eat rather than continue exploring. That is the secret to obtaining an eating score of zero in non-deprived mice. This point became obvious when freely fed or 24 h deprived mice were placed in an open field where food pellets were scattered on the floor (Figure 11.8). Shortly before the open field test, the same mice were given a 2 min eating test in a holding cage, as was done for mice in Figure 11.3. Each mouse was first tested before deprivation and after 24 h deprivation, then given a day of free feeding to recover, then another 24 h deprivation, and finally another test after free feeding. The consistency of behaviors in the 2 min eating test and eating of pellets in the open field was very good for all strains. Eating in the open field by non-deprived mice was very low, even lower than in the 2 min eating test, which suggests that the tendency to explore the novel open field was considerably stronger than any urge to nibble food. The 2 min test and the open field food consumption test showed some correlation across strains, but the result for DBA/2J and to a lesser extent C3H/HeJ and BTBR were different for the two measures. Activity in the open field was greatly reduced for strains BALB/cByJ, C3H/HeJ, and FVB/NJ when they were hungry and busy eating food pellets, whereas BTBR mice managed to travel widely in the open field and gulp many pellets at the same time. These data show clearly the strain-dependent nature of competing responses in any test of eating.

The data with the 2 min test in Figure 11.3 look so clean and neat. Unfortunately, there was a fatal flaw that became apparent only when the test was applied with a wider range of 21 strains (Figure 11.9). The problem was glaringly obvious with three of the four wild-derived inbred strains. Those strains did not care for human company and were very hard to handle (Wahlsten et al., 2003a). All but one strain lost considerable weight after 24 h deprivation, but three of them would not eat at all during the 2 min test in a holding cage when an

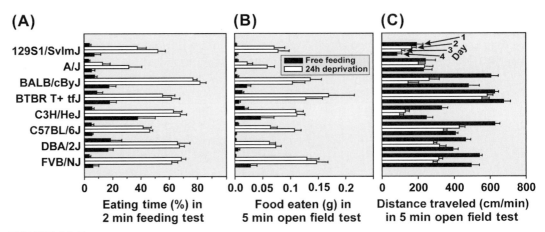

FIGURE 11.8

Mean scores for mice of 8 inbred strains from the original Priority A list of the Mouse Phenome Project that were assessed for one day prior to deprivation, then for 2 cycles of 24 h deprivation, and finally after a day of recovery. (A) Time spent eating in a 2 min test was strongly influenced by hunger. (B) Amount of food eaten in an open field test when a small food cube was placed in each corner showed the same pattern. (C) Motor activity was considerably lower in most strains when they had food available in the open field, with the exception of A/J, which was always low in activity and BTBR, which was exceptionally active and must have been chewing while moving. *(Based on data from the author's lab.)*

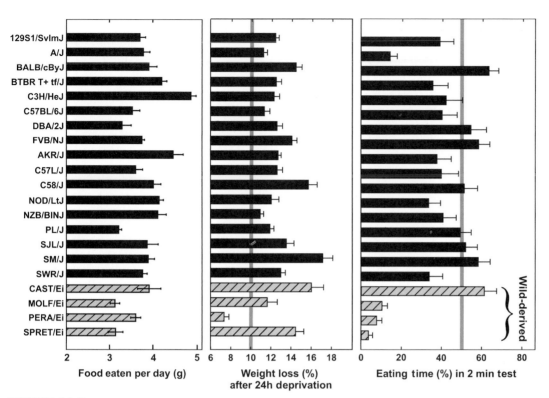

FIGURE 11.9

Data (mean ± standard error of the mean) for 21 inbred strains from the original Priority lists A and B of the Mouse Phenome Project, including 4 strains inbred from wild species of *Mus*. Average food consumption prior to deprivation differed substantially across strains. Weight loss after 24 h food deprivation also differed across strains, being remarkably low in PERA/Ei. Time spent eating during the 2 min test of hunger was very low in the very timid and hypoactive A/J strain as well as three of the wild-derived strains. *(Based on data from the author's lab.)*

experimenter was watching them. The test reflected a balance of three tendencies, not two: motivation to eat when hungry, tendency to explore a clean holding cage, and reluctance to eat when watched by a human. That stubborn fact implied that some of the non-wild derived strains that ate relatively little in the 2 min test might also have been influenced by the human observer. The strain BTBR, for example, lost considerable body weight but ate much less than BALB or FVB. The test was not a dependable indicator of hunger alone.

Revised eating test

Accordingly, a new method was sought to overcome complications in the 2 min eating test (Table 11.3). It was based on weight of food consumed and did not require a person to watch the mouse. The food was lab chow from the hopper of the mouse's own cage, and two blocks of that chow were carefully weighed and presented in a glass dish in that cage. Two mice were housed in each cage, so one was removed briefly to a holding cage during the test of its cage

TABLE 11.3 Two Tests of Eating and Hunger Where Lab Chow is Presented in a Clean Glass Dish

Feature	Version 1	Version 2
Trial duration	2 min	10 min
Test location	Clean holding cage	Home cage with cage mate removed
Food source	Fresh lab chow	Lab chow from home cage
Method of detecting eating	Observation by human	Weight of food consumed
Housing of mice in colony	Several per cage	Two per cage

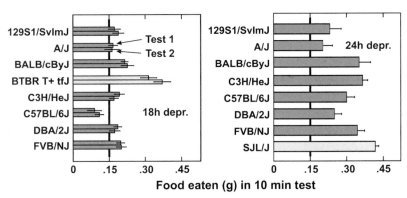

FIGURE 11.10

Amount of food consumed in 10 min in the home cage (see Table 11.3) after 18 h deprivation for 8 inbred strains of the original MPP A list and after 24 h for the next version of the MPP A list. Strains differing between the two lists are shown as light gray bars. For the 18 h deprivation condition, mice were allowed 6 h of free feeding after the first eating test. The two tests yielded remarkably similar results. Comparing the 18 h and 24 h data, results were very consistent for all but strain C57BL/6J. *(Based on data from the author's lab.)*

mate. After both mice had been tested, they were returned to their home cage and then resumed free feeding. Preliminary tests with different feeding times found that a 10 min test yielded small standard errors with four strains, and the amount of food eaten (0.15 to 0.25 g) was a small fraction of the total body weight gain of about 2 g in the first 4 h after returning to free feeding.

The 10 min eating test was then extended to eight inbred strains of the Mouse Phenome Project (MPP) A list. In one study done after 18 h deprivation, results on two tests conducted one week apart were very similar indeed, which demonstrated the high reliability of the test (Figure 11.10A). In a second study with 24 h deprivation, amounts of food eaten were higher than with 18 h deprivation and strain rank orders were very similar, with the marked exception of C57BL/6J. Especially noteworthy was the higher food consumption of the A/J strain that was so timid in the 2 min eating test (Figure 11.9) and the open field test (Figure 11.7).

Clearly, amounts of eating in the 10 min test differ between strains, but they cannot be considered direct measures of hunger. While all strains are hungry and eager to eat after 24 h of deprivation, it cannot be said that they are equally motivated. Factors such as metabolic rates, overall levels of activity, and vigor of behaviors appear to be involved. A/J mice are notoriously low in motor activity in novel environments, and they eat about half as much as SJL/J in the 10 min feeding test. Assessing the degree of hunger remains a challenging task for the researcher.

ELECTRIC SHOCK

Electric current from a high voltage source can be very unpleasant, as most people know from direct experience with sparks of static electricity in the winter or frayed appliance cords. Humans and virtually all other animals, even fruit flies, will react promptly to shock and avoid a situation where shock is given repeatedly. The stimulus is rarely encountered by animals outside of a laboratory and therefore might be termed "unnatural." For research on learning, however, shock is convenient to deliver and the mouse's reaction to it does not fade away. Telling friends and family that someone gives electric shocks to animals may evoke images of sadistic torture and Mengele, but shock can be given so that it is rather mild, just sufficient to motivate action. Traumatic shock is neither necessary nor ethically acceptable.

As physiologists such as Sherrington and Pavlov learned long ago, electric current stimulates the contraction of muscles. A shock to an animal's foot elicits an almost instantaneous contraction and withdrawal of the leg. The stronger the electric current, the stronger the reflexive response of the muscles and the experience of pain. The voltage needed to stimulate a single muscle fiber is very small, just a few millivolts (mV), whereas painful shocks usually occur when the voltage is 100 VAC or more. People are not harmed by touching the two poles of a flashlight battery when the voltage is so low (1.5 VDC), but electric current will still flow through the fingers. Because skin resistance is so high, the current will be very small, below the threshold for sensation. In discussing how to apply shock to a mouse, the basic principles of electricity are relevant but no consideration of physiology per se is required.

Electricity consists of electrons. If one place has a great abundance of electrons and the other has relatively few, there is a difference in voltage (V) or electric potential, measured in volts. If one pole is always high (+), the current will flow in one direction (direct or DC flow), whereas voltage from a generator reverses many times each second (60 Hz alternating or AC flow). When the two places are joined by a conductor of electricity, a current consisting of vast numbers of electrons will flow through the conductor. The quantity of current (I) is measured in amperes or amps (A). The resistance (R) of a conducting object to flow of electricity is measured in ohms (Ω). The current is jointly determined by the voltage across an object and its resistance; higher voltage causes higher current to flow, whereas higher resistance reduces the flow. The relation (Ohm's law) is summarized by the formula $I = V/R$. Thus, when current flows through an object, the voltage across the object must be $V = IR$. The electric power delivered to an object, measured in watts, is I^2R, a quantity that is important in choosing a light bulb but not in motivating a mouse.

The levels of current that animals find aversive are quite low. A shock of 0.001 amp or one milliamp (1 mA) to a mouse would be extremely intense, whereas a tiny trickle of current at 10^{-6} A or one microamp (1 μA) would be far below the threshold for sensation. Dry skin can have very high electrical resistance, more than 100,000 ohms (100K Ω) or even 1,000,000 (megaohm; 1 M Ω). If skin has resistance 10^6 Ω and current is 1 mA, the voltage must be $V = IR = 10^{-3}$ A $\times$ 10^6 Ω = 1000 V.

189

Basic features of a shock source

Shock sources used to motivate mice generally involve high voltage and can be dangerous to the people working with shock as well as to the mice. Great care is warranted at all phases. The shock source should be obtained from a reputable supplier or built by a qualified technician, and the installation of the device in the lab should be checked carefully by someone with sufficient expertise. The device and the behavioral test apparatus should be grounded, so that all exposed metallic parts of the entire system (except the shock grid) are at the same voltage level (0 V), termed "ground," which is the level at the third prong on an electrical socket that quite literally is connected to the ground of the earth.

A general schematic for a simple shock source suitable for work with mice is provided in Figure 11.11. This could be built in a local electronics shop. Commercial shock sources usually

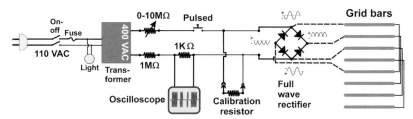

FIGURE 11.11
Diagram of a simple device for controlling electric shock delivered to a grid. The 1 MΩ resistor limits the maximum current to 400 μA and the variable resistor regulates the nominal shock intensity delivered to the mouse. The calibration resistor should have a value close to the resistance expected for a mouse (see Box 11.2). The oscilloscope shows the actual current flowing through the animal. The full-wave rectifier ensures that any possible pair of bars among a contiguous set of four differs in voltage.

involve more sophisticated electronics intended to regulate the current under diverse conditions. The user is cautioned that many shock sources were designed for work with rats, and currents suitable for motivating a rat are higher than those needed for mice.

Several parts of the simple system are essential. At the input, there must be a switch to turn the device on and off along with a bright light to warn the experimenter that the shock source is active. There should also be a fuse, just in case there is a short circuit. The wires then go to a transformer that increases the voltage considerably. In work with mice a 400 VAC transformer is sufficient and poses less danger than a 1,000 VAC model. On the output or "secondary" side of the transformer, one wire should be connected to a fixed, current-limiting resistor that determines the most intense current that could possibly flow through the circuit. With a 1 MΩ resistor and a 400 VAC transformer, the limit will be 400 μA, not enough to motivate a rat but quite sufficient for work with mice and low enough to pose little danger to humans. Then there is another resistor of low ohms connected to an oscilloscope to read the voltage across it. If the resistor is 10^3 Ω and the voltage reading is 0.1 V, the current is $I = 0.1/1000 = 0.1$ mA or 100 μA. On the other line from the transformer is a variable resistor that sets the current level to be used in a study. There should also be some kind of switch that can be pulsed on and off quickly, driven either by a motor and microswitch or an electronic timer. Finally, there is a full-wave rectifier with four nodes, each of which is connected to a bar in the shock grid. Within every group of four grid bars, there will be a strong voltage difference between every possible pair of bars. Joining every fourth bar electrifies the entire grid. Finally, there should be a socket to plug in a calibration resistor.

MONITOR THE ACTUAL CURRENT

Many devices have a dial for setting the current intensity. Never trust such a dial. It is no more than a rough guide to intensity. The investigator can monitor the current that flows through the mouse via the shock grid by recording voltage across a resistor in series. The monitoring should be done with an oscilloscope or chart recorder that can follow very fast changes in voltage as the mouse hops across the grid. There will be short periods when no current flows through the animal if it jumps off the grid. When current is read from an ammeter with a moving coil and needle indicator, the current will be an average, and the average will be lower when there are many brief periods with no current. Shock intensity should be reported as the *maximum* current flow through the mouse when it is fully in contact with the grid.

USE PULSED SHOCK

Electric current causes involuntary muscle contraction, which makes coordinated escape movements difficult. If the shock is delivered in pulses, on the other hand, the mouse can make its escape more effectively during the shock-free interval. The intensity of the shock can also be set lower when pulsed, because the initial onset of shock is perceived as the most aversive event, involving both surprise and a sharp pain. A reasonable pulse rate is 0.5 sec of shock every 2 sec.

CALIBRATE THE SYSTEM BEFORE USE

Points on a scale on the variable resistor can aid in determining the correct shock level, but the numerical value on a scale is not a dependable indicator. The best practice is to calibrate the system before use on a mouse. For this purpose, plug a calibration resistor into a socket on the shock source when no mouse is present. The resistor should have a resistance close to that expected for a mouse. Determine the actual current by monitoring voltage across the 1 KΩ resistor. It is also wise to note any current flowing when the calibration resistor is not inserted. If the oscilloscope shows any ripple of voltage at 60 Hz, there must be some leakage somewhere in the circuit, most likely at the shock grid. This can happen if the grid has not been constructed and cleaned properly.

BE ALERT FOR DEVIOUS ADAPTATIONS

It is important to observe mice during any training procedure that involves electric shock because there are so many ways that an animal can adapt to the situation. Some of these adaptations conflict with the goals of the study. The classic case is a mouse "bridging" the grid by spreading its legs to stand on bars with the same voltage. This is not difficult if the two AC lines from the transformer are simply wired to alternate bars. The full-wave rectifier circuit with four diodes eliminates this problem. Every fourth grid bar is wired to one node, so that the mouse can reach bars of the same voltage only by bridging four bars — an unlikely feat. Commercial shock grid scramblers also prevent bridging. Even then, a mouse can find a way to foil the experimenter. If it stands on just one bar and leans against a plastic wall, it will receive no shock. This is a rare event in tasks that are learned fairly quickly. If the task is too difficult, as with shuttle avoidance, devious adaptations are more likely to occur.

Variables that influence intensity

SKIN RESISTANCE

Skin is a living tissue; its electrical properties change with the animal's emotional state. People are familiar with the fact that skin becomes moist when a situation evokes fear, as in the so-called lie detector test or the galvanic skin response. Moist skin has much lower resistance than dry skin. After one or two experiences with electric shock, the mouse will certainly have lower skin resistance than when it was naive. Skin resistance ranges for four inbred strains are shown in Box 11.2 for different conditions (Wahlsten, 1972b). Whereas resistance of the dry feet of an anesthetized mouse is very high, resistance of wet feet and the entire mouse receiving shock on a grid is much lower and spans a narrow range. The calibration resistor shown in Figure 11.11 should be about 100 KΩ, close to what is expected during training.

11.2 SKIN RESISTANCE OF 4 MOUSE STRAINS

Dry feet	6.1 to 9.7 MΩ
Feet wet with saline	110 to 140 KΩ
On grid before shock	290 to 480 KΩ
On grid during shock	110 to 130 KΩ

Source: Wahlsten, 1972b.

URINE ON GRID BARS

After receiving one or two shocks, a mouse is almost certain to urinate in the apparatus. Some of this urine will get on the grid bars and then the animal's feet, lowering skin resistance even further. If this substance is not removed from the apparatus before the next mouse is tested, it can affect the next mouse in many ways. Odor from urine derived from a mouse that has been shocked will alarm another mouse. Residue from urine and feces can also provide a route for electric current and reduce the intensity of shock reaching the mouse. This problem can be greatly reduced by building apparatus so that the grid bars do not touch the walls of the enclosure and the grid can be easily removed for cleaning (Figure 11.12).

SURFACE TEXTURE OF THE GRID BARS

Pain from an electric shock depends on the density of current flowing through the skin. If a particular current is delivered to a small area of skin, it may cause a very unpleasant sensation, whereas the same current spread over a much larger patch of skin may not be felt

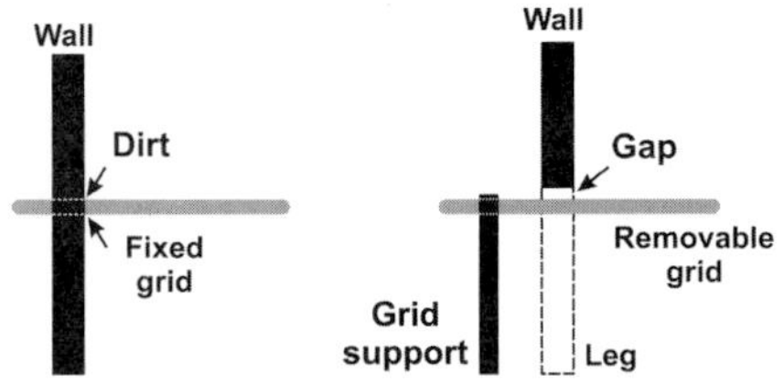

FIGURE 11.12

Detail of shock grid bar in relation to plastic wall of the test chamber. If the grid bar extends through a hole in the wall (left diagram), dirt can accumulate and cause leakage of current. Making the grid separate from the chamber so that the grid bar does not touch the wall (right diagram) prevents short circuits and facilitates cleaning of the apparatus.

at all. This same phenomenon occurs on a smaller scale when there is a rough metal surface versus a very smooth surface. A rough surface will deliver the electricity to just a few small points of skin, thereby generating more pain for a given level of current. It is therefore important not just to clean the grid bars but also to sand them periodically with a fine grit sandpaper and use the same grit throughout a series of studies.

Determining the correct level of shock

Because of the many kinds of metal grids and surfaces as well as shock source voltages and resistances in series with a transformer, it is good practice to calibrate the equipment to be used in any lab before commencing a study of learning. Genotypes differ in sensitivity and modes of reaction to shock (Kazdoba, Del Vecchio, & Hyde, 2007), and preliminary work should involve all genotypes to be used in a study. Two quantities are of primary concern: the threshold level of shock current that can be sensed by a mouse, and the most effective level for use in a particular kind of training.

Threshold of sensation

Mice do not like electric shock and will move away from it when they sense it. Thus, a simple test of sensation is a box with two grids, either of which can be electrified. The grids should be the same as those used in training. The grids should not be visibly different or discriminable by texture. An escape paradigm should be used in which the mouse is placed onto the grid that will soon be electrified. One second later the shock is turned on for 30 sec. If the mouse can sense it, it will become agitated and walk or run on the grid until it reaches a location on the other grid where there is no shock. It may then return to the shocked grid but will soon leave it. The measure of sensitivity is the percentage of time spent on the shocked grid, which will be close to 50% when the mouse cannot feel the shock. The test should be done with an ascending series of intensities, beginning at a very low level that is definitely below threshold and progressing to a level that is clearly sensed, then reversing the process with a descending series. Trials should not always be started on the same side, because the mouse will learn that contextual cues are associated with shock. A pseudo-random sequence of start locations should be used. The threshold level determined from this kind of preliminary study is not the level that should be used in training on a complex task. A higher level will be necessary, and it must be above the threshold for all genotypes.

Threshold of overt reaction

In training to escape or avoid a shock, most tasks require the animal to perform some kind of directed action. An escape response occurs while the animal senses the shock, whereas avoidance occurs before there is shock and may prevent its onset. To motivate an avoidance response, the shock must be sufficiently intense to evoke an escape reaction. Thus, the appropriate level of shock in a study of learning is one that will consistently provoke escape-related behavior in all genotypes. The successful escape response may require learning *where* to

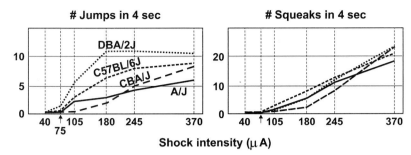

FIGURE 11.13

Reactions of four inbred strains (average of males and females) to foot shocks of different intensities. Each mouse received all intensities in an ascending series. Jumping was strongly strain dependent, while squeaking was similar for all strains. Jumping occurred at a lower shock intensity than squeaking for most mice. *(Redrawn with permission from Wahlsten, 1972b)*.

run or jump, but the immediate reaction of running or jumping must be sufficiently vigorous. The appropriate shock level should be judged according to the immediate reaction to the shock. An ascending series of intensities should be used, beginning below threshold. The descending series is probably not worth testing because of the powerful role of learning in the escape situation.

The example shown in Figure 11.13 involved four inbred strains assessed for jumping and squeaking behaviors (Wahlsten, 1972b). Most strains were silent at 105 μA, an intensity sufficient to elicit considerable jumping in three of the four strains. It is evident that 105 μA would not be a good choice for escape or avoidance training because one of the strains showed very little reaction. The next higher level of 180 μA clearly evoked responding from all individuals of all four strains. There was relatively little squeaking at 180 μA, which indicates that the shock was not intensely painful. Thus, 180 μA was a good choice for escape and avoidance training. There was no need to apply anything more painful to motivate the mice.

Kinds of responses and task requirements

Electric shock typically elicits withdrawal of paws from the offending surface at lower levels and jumping or running at higher levels. Because of the reflexive nature of the reaction, it is poor practice to require skilled motor performance to escape shock. Here is one area of study where anthropomorphism may engender wisdom. Humans react to shock almost the same way as mice. Shock is more likely to initiate panic than calm deliberation or careful planning of action.

Of the two principal reactions of mice, jumping and running, some strains show a marked tendency to perform one but not the other. DBA/2J mice in particular are likely to jump, even at relatively low levels of shock, whereas A/J mice usually run across the grid. If the task requires the animal to jump out of a box, this will be a difficult maneuver for A/J mice, whereas DBA/2J will encounter difficulty if the box has a high ceiling and the mouse must run through a small door to escape or avoid. Those tendencies were apparent on the first escape trial in three different tasks (Wahlsten, 1972b), all using identical shock grids (Figure 11.14). When the mouse had to jump out of a 10 cm high box to escape, DBA/2J mice achieved this very quickly while A/J mice struggled for quite a long time and sometimes did not escape within the limit of 30 sec. If there was a lid on the box and the mouse had to run through a 5 × 5 cm door to escape, A/J mice were actually a little faster than DBA/2J on the first escape trial. If either response was allowed, all mice escaped quickly using their preferred mode of action. Learning to avoid the shock that came on 5 sec after being placed on the grid showed a different pattern. Jump-out avoidance was generally easier for all strains and was similar to the either-way task, whereas running one way through the small door was more challenging, regardless of the kind of reaction on the first escape trial.

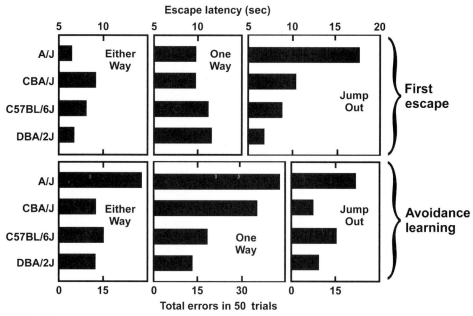

FIGURE 11.14

Performance of four inbred strains on three shock avoidance tasks that required the mouse to jump out of a box, run through a hole in its wall, or take either route out of the box. Latency of the first escape in the jump-out box was much shorter for strains shown in Figure 11.13 that jumped reflexively when shocked, whereas there was little strain difference in the one-way condition. Speed of learning to avoid the shock was not related to the initial reaction to the shock. Strain rank orders differed for the one-way and jump-out tasks, while they were similar for either-way and jump-out. Strain A/J was poor at any avoidance task, despite its good escape behavior on the first trial in the tasks that did not require jumping. *(Redrawn with permission from Wahlsten, 1972b).*

194

WATER ESCAPE

Mice do not like to be in water and will quickly climb out of it when given an opportunity. The motive to escape is strong and never wanes, and most mice will swim vigorously in a tank or maze. The configuration of the tank or maze varies greatly among labs (it will be discussed in depth in Chapter 13). The motivation to escape water is fairly simple, involving only a few parameters: depth, temperature, number of trials, opacity, and cleanliness.

Depth

When the Morris water maze is employed, almost all labs use a deep tank (Table 13.2). This is not necessary. Mice never dive to the bottom of a tank the way rats do. If the water gets beyond a certain depth, all mice will swim and never touch the bottom of the tank. One exception is strain C58/J that sinks even on the first trial and is not suitable for studies with water escape motivation. A series of trials was run in the author's lab with 8 inbred strains in a 4-arm water maze in which depth of the water was gradually increased from 1 to 10 cm. Many mice first began to swim when the depth reached 5 cm, but a few stood on their hind legs and did not swim until the depth was 7 to 9 cm. All mice spent the entire time swimming and never touched the bottom of the tank when water depth was 10 cm or more. Thus, a maze with perhaps a 12 cm water depth should be sufficient for training on almost any water-motivated task.

Temperature

Room temperature water is certainly the easiest to use because no heating or cooling is required, but room temperature varies considerably among labs. Figure 13.5 indicates that water temperature ranged from 15 to 27°C in different studies of mice in the Morris maze.

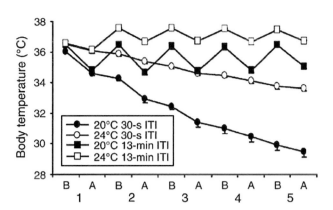

FIGURE 11.15

Body temperature determined by a rectal probe before (B) and after (A) five trials of swimming in a water tank with two water temperatures and two intertrial (ITI) intervals. With a long recovery interval, mice were able to regain most of the lost heat between trials, whereas those receiving a 30 sec interval lost almost 7°C over the five trials and were markedly hypothermic, especially when trained with the cooler water. *(Reprinted with permission from Iivonen, Nurminen, Harri, Tanila, & Puoliväli, 2003).*

Cold water can perhaps intensify the urge to get to dry land, but cold can also impair muscle function and even alter memory processes in some mice (Nowakowski, Swoap, & Sandstrom, 2009; Roder, Roder, & Gerlai, 1996). Water temperature of 20°C or less can induce hypothermia in mice, especially for certain genotypes (Stasko & Costa, 2004), and when the intertrial interval is so short that the animal cannot warm itself (Iivonen, Nurminen, Harri, Tanila, & Puoliväli, 2003). In one study of C57BL/6J-derived mice, core body temperature declined by about 7°C over five trials with a 30 sec intertrial interval (Figure 11.15), a common value in studies of the Morris water maze. Hypothermia was greatly reduced by using either 24°C water or a 13 min interval. Thus, water temperature in the range 24 to 26°C is a good choice for studies of learning. This can be achieved easily when room temperature is 21°C by placing a circulating heater in the tank between trials.

Number of trials

The oils in the mouse's fur help to maintain buoyancy but they become less effective with repeated trials, and the animal becomes increasingly waterlogged. The mouse will groom itself and remove some of the water between trials. Nevertheless, it becomes increasingly difficult for the animal to swim high in the water. Generally speaking, fewer trials per day will reduce this problem. Good results can be obtained with three or four daily trials for most strains, but two trials per day is an even better option for reducing waterlogging. The training sequence may then span two weeks. If the Morris version of the water escape task is used, one day at the end of the second week can be devoted to a probe trial with no platform present. That regimen would amount to 18 training trials and one probe trial. Alternatively, three trials per day with a test trial at the end of the fifth day would involve 14 training trials and one probe trial, and learning may not be sufficiently robust to yield good directed search for the platform for the less capable strains (Wahlsten et al., 2005). It is wise to choose task parameters adequate for the least competent genotypes (Stasko & Costa, 2004). Otherwise, some animals may appear to fail the task totally, when they would show some signs of competency with a better choice of parameters.

Opacity

For submerged platform water escape tasks, water rendered slightly opaque can prevent the mice from seeing the platform. Probably the best option is to add white tempera paint at a ratio of about 1 mL paint to 1 L of water. Tempera is sometimes derived from egg yolk and is

195

high in protein. It is non-toxic and can be eaten by children who use it for finger painting. Edibility and lack of toxicity are important features because mice will consume the liquid when cleaning themselves after a dip in the tank. Milk has similar properties to dilute tempera, but it is an excellent medium for the growth of bacteria at 26°C, especially when mixed with mouse excretions. Lime is not recommended for use with swimming mice; when added to water, it forms calcium hydroxide that is toxic when ingested. Some chalk has the same properties, although forms intended for children are usually safe.

Cleanliness

How would you like to take a bath in a tub of water that has been used by 40 other people the same day? Humans would find the prospect repulsive and a threat to health, but mice are routinely immersed in that kind of soup. Given the volume of a large water maze, it may seem impractical to change the water between mice. A filtration system used for large fish tanks could be very effective, but it would also remove the tempera paint. The possible importance of water cleanliness has not been assessed in formal studies. Mice are small relative to tank volume and the accumulated urine never becomes highly concentrated. The problem can be reduced by exchanging a few liters of dirty water for fresh opaque water after each mouse. Boli can be swept from the tank between trials using a fine mesh tropical fish net. By all means, the tank must be drained and thoroughly disinfected at the end of each day. The platform should also be disinfected.

AIR PUFF AVERSION

One of the most effective ways to provoke an alarm reaction in weanling mice is to blow into their cage. The wild jumping that ensues ("popcorn" behavior) suggests mice do not like to have air blown at them, but there are also odors in human breath that are offensive to mice. Puffs of clean air are mildly aversive to mice and may provide a gentler motivating stimulus in some tasks (Clark et al., 2003). The IntelliCage, for example, uses mild air puffs to teach mice which drinking spout to use and which to avoid (Galsworthy et al., 2005). Research with compressed air in the author's laboratory indicates that it may be very effective for simple inhibitory avoidance learning but not for active avoidance.

Parameters of air puffs

SOURCE OF AIR

Compressed air can be purchased in cylinders and delivered to the animal via a two-stage regulator that allows adjustment of air pressure at its outlet (Figure 11.16). The flow of air can be calibrated with a flow meter and turned on and off with a solenoid. Commercial compressed air and attachments are usually made for industrial use and may have traces of machine oils that are noxious to mice. It is therefore essential that scrupulously clean, medical grade air intended for use with humans be delivered to the lab mice. The regulator and valves should also be free from volatile oils, and all connecting parts and tubing should be cleaned with acetone and dried before use with mice. Many arrangements of the parts of the system are possible, but two design features are important. First, the length and volume of the path from the solenoid to the nozzles should be as small as possible, so that the onset of air at the nozzles is abrupt. A smooth onset that occurs when the volume in the path is gradually compressed will encourage habituation to the stimulus. Observations of mice receiving air puffs reveal that the onset of the air is by far the most noxious aspect of the stimulus, unlike electric shock where the entire experience from beginning to end is aversive. Second, the volume of air between the solenoid and the pressure regulator should be relatively large so that a sudden increase in flow through the nozzles will not rapidly deplete the supply of air in the tubing. A pressure gauge in the path near the flow meter can reveal a sudden drop in pressure.

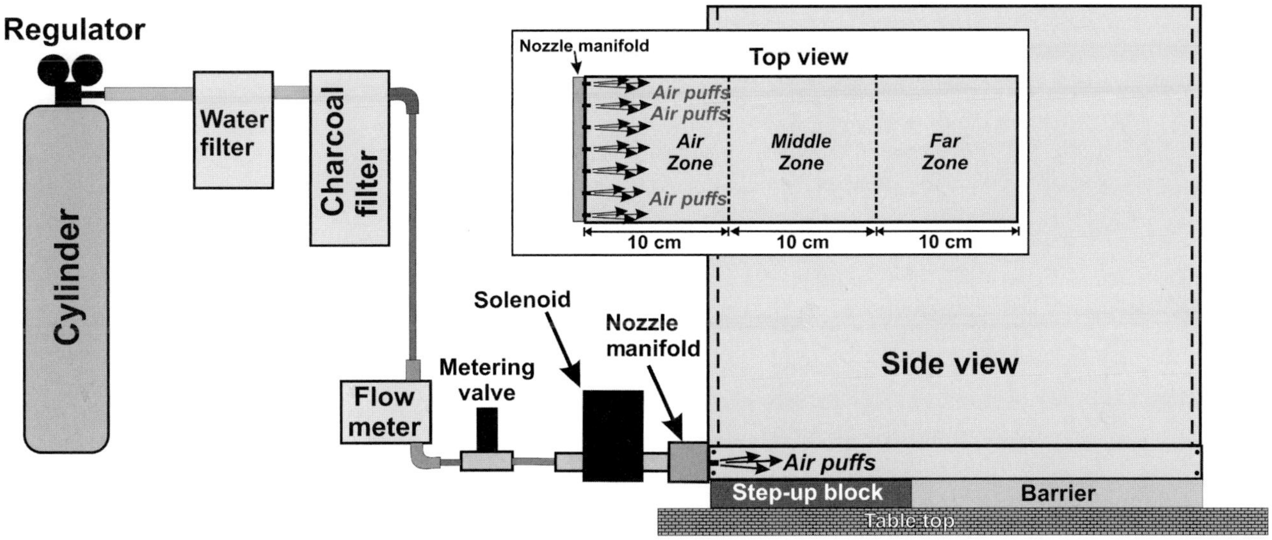

FIGURE 11.16

Diagram of apparatus to deliver air puffs to mice. Clean air from a cylinder is reduced to 207 kPa (30 psi) by a regulator, purified further by two filters, and passed through a flow meter and metering valve that allow fine adjustments to flow rate. A solenoid attached to a manifold turns the air flow abruptly on and off. Seven nozzles protrude from the manifold so that mice in the air zone experience high air velocity at any distance from the side walls. Air flow in the far zone is very mild and non-aversive. For the inhibitory avoidance test described in Figure 11.18, mice had to step up a 2 cm high block to enter the air zone (see also Figure 3.3).

FLOW RATE

Mice are not bothered by a gentle breeze. A rather high velocity, directional flow of air is needed to motivate an escape or avoidance response. Three variables govern flow rate: pressure at the air source (outflow of the regulator), diameter of the orifice nearest the mouse, and resistance to the flow of air along the path through valves and tubing. Higher pressure causes faster flow, a larger orifice allows faster flow, and high resistance attenuates flow. A hydraulic engineer might be able to compute the resistance of all the twists and turns in the tubing and valves, but in the mouse lab the best option is to monitor air flow with a flow meter. Two kinds of flow meter can be used. An inexpensive device with a ball in a cone-shaped tube is good for determining steady flow rate, but the ball will shoot to the top of the tube during the initial surge of air. A meter based on cooling of a hot wire can transduce rapid changes in velocity. That velocity, however, will show what is happening at the meter, not at the nozzle near the mouse.

Flow rate at the meter is given as volume of air per unit time (L/min). Suppose flow rate F is the volume of air that passes through a nozzle in one minute. If nozzle diameter is D, the cross-sectional area of the nozzle is $\pi(D/2)^2$. The length of the column of air that flows in one minute is thus $F/[\pi(D/2)^2] = 4F/(\pi D^2)$. Resistance is not a separate factor in the equation; it is an unmeasured determinant of flow rate. If there is more than one nozzle, flow through each nozzle is the total divided by the number of nozzles, more or less. Speed of air flow at the nozzles in one second is then $4F/[60(\pi D^2)]$. Values for different nozzle diameters are given in Box 11.3.

NOZZLE DESIGN

For most applications involving mouse behavior, air will be delivered through small nozzles attached to a hollow manifold (Figure 11.16). The nozzles can simply be holes drilled in the walls of the manifold. If the researcher wants to vary nozzle size, it is better to use small screws with holes drilled down their centers. Air speed falls rapidly with distance from the nozzle, as air expands from system pressure to atmospheric pressure. The speeds in Box 11.3 are those

11.3 NOZZLE DIAMETER (D) AND AIR SPEED (M/SEC) WHEN FLOW VOLUME F = 1 L/MIN

D (mm)	D(inch)(Drill#)	Speed
0.4	0.0157″(#78)	132
0.8	0.0315″(#68)	33
1.0	0.0394″(#60)	21
1.2	0.0472″(#56)	15
1.6	0.0630″(#52)	8
2.0	0.0787″(#47)	5

experienced by an animal that places its snout close to the nozzle. How far from the nozzle a mouse must be to experience no aversion must be determined empirically.

PULSED AIR

If the mouse does not flee soon after the onset of air flow, it may habituate rapidly. Air is far more noxious when delivered in short puffs, each with a rapid onset. An electronic timer can be used to operate the solenoid and control puff duration and spacing. In the author's lab, 0.5 sec of air alternating with 0.5 sec of no air has been effective, but temporal parameters need further evaluation.

SOLENOID NOISE

The most abrupt onset of air flow occurs when the solenoid is mounted directly on the manifold. Because most solenoid valves emit a sharp, metallic click when actuated, the noise may elicit a startle response. Some noise at air onset cannot be prevented because the air will generate a hiss at the nozzle that can alarm a mouse. Solenoid noise will augment the startle response to some extent, depending on features of the specific device. The contribution of that noise to escape or avoidance behavior can be determined with a series of preliminary trials with solenoid snap but no air flow. For the equipment used in the author's lab, the solenoid noise added nothing to the unpleasantness of the air.

DIRECTION OF AIR FLOW

In preliminary tests in the author's lab, it was discovered that mice often will not run away from puffs of air that come from beneath the animal. Instead, they hold fast to a grid or hunch over a plate with many nozzle holes. Air puffs are most effective at provoking withdrawal when delivered to the animal's face and directed horizontally, so that the direction of escape is obvious.

Determining optimal air flow rate

For use in studies of learning, there must be a zone where air puffs occur with considerable intensity and a safe zone where air flow is minimal. The configuration of an apparatus 30 cm long and 10 cm wide is shown in Figure 11.16. Air is delivered to the air zone through seven nozzles. For the data shown in Figure 11.17 for four inbred mouse strains, each trial began with the mouse in the middle zone. Video tracking determined the time spent in each of three zones (air, middle, far) during a 30 sec or 60 sec trial. No flow and then an ascending series of four flow rates were given to each animal in a single session. Even the 5 L/min rate had a deterent effect on many mice but there was considerable variability, whereas all mice exposed to 20 L/min stayed away from the nozzles and most of them preferred the far zone in the chamber, with the exception of DBA/2J. There was little difference between 15 and 20 L/min, but subsequent studies of avoidance behavior showed that the higher flow was required to deter most mice.

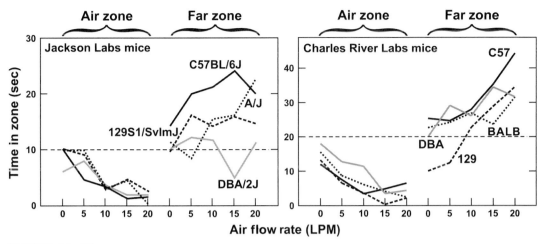

FIGURE 11.17

Time spent in the air and far zones (Figure 11.16) as a function of air flow rate through the seven nozzles. Rate through any one nozzle was one-seventh of the indicated rate. At the higher flow rates, most mice avoided the air zone and spent most of the time in the far zone. Trial duration was 30 sec for the four Jackson Lab strains and 60 sec for the strains from Charles River Labs. *(Based on data from the author's lab.)*

Avoidance of air puffs

The promise of compressed air as an aversive stimulus is that it can motivate mice to learn about risky zones in an environment without the pain of electric shock (Clark et al., 2003). Accordingly, a series of trials was run with eight strains using an inhibitory avoidance procedure. The mouse began each trial in the far zone and was allowed 60 sec to explore the box. In preliminary tests, it was found that better results were obtained when the air zone was made distinctive by raising it 2 cm above the floor of the apparatus so that the mouse had to step up a short distance to reach it. Almost all mice did indeed approach the nozzles on the first trial and approached them two or three times, receiving air puffs on each occasion. Over the course of eight trials given two per day, most mice learned to stay away from the air zone (Figure 11.18). After two or three trials, mice of six of the eight strains rarely approached the air zone, but DBA/2J mice persisted and often required a "reminder" puff to send them scurrying to the far zone.

It is clear that air puffs, as delivered in these preliminary investigations, are a mild form of punishment but are sufficient to motivate behavior in most mice. DBA/2J mice show less aversion to air puffs than other strains, in contrast to their great sensitivity to electric shock (Figures 11.13 and 11.14). Further evaluation of this form of motivation is clearly needed. It will be important to try air puffs with a wider range of mouse strains, and it is probably worth using even higher air flow rates or different pulse parameters to make the air puffs more aversive.

MOTIVATION AND LEARNING

Many studies of learning in mice have found marked strain differences (Nguyen, 2006; Wahlsten, 1972a), and numerous targeted mutations have been created that impair learning and memory. The intricate details of molecular and electrophysiological events forming the basis for information storage and retrieval have been well documented in mice (Nguyen, 2006; Powell, 2006; Wehner, Bowers, & Paylor, 1996). What is less well investigated is the possibility that the various strains and mutations also differ in the motivating properties of hunger, shock, water immersion, or air puffs. A strain difference in rate of learning can be attributed solely to associative processes only when the animals are equally motivated. Proving that identical operations yield identical degrees of motivation is a daunting task. There are many ways to

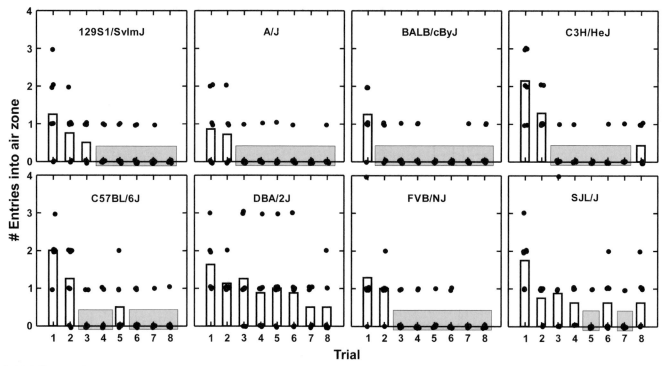

FIGURE 11.18

Mice of eight inbred strains (the second MPP A group) were given two trials per day for four consecutive days of inhibitory avoidance training; they were started in the far zone of the device shown in Figure 11.16 and then exposed to 20 LPM air puffs pulsed 0.5 sec on:0.5 sec off whenever they entered the air zone with all four paws. After one or two trials, most mice avoided the air zone altogether or received just one reminder puff. Data are jittered to show overlapping points for individual mice. Open bars show means where more than half of the mice entered the air zone, while gray zones show trials when fewer than half entered the air zone even once. DBA/2J mice often returned to the air zone and received more puffs than other strains. *(Based on data from the author's lab.)*

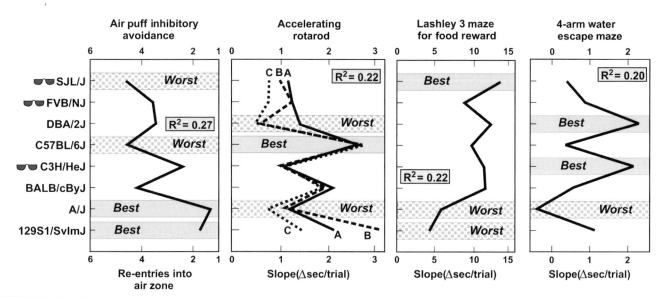

FIGURE 11.19

Relative performances of the same eight inbred strains (second MPP A list) on four refined tests of learning that involved different sensory and motor requirements as well as sources of motivation. Each task was administered with two trials per day for 10 days. Effect sizes of strain differences, indicated by adjusted R^2 values, were very similar for the four tasks. Striking variations between tasks in strain rank orders were apparent. Strain means are joined by lines to show the strain profile. For the accelerating rotarod, values for three studies (A, B, C) done with the same parameters are shown. The 4-arm water escape maze had an escape platform in every arm so that room cues were not important for navigation. *(Based on data from the author's lab collected with the assistance of Elizabeth Munn and Emily Marcotte.)*

manipulate motivation, as outlined in this chapter, but there are few ways to measure the level of motivation itself. Without a valid measure, there is no way to equate levels of motivation.

This becomes an important issue when making inferences about relative abilities of mice to learn. Claims that mice differ substantially in general learning ability (Matzel et al., 2003), termed "intelligence" in humans, are buttressed by correlations of scores on different kinds of learning tasks. Whether levels of motivation on those tasks are equivalent across genotypes has not been determined, therefore the distinction between associative and non-associative processes that give rise to correlations cannot be fairly assessed. It is apparent that variants of task parameters can alter strain rank orders in a major way (Chapter 13). Likewise, sensory and motor requirements of many tasks are disparate. When separate samples of eight inbred strains are compared across several refined tests of learning that involve rather different task requirements, several striking shifts in strain rank orders are apparent (Figure 11.19). No strain is consistently good or poor across all tasks, and almost every strain excels at something. On the other hand, when individuals from a genetically segregating population are evaluated with several different tests per mouse, the same kind of processes could be influencing the data, but this will not be apparent from a matrix of modest correlations.

Qualities of Behavioral Tests

To decide whether a test of behavior is a good test, many attributes need to be considered. A test might yield very accurate numbers but be almost impossible to interpret. Elaborate apparatus with sophisticated electronics might yield copious amounts of data automatically, even though behavior in the device is so highly unstable that precision of the measures is low. Some of the most important qualities of any test are discussed by Wahlsten & Crabbe (2007) and summarized in Table 12.1.

RESOLUTION, ACCURACY, AND PRECISION

Most data from a test of behavior are numbers. Even if the behaviors themselves are categories such as rearing or grooming, the data are usually frequency counts, time spent doing something, or latency to the first action of each kind. When data are collected by an electronic device, the numbers may be quite accurate, much more than any judgment made by a human observer, but they may not be very precise because the object of measurement, a living animal, is constantly changing.

Resolution of an instrument is sometimes termed *readability* of a scale or the smallest scale division. There are two parts to this quality: the number of significant digits and the number of decimal places or smallest scale division. For example, in a computer there is a clock that ticks very rapidly and a timer that keeps track of clock ticks. If the processor works with 64-bit numbers, one number can keep track of up to $2^{64} = 184,467,440,737,096$ clock ticks, and it can tell the difference between $184,467,440,737,091$ and $184,467,440,737,092$. The resolution or small-scale division is one tick, and there are 15 significant decimal digits. In a global positioning system (GPS), the distance from the equator, termed the northing, might be measured by a high-quality device as $3,690,582.18$ m, so that there are 9 significant digits and an astounding resolution of 1 cm. A fine electronic balance might be able to read up to 100 g with a resolution of 0.001 g or 1 mg, which would entail five significant digits when mouse body weight is recorded as 27.661 g.

The marvelous readability of electronic devices is sometimes defeated by the behavior of live mice. It would not do much good to know that a mouse at one moment weighs 27.646 g,

TABLE 12.1 Definitions of Terms

Term	Definition	Example
Resolution	The smallest difference in a measurement that can be detected	1 pixel in 1,600 for a monitor 0.1 mg for a balance
Significant digits	The number of meaningful digits in a measurement	17,348.4 turns of a running wheel; 0.4528 g brain weight
Accuracy	The agreement of a measurement with a known standard; degree of bias	Calibration mass = 20.000 g
Precision	Repeatability or similarity of measures of one animal	Grip strength = 42 g, 57 g, then 39 g; brain weight = 452.7 mg, then 452.4 mg
Consistency	Similarity of measures of the same animal over a period of time; predictability of later values on the basis of earlier measures	$R^2 = 0.67$ for line of best fit to rotarod fall latencies for 1 mouse
Reliability	Correlation of measures on several animals on different trials or days	$r = 0.73$ for accelerating rotarod
Replicability	Similarity of results of two experiments in same or different labs	Strain x lab interaction $P < 0.00001$
Validity	Does the test measure the construct it is designed to assess?	Diazepam increases exploration of open arms of elevated plus maze

because a measurement taken a few moments later may say that the mouse is now 27.691 g. Although the device can resolve differences of 0.001 g, two of the decimal places are effectively meaningless numbers because motion of the animal on the balance pan makes the *repeatability* or *precision* of a measurement much worse than the resolution of the measuring instrument. Even if the mouse is asleep on the balance pan, weight will fluctuate at the third decimal place as it breathes and twitches, unless readings are averaged over a long period of time. Furthermore, the animal is continuously losing small amounts of weight as it breathes water vapor into the room. Because of the moment-to-moment fluctuations in mouse behavior, resolution of measurements is rarely a limiting factor for a behavioral test. Little is gained by measuring something with considerably greater resolution than the repeatability or precision of the measurement when performed on the same individual in a narrow period of time. The net result is that a cheap balance with a resolution of only 0.1 g is perfectly adequate for measuring mouse body weight in most situations.

For mass of a lifeless fixed brain, on the other hand, the 0.1 g resolution of a triple-beam balance would be inadequate, and a good electronic balance that reads to 1 mg would be better. Precision of brain mass would be very good when readings are taken without removing it from the balance pan, but in reality a person must blot liquid from the surface of the brain before placing it on the pan, and variations in technique can change the measured mass by a few milligrams.

Repeatability may be lower than desired because of the way the identical mouse interacts with the measuring instrument or because the animal changes. Short-term repeatability can be assessed by taking the second measurement soon after the first (Figure 12.1A), whereas longer term repeatability entails a greater period of time between measurements (Figure 12.1C).

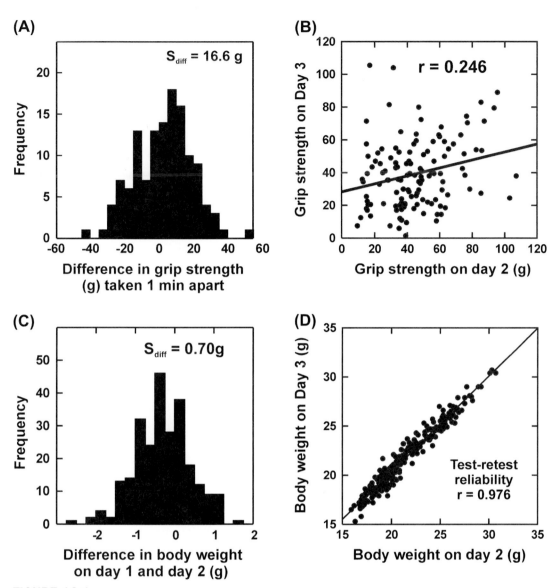

FIGURE 12.1

Body weights and grip strength of mice of the strains 129S1/SvlmJ, A/J, BALB/cByJ, C3H/HeJ, C57BL/6J, DBA/2J, FVB/NJ, and SJL/J. (A) Two measures of grip strength for 128 mice were taken 1 min apart, and the consistency or repeatability of the measures was low. (B) Test—retest reliability of mean grip strength on two trials on each of two days was also low. (C) Repeatability of body weight on two days was quite good, and (D) reliability was very high for 253 mice from the same eight strains. All mice were tested in the author's lab at the University of Windsor.

In both situations, repeatability is represented by the difference between measures of the same individual.

Shortcomings in repeatability do not always arise from changes in a live animal. Video tracking (Chapter 14) can yield measures of distance traveled in a maze that seem to be remarkably precise. For example, the path length might be reported by one system as 12.363360760 m and 14.115 m by another. When an inanimate object that always follows the same path is measured, however, the path length can be 617.336869345 cm on one trial and 624.566319866 cm on the next, which reveals that repeatability is not even 1 cm out of 600 cm (Bailoo et al., 2010).

Thus, less than perfect repetition of a measurement can arise from two sources: the measuring system and variations in the mouse. The former can be assessed by using some kind of inanimate object of measurement that does not change from one occasion to another. If repeatability when working with a live animal is substantially inferior to variations arising from the instrument, it is behavior that makes the measurement relatively unstable. Whether those behavioral variations reveal important facts about the nervous system or are just a hindrance to understanding depends on the specific situation and application.

Accuracy denotes the agreement of a measurement with some well-defined standard. A device such as a thermometer might measure temperature with precision of 0.1°C, while it consistently underestimates true temperature by more than one degree. Similarly, cheap digital bathroom scales give body mass to the nearest 0.1 kg but commonly deviate from the true value by more than 1 kg. To be very accurate, an instrument must be calibrated with respect to some standard. For an electronic balance, special calibration masses are used. Class S masses, as defined by the National Bureau of Standards in the United States, are guaranteed to be within 0.1 mg for a 20.000 g mass. Standards for many kinds of measures have been established by the International Organization for Standards (ISO). The National Science and Technology Committee (NSTC) in the United States has established a system for ensuring the proper calibration of many kinds of instruments, and the research laboratory can purchase thermometers, balances, and other useful devices that come with an NSTC-traceable certificate that traces the calibration of the instrument to a known standard. It is good practice for the mouse lab to use instruments with traceable calibrations to ensure good accuracy.

The accuracy of video-tracking systems used with mice would appear from the path traced on the computer screen to be outstanding, but careful measurements of paths of a ping pong ball moving in a perfect circle reveal some noteworthy deviations from the true path length (Chapter 14). Even under the best lighting conditions where there is high contrast of the animal with the background, the measured path length often deviates from the true value by 1% or more, while under less than optimal conditions the deviation can be far larger (Bailoo et al., 2010).

An example of a test device is shown in Figure 12.2. Running of a mouse on a wheel is detected by photocells using 18 reflective strips along the perimeter of the wheel. Thus, movements that change the wheel by less than 1/36 of the circumference or about 10 degrees cannot be detected. This will not be a major issue when the wheel turns rapidly in one direction. The device actually counts the number of strips that move past the photocells, and this value is then transformed by a computer into distance traveled. The radius of the wheel at the surface of the rungs on which the mouse stands is 8.33 cm, as determined by a caliper. Repeating the measurement might yield 8.32 cm, which makes precision 0.1 mm. One complete revolution of the wheel amounts to $2\pi r = 52.320084$ cm according to a calculator, but this is actually just 52.3 cm because radius is measured with only three significant digits. The accuracy of the measure is limited by the part of the system having the least precision. On the other hand, inaccuracy of the measured distance run will be far worse than its precision if the caliper was not calibrated correctly. If an animal registers 317,569 strip hits in 24 h, the number will be precise to about ± 1 strip hit. That many strip hits corresponds to 17,642.72 revolutions, the equivalent of 17,642.72 (0.52320084 m) = 9,230.685924 m or 9.23069 km in one day, a rather good marathon for a little mouse. The measurement really has only three significant digits, 9.24 km, because radius has only three. The researcher wants to report distance traveled, not photocell counts, because distance is understood the same way in all labs using all kinds of wheels.

CONSISTENCY

Consistency of a measure of behavior is high when two or more measures of the same individual are very similar within a measurement occasion, or when the measure early in the test can

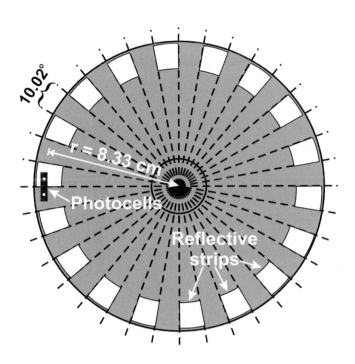

FIGURE 12.2
Diagram of a running wheel where motion of the wheel is detected with 18 reflective strips attached to a disk (Schalomon & Wahlsten, 2002). Two photocells that detect a strip allow direction of rotation to be determined. The resolution of the measurement of radius with a caliper (0.01 cm in 8 cm or about 1 part per thousand) limits the resolution of the measure of distance traveled.

accurately predict the value later in the test for that one individual. Consistency depends only on things that occur during a short trial or session, whereas repeatability can be influenced by myriad events that occur between one testing occasion and another. It is not the test per se that is consistent. The individual animal behaves consistently or shows large fluctuations from moment to moment. To some extent the fluctuations do tell something about the test situation, because behavior in some kinds of tests tends to be inconsistent for many of the mice.

Figure 12.3 shows the percentage of time spent moving during 1 min blocks in an open field, as assessed by video tracking. Averaged over all 200 mice from 21 inbred strains tested in the Edmonton lab, there is a smooth decline, but the small standard error bars conceal large variations within and between individuals. Four examples of mice with high consistency across time show how this can occur with no average change across time or consistent increase or decrease. If a behavior exhibits a smooth decline during a test session owing to habituation or fatigue, repeatability of the measurement over time will not be very good, even though consistency of the behavior is high.

Generally speaking, when there is less consistency, it is important to run the test with a longer trial or more trials in a single session to obtain a good estimate of average activity level, if that is the goal of the study. Even if consistency is high, a short trial may not be the best choice when there are markedly different trends of change across time for different individuals, as seen in Figure 12.3B. A test lasting only 1 min could not distinguish between mice that start at similar activity levels and then diverge. Furthermore, ostensibly the same behavior can be sensitive to different processes early versus late in a test session (Blizard, Takahashi, Galsworthy, Martin, & Koide, 2007), which allows important information to be extracted from a longer trial that might otherwise be obscured by taking an average.

Especially when working with a new kind of test, it is important to inspect individual data across time or trials in order to make an informed choice of test duration. Although there are

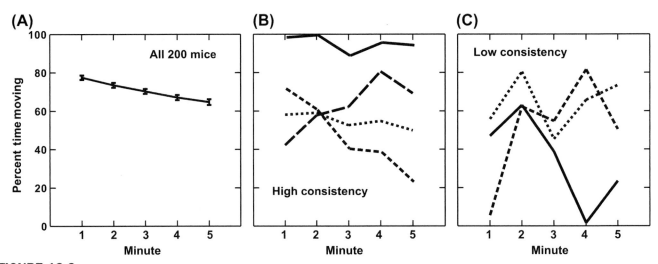

FIGURE 12.3

Percentage of time spent moving during 1 min blocks in an open field, as assessed by video tracking. (A) Means of all 200 mice from 21 inbred strains tested in the Edmonton lab show a smooth decline with very small standard error bars. (B) Some individual mice show great consistency or predictability over time, while others (C) are highly variable from minute to minute. *(Based on data reported in Wahlsten, Bachmanov, Finn, & Crabbe, 2006.)*

a number of abstract statistics such as Cronbach's coefficient alpha that summarize consistency for a group of diverse individuals, close inspection of data from individual mice has much to recommend it because of the wide variety of patterns than can be seen.

RELIABILITY

Test–retest reliability (r_{tt}) of a test expresses the similarity of individual scores on two successive measurements relative to the span of differences among individuals in the sample. It is indicated by a simple Pearson correlation of two sets of test scores on the same mice (Figure 12.1B). Reliability tends to be higher when: (a) repeatability and consistency of a measurement are very good for an individual, and (b) the range of individual scores in the sample is wide. In work with mice, a critical factor is the choice of genotypes to be examined. The wider the range of genotypes and therefore phenotypes, the higher the test reliability. Thus, reliability is a property of both the test and the populations that are sampled. Consequently, comparisons of reliabilities for different kinds of tests are meaningful only if they are based on the same population of mice.

The example in Figure 12.4 is based on data for 21 inbred strains tested on the accelerating rotarod at two laboratories as part of the Mouse Phenome Project (MPP; Rustay et al., 2003a). Test–retest reliability based on means of three trials for all n = 383 mice was 0.53 when comparing days 1 and 2 (Figure 12.4C) but a much higher 0.76 for days 2 and 3, after behavior had approached an asymptote for many animals. Correlation was considerably lower for the three trials before and after the ethanol injection on day 3, even though they occurred close together in time, most likely because the injection had disparate effects on various mice. The range of scores after ethanol injection was considerably reduced because so many mice were substantially impaired, and a reduced range of scores usually reduces the correlation. Reliability was higher if only four extreme strains were considered (A/J, FVB/NJ, NOD/LtJ, PL/J). Reliabilities for day 2 versus 3 within each strain ranged widely and had a median value of r = 0.55 when data were combined for the two labs, but the median was only r = 0.45 when the substantial lab difference was removed. Thus, test–retest reliability depends strongly on the range of scores and the kinds of treatments applied as well as the stability of individual measures across days.

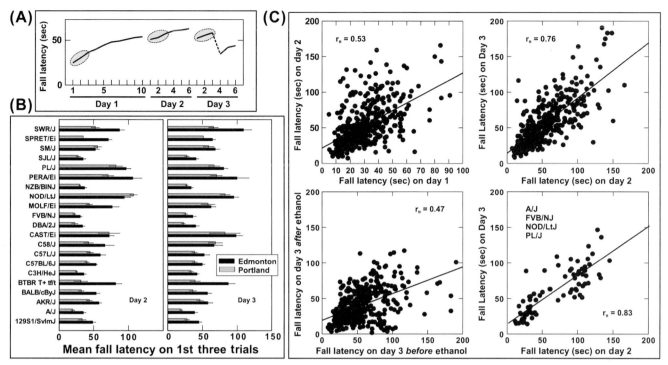

FIGURE 12.4

Performance of 21 inbred strains on the accelerating rotarod evaluated at two sites using the same equipment and test protocols. (A) Mice were given 10 training trials on day 1, then another six trials on day 2 without any injection. On day 3 there were three trials prior to injection and then three after an ethanol injection. (B) The patterns of strain means for the two labs were remarkably similar on day 2 and day 3 prior to injection, although fall latencies were generally lower in Portland than Edmonton. (C) Correlations of the scores for all mice were determined for the three trial blocks highlighted in panel A and the three trials after the injection. Correlation for the two highest and two lowest strains was quite high. *(Based on data reported in Rustay, Wahlsten, & Crabbe, 2003a.)*

There is no generally accepted criterion for adequate reliability of a test of mouse behavior. Considering many kinds of tests, $r_{tt} > 0.5$ is respectable when several inbred strains are involved. It would be most unusual to find a behavioral test with a reliability exceeding 0.95, a value typical of body weight or two measures of fixed brains on successive days. To establish a widely accepted criterion, it will first be necessary to decide on a standard set of inbred strains to be measured. At one time the eight strains of the A list of the Mouse Phenome Database (MPD) were used by several research groups as a standard reference set, but the A list was recently replaced by a new tier structure at the Jackson Laboratory that has 16 strains in Tier 1 and does not include one (SJL/J) in Tier 1 that was formerly on the A list.

Reliabilities for several kinds of tests based on 21 inbred strains, representing all of the MPP priority lists A and B plus the strain BTBR T + tf/J, are shown in Figure 12.5. Each mouse was tested twice on successive days, and some measures taken during a test showed much higher reliability than others. Distance moved through an apparatus tends to have high reliability when extreme inbred strains such as C57BL/6J and A/J are included in the study. Some indicators, such as percent alternations between arms in a Y maze, are much less stable and reliable.

When reliability of a test appears to be quite low (e.g., $r_{tt} < 0.3$ or 0.4), variation in scores within a group tend to be large compared with between-group differences, and sensitivity of the test to a wide range of treatment effects will be reduced. There are exceptions, such as the grip strength test (Figure 12.1B), which shows large fluctuations from trial to trial but is nevertheless very sensitive to disruptive effects of ethanol. Low test reliability, as assessed with untreated control animals, does not undermine results if treatment effects are very large. For

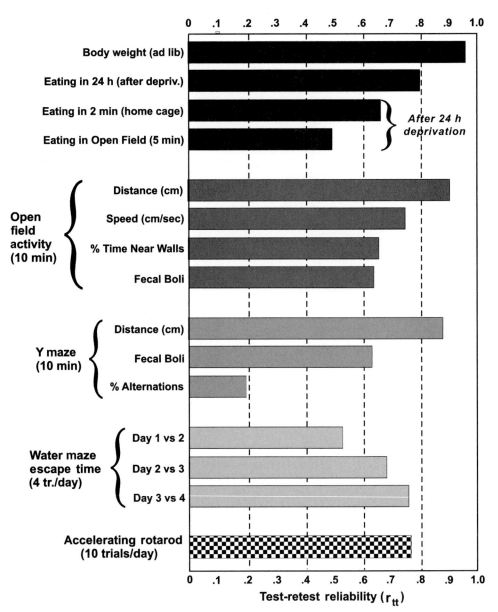

FIGURE 12.5

Test—retest correlations for 96 mice (6 of each sex) of 8 inbred strains tested twice on the same apparatus on successive days. The strains were from the original MPD A list (129S1/SvlmJ, A/J, BALB/cByJ, BTBR T+ tf/J, C3H/HeJ, C57BL/6J, DBA/2J, FVB/NJ). Reliabilities for measures of motor activity were almost as high as for body weight. All but alternations in the Y maze exceeded 0.5. *(Reprinted with permission from Wahlsten and Crabbe, 2010.)*

a test that is to serve a more general purpose in studies with many kinds of treatments, however, high reliability is a desirable feature.

REPLICABILITY

An entire experiment can be replicated, and a study with good replicability yields similar results on different occasions in the same lab or across different labs. This question is discussed at length in Chapter 15. For replicability, the issue is not the similarity of scores of individual mice; instead, the concern is with the patterns of group mean differences. When a study is replicated, the mice in the two experiments are usually different animals. The test of replicability is commonly done with a factorial analysis of variance (ANOVA). If two independent

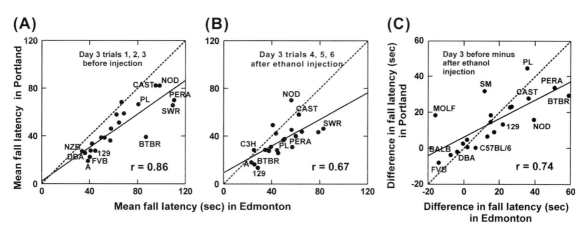

FIGURE 12.6
Correlations of the mean scores of 21 inbred strains on the accelerating rotarod evaluated at two sites using the same equipment and test protocols. Only the strain root name is shown for extreme strains, without substrain symbols. The full strain names are shown in Figure 12.4. Dashed lines denote equal values in both labs. (A) The pattern of strain means was very similar on the three trials on day 3 prior to an ethanol injection, even though fall latencies were considerably longer in Edmonton (63.6 sec) than Portland (43.9 sec). (B) The strain correlation was reduced after the ethanol injection that reduced the range of scores. (C) There was good agreement between labs on the magnitude of the ethanol effect across the strains. *(Based on data reported in Rustay et al., 2003a.)*

studies are compared with ANOVA and the study × treatment or study × genotype interaction term is not statistically significant, then results are well replicated, even if the overall mean scores are higher in one study than the other.

It is meaningful to compare results of an entire experiment on two successive days in the same lab or two labs (Rustay et al., 2003a). Figure 12.4B shows strain mean fall latencies on the rotarod in two labs on days 2 and 3. The patterns of strain differences and lab differences are remarkably similar on the two days. Interaction effects can be evaluated with repeated measures ANOVA because the same mice were tested both days. Averaged over the two days, the strain × lab interaction was significant ($P = 0.006$), whereas the lab × strain effect itself did not differ significantly between days (day × strain × lab effect, $P = 0.04$). Likewise, the ethanol effect on behavior on day 3 was very large, and strains differed considerably in the disruptive effects of ethanol ($P < 0.00001$), but the strain × ethanol interaction effect did not differ significantly between labs ($P = 0.11$). The correlations of strain means between the two labs was $r = 0.86$ for mean fall latency on the first three trials of day 3 (Figure 12.6A) and $r = 0.67$ for mean latency under the influence of ethanol (Figure 12.6B). The strain correlation for the two labs for change caused by ethanol was lower, but there was still good agreement between labs about which strains change the most or least (Figure 12.6C). Thus, the two labs obtained remarkably similar results and replicated their main findings, even though the mean fall latency was about 60 sec in Edmonton and 40 sec in Portland. This issue is discussed further in Chapter 15 (see Figure 15.5).

VALIDITY

Test validity is good if the test measures the kind of behavior or the construct that it was designed to assess. This cannot be determined from internal consistency, reliability, or replicability. Those features could be outstanding and yet the test might be worthless to the investigator because it does not measure anything of interest. For example, the researcher may want to measure memory for events associated in time, but the mice that do worst on the test may fail because they are blind and cannot see the stimuli (Brown & Wong, 2007). It may be a good test of memory for some mice but not others. If blind mice with retinal degeneration (Chapter 2) are to be studied, a task should not rely strongly on visual cues, unless it is designed to be a test of visual ability.

There are several kinds of validity that have been defined by psychologists who create tests for use with humans (Gregory, 2010; Kaplan & Saccuzzo, 2009), and some of these can be applied equally well with mice. The simplest is *face validity*, something that is apparent for tests where the entity measured is quite clear from the actual test and requires no external evidence to demonstrate validity. Tests that measure gross locomotor activity by photocell beam breaks or turns of a running wheel possess good face validity because the only way to achieve a high score is for the mouse to move extensively during a trial, while a low score denotes little movement. The test provides no clue about *why* a mouse may be very high or low in activity, and the test score therefore tells the researcher nothing more than what is apparent on its face, from the stream of behavior.

The two-bottle preference test used to compare drinking tap water versus a 10% solution of ethanol also has high face validity as a test of preference. This will be true, however, only if the positions of the tap water and ethanol solutions on the right and left sides of the cage are changed from day to day to compensate for position bias (McClearn & Rodgers, 1959). If ethanol is always on the left and a mouse drinks more solution from a tube on the left side of the cage, a bias in favor of drinking from the left tube might be misinterpreted as a preference for ethanol. Thus, validity depends strongly on the proper control for extraneous or confounding factors during the test.

Criterion validity of a new test can be assessed if the researcher has some other measure that is known to be a good indicator of a behavioral process. Scores on the new test should then correlate highly with measures of the same mice on the other test. If the criterion is so widely recognized as a good measure, why not just use it? Perhaps it requires invasive sampling of blood or brain electrical activity, or maybe it is based on a very long test with many items or trials. There may be an advantage for the research if an abbreviated and simpler test could be devised and substituted for the more demanding one. In human psychology a well-standardized test with many items, such as the Weschler Intelligence Scale for Children (WISC), might serve as the criterion when devising a shorter test of mental ability. If the correlation of the abbreviated test with the WISC is very high, the short test may be taken as a convenient indicator of intelligence. In work with mice, there are currently no tests that are widely accepted as a criterion or gold standard for measuring some process or construct (Crawley, 2008).

Construct validity can be established if some treatment is known to alter a psychological process or construct in a certain way. The treatment should then have a similar effect on scores of a test that purports to indicate that construct. This has been done in the study of anxiety in mice and rats. Defining anxiety in rodents is hazardous because the concept was derived from human experiences and cognitive structures. There is no guarantee of homology between human and rodent forms of anxiety. The indicators in mice are usually behaviors that show a preference for the less threatening parts of an apparatus (Henderson et al., 2004), whereas in humans they usually involve expressions of language. Validity on the mouse tests measuring anxiety is conferred when the behaviors respond to anxiolytic drugs such as the benzodiazepines, which are known to reduce anxiety in humans (Holmes, 2001). This approach has been used to demonstrate the validity of the elevated plus maze as a measure of anxiety-related behavior. Investigators prefer to call the exploration of open arms an anxiety-related behavior rather than a direct measure of anxiety. The test does not measure murine anxiety. Nobody really knows what anxiety means to a mouse. Instead, the test for mice yields a behavioral measure that is a good model for behavioral states in humans that do involve anxiety. Construct validity for the elevated plus maze is rather difficult to establish in any general sense because the drug effects that increase open-arm exploration on the first trial in the maze are often not evident the second time the mouse encounters the same maze, unless the two trials are longer than the usual 5 min (Holmes & Rodgers, 1999). The behavioral indicator of anxiety, avoidance of the open arms of the maze, may change very little from the first to the second experience, whereas the drug effect changes greatly.

Establishing test validity is probably the most challenging aspect of work with any behavioral test. Validity is a major concern when engaging in task refinement (Chapter 13). Many of the changes that might be recommended in common tests serve to enhance validity rather than reliability.

Task Refinement and Standardization

Even the simplest test of mouse behavior can be done in many ways. It would be very helpful to know if one of these ways is better than the others. If it is, perhaps the task could be improved and refined to yield better data. The field might then adopt the improved version as a standard version in different labs. Whether tests should be standardized at all has been debated for several years, and there currently is no apparent trend toward greater homogeneity of test parameters across labs. On the contrary, one group of researchers recently lamented that in work with mice "laboratory behavioral science remains simplistic, unstandardized, and underfunded" (Blizard et al., 2007). This chapter discusses the most promising approaches to refining behavioral tests. If there is to be improved standardization, hopefully the field will adopt a good set of parameters that eliminate all major flaws in a test.

FLAWED TESTS AND MISSTEPS IN THE RESEARCH PROCESS

There are many excellent behavioral tests available for research with mice but there are no perfect tests. Many have flaws or are applied in less than optimal ways. The need for improved tests arose at the beginning of behavioral research on animals and continues today. The title of one recent review expresses the situation well: "How Many Ways Can Mouse Behavioral Experiments Go Wrong?" (Schellinck et al., 2010). Judging from the length of that review

(110 pages), there are many indeed, even after more than 100 years of experience with testing mouse behavior.

One of the earliest studies of rodent behavior in the laboratory trained rats to find food in a maze (Small, 1901). A prescient comment by that author deserves repeating: "the experiments must conform to the psycho-biological character of an animal if sane results are to be obtained." Since then, several kinds of tests in the published literature have had major flaws that generated insane results. Most mouse researchers have experienced the frustration of running mice through a seemingly clever test using elegant apparatus, only to discover that the data are worthless. In a real debacle, the defects are obvious and may even be beyond repair. The apparatus will end up in the garbage bin and only close colleagues learn from the experience. For other tests, the defects may be more subtle and can be remedied if they are detected early in the process of designing and refining a test.

The steps in successful research are well known, but much can be learned from the missteps too. What begins as a misstep may develop into an important discovery. The history of science is replete with accidental discoveries that led someone out of a well-worn conceptual rut. There can be no systematic science of the unanticipated. Missteps can only be presented through specific examples. Several of the examples presented here involve the author, who has made more than two mistakes and watched others stumble too.

The "lab cat"

Many of the basic procedures in behavioral testing that we now take for granted were lacking in early tests. The work of Small provides several examples of now extinct methods. His maze, amazingly complex by today's standards, was copied from the famous Hampton Court maze in England that challenged the spatial abilities of many humans (Figure 13.1). It had a wooden floor and wire mesh walls that could never be properly cleaned to remove odor trails. Even a blind rat was able to learn the maze. The route to the goal was the same on every trial for every rat. Two or three rats, either wild-caught or a white lab variety of unknown origin, were run through the maze at one time, and they remained there overnight, making many reverse trips back to the home cage from the food source, carrying most of the food back to the home

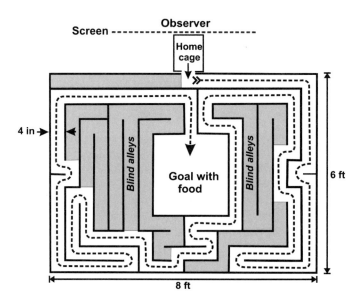

FIGURE 13.1

Maze for rats modeled after the Hampton Court maze for humans in England. Two or three rats spent the entire night in the maze. The correct path was the same for all animals on every trial, and the maze was made from wood and wire that could not be properly cleaned to remove odor trails. *Redrawn from Small (1901).*

cage. Sometimes a rat's attention was said to be distracted by unspecified "foreign matter." Then on one trial there was a "slight difference" in protocol: "The laboratory cat jumped up onto the maze and showed friendly interest in the rats. They were startled … the wild rats were stiff with terror for 15 minutes." Descriptions of the animals included obvious instances of anthropomorphism: "indecisiveness is a constant trait of the rat character" … "general appearance of *lostness*" … "rats tend to let loose the play instinct" … "strolling nonchalantly into blind alleys" … "primitive craving for knowledge."

It is a fascinating fact that rats and mice can indeed learn a path through a very complex environment. Anyone with murine visitors in the kitchen knows this all too well. Nevertheless, the technology of behavioral testing was eventually simplified to control many influences on the learning process and compare discrete steps in the process in different groups of animals. Apparatus is now made from material that can be thoroughly cleaned. Cats are still found in a few mouse labs but only as experimental treatments (Adamec, Head, Blundell, Burton, & Berton, 2006), because mice are greatly alarmed by the odor of cats (Arakawa, Blanchard, Arakawa, Dunlap, & Blanchard, 2008; Blanchard, Griebel, & Blanchard, 2003). Contrary to the colorful prose of Small, imputing human motives and behavioral proclivities to mice is no longer tolerated. The charm of early research has been replaced by austere abstraction. Behavioral genetics has adopted the model of controlling variables that is so successful in physics.

Mouse shuttle avoidance

An influential paper that applied mouse genetics to the scientific study of memory was published as a lead article in *Science* in 1969 by Bovet, Bovet-Nitti, and Oliverio. Bovet had already won the Nobel Prize in 1957 for his work on the pharmacology of the autonomic nervous system, and many researchers took note when he began to study mouse behavior. The study compared three inbred strains of mice on several learning tasks. The data in the article were incredibly clear; all mice within a strain produced almost identical learning curves, while strains differed greatly. Genetic differences appeared to explain almost all variation in learning, which made mice prime candidates for genetic analysis, far superior to outbred rats. Most remarkably, the C3H strain of mice appeared to have good long-term memory but poor short-term memory, whereas DBA mice seemed to retain memory well in the short term but not 24 hours later. This impressive result prompted several other memory researchers to order the same strains of mice, hoping to repeat the results and discover memory genes. One of those researchers was James McGaugh at the University of California Irvine, where the author happened to be doing Ph.D. studies on the psychology of learning. McGaugh visited Bovet's lab in Rome and made detailed measurements of his apparatus, then returned home and had a duplicate built.

The apparatus was automated and could test several mice at once on shuttle avoidance learning (see Figure 3.3, page 221). A shuttle box has two identical chambers, each with a grid floor for giving electric shock and a signal light. A small doorway separates the boxes. The mouse starts in one side (A), the light comes on, and 5 or 10 sec later the shock comes on, whereupon the mouse quickly scampers to the other box (B). After an intertrial interval of 30 sec or more, the light comes on in box B and the mouse must run back to box A to escape or avoid the shock. It all looks so efficient, until a mouse is asked for its opinion on the task.

The author and another graduate student were nominated to train mice from the same strains used by Bovet. The shuttle boxes were an unmitigated disaster. Many of the mice never learned at all, or at least they did not learn what the experimenters wanted them to do. Several found ingenious (for a mouse) ways to avoid the shock without running back and forth at all. For example, one mouse never escaped from box A to B because it discovered a way to spread its legs far apart and stand on alternate grid bars that happened to have the same polarity (see Figure 11.11). It never got shocked, provided it remained motionless trial after trial. Another

217

managed to perch on one bar in the doorway between the two boxes. A mouse that jumped vigorously found a thin gap between the lid and wall of the shuttle box, and it hung suspended off the grid for many trials. Others froze in the far corner of a box until prodded to escape when the shock came on. Differences between mice of the same strain were so great that the data were worthless for exploring the genetics of memory.

Other researchers tested mice on a much simpler task, inhibitory step-through avoidance, where the animal learns not to re-enter a chamber where it previously received a shock. Both rats and mice were usually quite good at this task, and a durable memory was generated after only one bad experience in the shock compartment for most mice (Flood et al., 1986; Geller et al., 1969). The task was used to demonstrate time-dependent consolidation of memory by giving different amounts of time between the initial experience and a treatment that impairs memory storage. In J. L. McGaugh's lab, the memory curves for C3H and DBA were virtually identical (personal communication, 1972).

A review of the literature on learning later identified the Bovet et al., study (1969) as an extreme outlier in the world literature on inbred strains (Wahlsten, 1972a). The full story of why such beautiful data emerged from one lab but not another that used the same genetic strains has yet to be written. What is clear is that small details of a task can make a huge difference to individual animals and the variability among animals can become excessive if the task requires them to do things that are contrary to their more common tendencies. The mice conveyed to the experimenters that there was something wrong with the task, not their memories.

Bar press avoidance

New behavioral tests are sometimes designed according to theory that explains why they *should* yield good data. An apparatus is built and training begins, whereupon it is instantly obvious that something important is missing from the formula. Such was the case with bar press shock avoidance. The task promised high throughput because it could be automated with simple technology. In an avoidance paradigm, a neutral signal such as a tone or light warns that shock to the feet is imminent. According to theory, when the tone is followed a few seconds later by shock, the tone should evoke fear via Pavlovian conditioning, and a bar press that turns off the tone should be reinforced by the reduction in that fear.

Researchers were surprised to discover that the animals would quickly learn to press the bar to turn off or escape the shock, but they simply would not learn to press it *before* the shock came on, even though that simple act would avoid the shock entirely. In one study, most of the rats failed to learn the avoidance response even after 1,000 trials over many days (D'Amato & Schiff, 1964). Behavior in that apparatus seemed quite maladaptive. Several investigators then tried modifications of the task that theory or intuition told them might make a difference. One promising approach was based on the idea that the fear was not evoked solely by the tone or light but was associated with the contextual cues of the chamber. Accordingly, a new apparatus was built that had two distinctive chambers: the usual box with a grid floor to provide shock and a bar to be pressed, and a second box with no bar and a smooth floor where there was never any shock (Wahlsten, Cole, Sharp, & Fantino, 1968). The two chambers were joined by a door that could be raised by a motor, such that a bar press opened the door and allowed the animal to leave the accursed shock box altogether. Fear would then be greatly reduced following a bar press and the animals should learn better. So they did, but there was one little hitch to this paradigm: once the rat ran into the safe box, it had to be returned to the shock box by the experimenter before the start of the next trial. This entailed handling the animal shortly before the start of the trial, so a control group was added that got the same handling as the experimental group but was never allowed to escape from the shock box. Surprisingly, the handled group also learned much faster and was just as good as the animals that were allowed to escape into the safe box. Escape from the contextual cues was not essential to reinforce bar pressing.

Pairing contextual cues with an electric shock evokes freezing, a species-specific defense reaction (Bolles, 1971). When the animal is returned to that context, freezing antagonizes bar pressing. Handling the animal disrupts the freezing response enabling the animal to learn the bar press connection. Eventually, researchers found it more meaningful to study the freezing response, as in the contextual and cued fear conditioning tasks (FCA, FCC, Table 10.2). One other salient fact contributed to the disuse of handling to aid avoidance learning: rats and mice do not like to receive electric shock, and rodents sometimes try to bite the experimenter soon after being shocked. As Hymie Anisman, then at the University of Waterloo, learned to his dismay, a stock of hooded rats bred at the Royal Victoria Hospital in Montreal became so vicious after just one shock that it was almost impossible to handle them at all.

Today the shuttle shock avoidance and bar press avoidance tasks have almost completely disappeared from the repertoire of mouse learning tests. Researchers who had never read the work of Small from 1901 relearned the value of his admonition about the task conforming to the psychobiological character of the animal. A major question for the field today is whether our current bevy of tests also harbors flaws and foibles that need to be purged.

TASK COMPLEXITY

One of the greatest challenges when seeking to evaluate and improve a test is the sheer complexity of even the simplest situations. There are three broad categories of features of a test situation.

1. Several aspects of the environment impinge on the animal during a test.
   a. Certain of these may also be part of the global lab environment (Chapter 15), such as air temperature, humidity, and odors that travel through the air conditioning.
   b. Others are specific to the test or may be shared by different kinds of tests given in the same room, including light quality and intensity, nearby noises and vibrations, and other animals awaiting their tests. The experimenter also belongs in this category.
2. The physical apparatus is unique for each test. It includes dimensions of structures, kinds of walls and floors, and motors or electronic devices to create motion or measure actions of the mouse.
3. Procedures involve where to start the trial, number and duration of trials, means of cleaning the apparatus between animals, and ways of recording specific behaviors.

The relatively simple open field

Among the tests presented in Chapter 3, open field activity appears to be rather simple and has relatively few variables. Nevertheless, mouse behavior in the open field "reflects the influence of multiple complex processes" (Blizard et al., 2007), including motivational processes, and it has defied all attempts at standardization. Twenty-three features of this task are listed in Table 13.1. To be included on the list, the feature must be something that often varies among labs and could reasonably influence the measures of behavior during the test, especially when it is administered to different strains. Comparison is made of values employed by two groups of researchers using substantially different task configurations (Kafkafi et al., 2003; Kafkafi, Benjamini, Sakov, Elmer, & Golani, 2005; Wahlsten et al., 2006). There are two major differences between them. First, the open fields are very different in size and stimulus surroundings, so different that one could reasonably view them as different tests. Second, the kinds of data collected and the manner of analyzing the video-based XY coordinates of the sequence of mouse centers are different.

Results of the two kinds of tests for the same inbred strains were remarkably different for certain strains (Figure 13.2), especially 129S1/SvImJ, BALB/cByJ, and FVB/NJ. The correlation of strain means between Windsor and UNCG labs using the same apparatus and almost identical protocols was r = 0.92, whereas the correlation of those two labs with the means from the Kafkafi et al. (2005) study were less than r = 0.5. There are at least eight differences

TABLE 13.1 Features of Two Open Field Activity Tests

Item	UNCG lab	Kafkafi et al.[a]
	Proximal features of the room environment during the test (not colony room)	
Room temperature/humidity	21°C, 30–80% RH	22°C
Occasional odors in room air	Coffee, popcorn, rats	?
Constant background noise	Vent fans; 40 dB	?
Occasional sharp sounds	Banging on nearby cage racks	?
Vibrations of the floor	Elevator nearby (45 dB)	?
Lighting at test apparatus	Fluorescent, 20 lux	2 × 40 W neon bulb
Light in waiting area	173 lux	?
Phase of light—dark cycle	Light	Dark
Specific experimenters	M. B., E. H.	3 or more
	Physical test apparatus	
Shape	Square	Round
Dimensions	40 cm L × 40 cm W × 30 cm H	2.1–2.5 m D × 50 cm H
Wall material	Clear 3 mm polycarbonate	Gray paint
Top or lid	None	None
Floor	Clean waxed paper	Non-porous gray
View from inside the box	White cubicle walls	Objects on walls of tub
	Procedures during trials	
Prior tests of same mouse	4th or 5th test in a battery	None; single test
#, spacing, duration of trials	1 trial of 5 min	1 trial of 30 min
Starting location	Center of box	Near wall
Handling method to start	Gloved hand grasping tail	Small black box lifted away
Postures scored by person	Rear, lean, groom, boli, urine	None
Automated measures	Video tracking (AnyMaze)	Video tracking (EthoVision)
Location of experimenter	Out of view; watches monitor	?
Cleaning of apparatus	Discard paper; wipe walls with protease	?

[a]**Sources**: *Kafkafi et al., 2003; Kafkafi, Benjamini, Sakov, Elmer, & Golani, 2005.*

between the tasks in the two labs that might reasonably contribute to discrepancies in results. For example, trial lengths were 2, 5, and 30 min in the studies at UNCG, Windsor, and the Kafkafi et al. (2005) consortium sites, respectively, and habituation tends to reduce activity in longer trials but in a strain dependent manner. Despite this variation in trial length, scores were very similar indeed for the A/J, DBA/2J, and SJL/J strains across diverse conditions. A systematic study within one lab of the impact of each of eight factors in activity levels would be a major undertaking if they were evaluated one at a time, whereas an immense multifactor experiment to detect all possible interactions would suck resources into a sinkhole. If eight strains, five mice of each sex, were evaluated with a square or round field having diameters of 40 cm, 1 m, or 2.5 m, starting the mouse in the center or at the wall, with a clean paper floor or plastic floor cleaned with alcohol each trial, the experiment would involve 1,920 mice.

The open field test is remarkably robust with respect to many kinds of task variations when the devices are square and reasonably similar in size (Wahlsten et al., 2006). Strain correlations among studies done four to five decades apart are greater than or equal to $r = 0.85$ (Figure 13.3), and the correlation between two labs (Edmonton, Portland) using mice tested at the same time with identical apparatus and protocols is very high. Comparing Figure 13.3 with Figure 13.2, it appears that there are some major variations in the open field test that have rather large effects on certain strains. The largest discrepancies in Figure 13.2 were for the 129S1/SvImJ and FVB/NJ. These were not included in the older studies because they did not then exist.

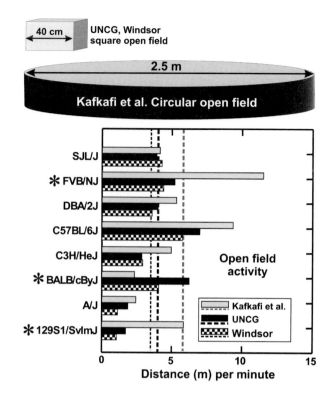

FIGURE 13.2

Distance traveled per minute in an open field by the same eight strains of mice in studies by two groups of researchers. The data for the 40 × 40 cm box are from the author's laboratories at the University of North Carolina and the University of Windsor, whereas those for the large round field are reproduced with permission from Kafkafi, Benjamini, Sakov, Elmer, & Golani (2005). Dashed lines show means of the eight strains for each data set. Strains with an asterisk were markedly different between the two kinds of open fields.

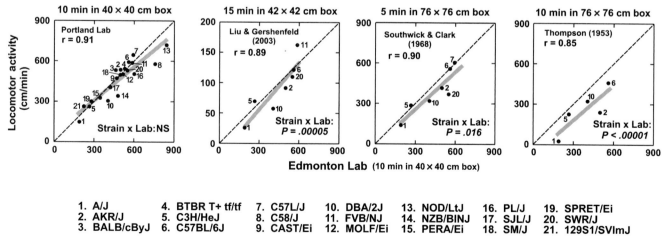

FIGURE 13.3

Correlations of distance traveled per minute in an open field in five laboratories. Data from the Edmonton and Portland labs were collected simultaneously in 2002 using identical apparatus and protocols, whereas data from the other three labs were obtained using somewhat different apparatus and procedures (Liu & Gershenfeld, 2003; Southwick & Clark, 1968; Thompson, 1953). *(Reproduced with permission from Wahlsten, Bachmanov, Finn, & Crabbe, 2006.)*

The submerged platform water escape task

The most popular method for assessing spatial memory in mice is the Morris water escape task (see Table 3.1). The mouse must find an invisible, submerged platform located away from the wall of the tank in order to escape from the water (Figure 13.4A). The featureless tank is located in visually complex room surroundings so that spatial location is the only clue. After several learning trials when latency and path lengths decline, a test trial is conducted with the platform absent and time in the formerly correct quadrant is noted. Mice that have learned the spatial location confine their search mainly to that quadrant, whereas poor learners swim widely in the tank and search in all four quadrants equally. The 29 salient features of this task are listed in Table 13.2. Combining them with the additional nine proximal features of the

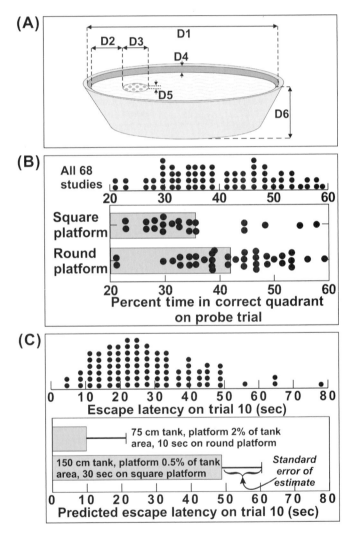

FIGURE 13.4

(A) Diagram of the Morris submerged platform water escape task showing diameter of tank (D1), distance of platform from wall (D2), diameter of platform (D3), height of rim above water (D4), depth of platform under water (D5), and depth of water (D6). Those and other task parameters were noted in 100 published studies that used the C57BL/6 mouse strain. (B) Percent time in the correct quadrant on probe trial ranged widely in studies where the specific value could be determined. The chance level for random search is 25%. Studies that used a square platform reported poorer performance than those with a round platform. (C) Escape latency on trial number 10 from 81 studies. Bars show the mean values from a regression equation that predicted escape latency from four task parameters (tank diameter, size of platform relative to tank size, time on escape platform, and platform shape). The four parameters accounted for 35% of variance in latency between studies. *Data collected and analyzed with the assistance of Emily Marcotte.*

TABLE 13.2 Features of the Morris Water Escape Task[a]

Feature	Range in Literature
Physical features of water tank	
Tank diameter	60 to 200 cm
Water depth	14 to 70 cm
Rim height above water	6 to 44 cm
Rim height/diameter ratio	0.025 to 0.60
Tank color	White, gray, black
Platform diameter	5 to 23 cm
Platform distance from wall	10 to 43 cm
Platform depth under water	0.05 to 2 cm
Platform as % of tank area	0.2 to 3.5%
Platform shape	Round, square
Platform material	Smooth plastic, mesh, cloth
Platform location across trials	Constant, variable
Water pigmentation	None, paint, milk, chalk, lime
Water temperature	15 to 27°C
Water change frequency	??; end of day
Sensory environment	
Intramaze cues	??; none, smooth tank
Extramaze cues	??; many at different distances
Location of experimenter	??; always at N end
Location of lights	??; 4 × 60 W bulbs + room light
Procedures	
Trials/day × days = # trials	1 to 12 × 1 to 66 = 3 to 176
Trial time limit	30 to 120 sec
Platform time limit	5 to 30 sec
Intertrial location	Holding cage; platform
Handling method	Gloved hand; plastic scoop
Start location	Near wall; center
Start on different trials	N S E W balanced or random
Probe trial #	9 to 75
Probe trial duration	20 to 120 sec
Rescue protocol	??; if submerged, stop trials

Note: ?? means not specified in the article.
[a]Based on 100 published articles using C57BL/6 mice compiled with assistance of E. Marcotte.

room environment of a test from Table 13.1 yields a total of 38 task parameters for the Morris task that might reasonably influence some measure of the behaviors. Thus, merely adding an escape platform to a bare water tank greatly increases task complexity. The apparatus looks so simple, but it is not.

Along with popularity comes diversity of methods when investigators with different scientific backgrounds seek to study a complex mouse behavior. The diversity was documented by tabulating parameters for 100 published studies that all used the C57BL/6 mouse strain, usually as a control for effects of a knockout or drug. The full list of studies and a massive table of parameter values is available from the author. Table 13.2 gives the range of published values, which is quite extraordinary and in some cases entails task parameters that are highly dubious. For example, a platform depth of 2 cm is typical of studies using rats, as is a tank diameter of 2 m. A mouse is still mostly submerged in water on such a platform. It appears that some investigators ran mice in the same tank used for rats. Some added lime, a mildly toxic substance, to make the water white.

The Morris task is designed to assess memory for spatial location in a complex visual field. It is somewhat surprising that so few articles described the visual cues available to the mice. Another potentially critical cue was rarely mentioned: the location of the experimenter who places the animal into the tank and retrieves it at the end of a trial. If the experimenter stands in approximately the same place on every trial, this will create a highly salient cue. The path traveled by the mouse is often documented with video-tracking software, and lighting is critical for obtaining good tracking (Chapter 14). The location and number of lights ideal for good tracking may not be the best condition for obtaining good spatial learning. It is difficult to judge this factor from the published literature because lighting is almost never described.

Do task parameters matter?

At first glance, the diversity of task parameters used in the Morris maze (Figure 13.5) suggests that those specific parameters must not be very important because all of the studies were passed by referees and then published. Fortunately, the diversity creates an opportunity for an empirical test of the importance of parameters in the literature. The studies gave results in terms of latency and distance to find the platform during training and then quadrant search times on probe trials (Figure 13.4B, C). Many of the parameters in Table 13.2 are continuous variables that can be used as predictors in a multiple regression equation, and categorical variables can be converted to effect coded predictors. It is then feasible to use forward stepwise regression to find variables that are associated with performance during learning and on the probe trial. A very conservative method can be used in which every study was counted as a sample size of one and assigned equal weight, regardless of the number of mice tested. That method can detect only variables that have a substantial impact on performance.

224

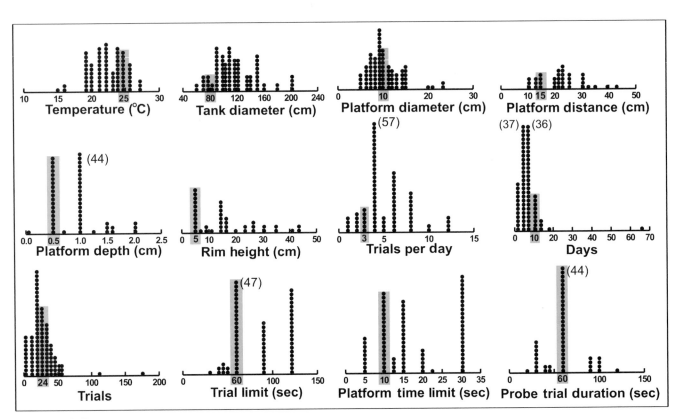

FIGURE 13.5

Frequency distributions of parameters used for various features of the Morris task in 100 published studies of C57BL/6 mice. Some values were not given for every study. The value highlighted in gray is proposed by the author for a standard version of the task.

Probe trial performance

It is generally believed that concentration of search for the platform in the correct quadrant of the water tank is the best indicator of spatial memory, because latency and distance can improve with practice through the use of non-spatial swim strategies (Gerlai, 1999; Wolfer, Madani, Valenti, & Lipp, 2001; Wolfer et al., 1998).The range of scores for percent time in the correct quadrant was very wide indeed (21 to 59%; Figure 13.4B), and this range for genetically identical (or nearly so) C57BL/6 mice demands an explanation. It is surprising that the regression analysis uncovered only one task parameter that significantly correlated with percent time in the correct quadrant: shape of the platform. As shown in Figures 13.4B and 13.6, the statistical size of the effect was not large, the range of scores for a given platform shape was very wide, and the vast majority of variance among studies could not be accounted for by the information in published reports. Platform shape was unrelated to improvement in latency or distance early in training ($r = -0.13$ and -0.09, respectively), but the correlation of platform shape with improvement in latency from trial 10 to the last trial was notably higher ($r = -0.33$). Thus, shape of the platform seemed to be more closely related to mastery of the task than to initial learning. Little or no information was provided in any study concerning the actual room cues available to the mice. It is conceivable that some labs used cue configurations that were highly salient and relevant to platform location, whereas others presented the mice with more subtle and difficult discriminations. Whatever the cause, there must have been important aspects of the test situation that strongly influenced spatial learning and memory but were not reported in the methods section of the published reports.

Escape latencies and rate of learning

In the analysis of latency and distance, all test parameters that accounted for a significant ($P < 0.05$) portion of variation were first included in the equation, and then the percent of

225

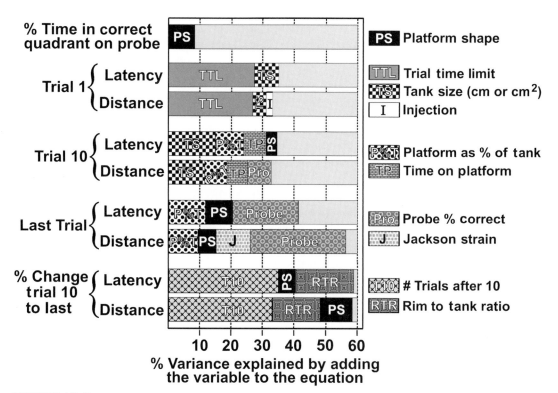

FIGURE 13.6
Results of a multiple regression analysis showing percentage of variance in each dependent variable that is accounted for by inclusion of predictors with a forward stepwise method. Further details of the analysis are given in the text and Table 13.3.

time spent in the correct quadrant during the probe trial was added as a predictor variable. Quadrant time on a probe trial occurs *after* training trials and is not a predictor in the usual sense. Quadrant time was used as a proxy variable for unreported task features, especially extramaze cues, in order to discover whether they affected escape latency and distance in a similar way.

PERFORMANCE ON TRIAL 1

Escape latency and distance on the first trial were strongly influenced by the length of the trial imposed by the experimenter, which implies that in many studies the C57BL/6 mice failed to escape on the first trial. Latency and distance were also somewhat longer in larger tanks. For distance only, mice that received a saline injection before the trial had 156 cm longer path lengths on average, but the statistical size of the effect was small (added $R^2 = 0.021$).

PERFORMANCE ON TRIAL 10

Mice improved considerably over the first 10 trials and trial time limit no longer exerted a significant influence on escape behavior. Both the size of the tank and the platform size as a percent of tank area were substantially associated with performance, which was better in smaller tanks and larger platforms relative to tank size. For every 1 cm increase in tank diameter, mean latency increased 0.2 sec and mean distance increased 3.1 cm (Table 13.3), which is roughly equal to the increase in tank circumference (πd). Increasing time on platform was associated with both longer latencies and distances. For latency only, shape of the platform had a small effect; latencies were about 7 sec longer when the platform was square. For distance only, adding percent time in the correct quadrant on probe trials to the equation improved the fit to the data substantially (added $R^2 = 0.075$).

For both latency and distance, more than 30% of the variance among studies could be attributed to the combined effects of four significant predictors. The size of the relation is indicated in Figure 13.4C for escape latency, where there was a difference of almost 40 sec between values of parameters that were expected to yield good performance and a set of four associated with inferior performance. The specific values were chosen for insertion in the regression equation so that they differed considerably but did not represent the most extreme values seen in the studies (Figure 13.5).

PERFORMANCE ON THE LAST TRIAL

When mice approached asymptotic mastery of escape from the Morris maze, overall size of the tank no longer made much difference, whereas a larger platform size as a percent of tank size was associated with superior escape behavior and a round platform was associated with better performance (Table 13.3, Figure 13.6). For both latency and distance, adding percent time in the correct quadrant on probe trials increased the fit of the equation by a large amount. This suggests strongly that there were unreported factors that were very important for both mastery of the escape task in the final phase of learning and concentrated search for the platform on probe trials that was also observed at the end of training. The most cogent explanation for this pattern is that memory for platform location based on spatial cues is most important for performance late in the learning process, whereas earlier during training the mice are learning motor skills and relying more strongly on non-spatial search strategies that are less efficient.

IMPROVEMENT FROM TRIAL 10 TO THE LAST TRIAL

When a change score was examined, a larger number of trials administered beyond trial 10 clearly improved performance. For every additional trial, latency declined by about 0.7 sec and distance by about 0.7 cm. The extent of improvement was about 16% lower with a square platform for both latency and distance (Table 13.3). The ratio of rim height to tank diameter contributed significantly to learning, such that for every 0.1 increase in the ratio, latency

TABLE 13.3 Multiple Regression Equations for Predicting Seven Measures of Performance in Morris Maze

Measure/Predictor	N	Coefficient	Adj. R^2	R^2 Add	Tolerance	P
Percent Correct						
Platform shape	57	−6.302	0.083	0.083	1	0.01715*
Escape Latency Trial 1						
Trial limit	82	0.369	0.351	0.275	0.953	0.000005
Tank area	75	0.001		0.076	0.953	0.000555
Escape Latency Trial 10						
Tank diameter	81	0.195	0.346	0.15	0.959	0.000225
Percent platform	75	−6.439		0.09	0.93	0.005435
Platform time	68	0.4		0.071	0.969	0.01719*
Platform shape	66	6.634		0.035	0.985	0.03959*
Escape Latency Last						
Percent platform	77	−12.211	0.414	0.118	0.984	0.000035
Platform shape	75	8.641		0.086	0.894	0.00166*
Probe % correct	51	−0.311		0.209	0.891	0.01118
% Lat. change Tr. 10 to Last						
# Trials after trial 10	87	0.707	0.59	0.357	0.855	0.000055
Platform shape	72	−19.903		0.053	0.969	0.0016*
Rim to tank ratio	35	−41.942		0.18	0.879	0.03095
Distance Trial 1						
Trial limit	82	6.993	0.331	0.269	0.939	0.00001
Tank area	75	0.018		0.042	0.949	0.00295
Injection	72	156.635		0.021	0.988	0.09442*
Distance Trial 10						
Tank diameter	81	3.141	0.327	0.116	0.831	0.005605
Percent platform	75	−203.549		0.07	0.911	0.00609
Platform time	68	7.429		0.066	0.968	0.05769*
Probe % correct	46	−7.645		0.075	0.877	0.01877
Distance Last						
Percent platform	77	−231.599	0.564	0.094	0.998	0.000525
Platform shape	75	173.304		0.058	0.813	0.00583*
Jackson strain	54	133.463		0.11	0.997	0.02132*
Probe % correct	32	−7.071		0.301	0.812	0.01009
% Dist. change Tr. 10 to Last						
# Trials after trial 10	87	0.724	0.585	0.339	0.855	0.00006
Rim to tank ratio	36	−56.959		0.158	0.879	0.00812
Platform shape	35	−16.441		0.088	0.969	0.00959*

Note: N is number of values available with complete data when the variable was included in the equation. This number always declined as variables with missing data were included. Adjusted R^2 is the goodness of fit for the best equation with all significant predictors included. R^2 Add is the increment in the multiple R^2 resulting from the inclusion of each additional variable into the equation. Tolerance indicates the correlation among the predictors being near 1.0 when they are independent. P is the one-tailed probability except when * indicates two-tailed.

improvement was 4.2% less and distance improvement was 5.6% less. For a tank where the height of the rim was half the tank diameter, improvement from trial 10 to the last trial was reduced by more than 20% compared to a tank with a very low rim. Thus, in studies where the low rim allowed maximum visibility of spatial room cues, improvement of latency and distance was considerably better. The fact that rim to tank ratio was significantly related only to improved performance at the end of training suggests that it was indeed related to utilization of spatial cues. When percent time in the correct quadrant on probe trials was added to the equation with only one predictor (number of trials beyond 10) already in the equation, it contributed significantly ($P = 0.007$) to explanation of both reduced latency and distance, but once rim to tank ratio was in the equation, percent correct on probe trials no longer made a significant addition to the story. This supports a role for spatial cues late in training.

Several conclusions are warranted from this survey and analysis of the water escape task.

1. The rim height of the tank above the water should be as small as possible to provide an unimpeded view of spatial room cues. Tests in the author's lab established that 5 cm is a good value: no mouse was ever able to reach that far up the wall while swimming.
2. The platform should be round, so that the mouse encounters the same tactile cues, regardless of the direction from which it approaches the platform.
3. The percentage of the tank occupied by the platform should be substantial, perhaps 2 or 3%, so that finding the platform is not too difficult.
4. The tank should not be so large that the mouse becomes waterlogged while searching for the platform. A tank with a 70 to 100 cm diameter is sufficient.
5. There should be sufficient trials properly spaced so that most normal mice approach mastery of the swim task before a probe trial is given. If the tank is not too large or the platform too small, four trials per day for five days should suffice.

Clearly, task parameters can have a very large impact on performance. As indicated in Figure 13.6, half or more of the variance among studies in a performance measure were associated with parametric variations in three to four features of the task. Which parameters matter most depends on the specific measure of behavior. The analysis was done for the Morris task because the large literature made it easy to locate many relevant studies, and there was equally great diversity of parameters reported in the literature. There is also a large literature on the elevated plus maze, but the range of parameter values is much smaller. For example, most studies use opaque plastic arms of 6×25 or 30 cm. The literature is growing rapidly, so it should be possible to conduct similar reviews and analyses for several other tests.

TASK REFINEMENT

Despite the wide range of parameter values evident for most published tests, there are usually one or two values that are used most often. Unfortunately, the most common choices are not necessarily the best choices. Some popular tests have stubborn flaws that undermine validity. It may then be desirable to identify and remedy shortcomings (Wahlsten et al., 2003).

The accelerating rotarod

A case in point is the accelerating rotarod test of motor coordination (Figure 3.1) that was originally devised for studying drug effects on rats (Dunham & Miya, 1957) and later adapted for mice (Jones & Roberts, 1968). The mouse must walk and then run faster and faster to stay on the rod, and those that have longer fall latencies are said to be better coordinated. Several manufacturers sell versions of the device, one with a larger diameter rod for rats and another with a smaller rod for mice. Some use a smooth rod and others craft it with grooves that provide a better grip. The grip may be so good that the mouse can hold onto and wrap around the rod, then rotate with it, rather than walking on it, which yields longer fall latencies. That kind of grooved rod is not measuring the same aspect of motor behavior as one that is smooth so that the mouse must walk on top of it.

The obvious shortcomings with the test have given rise to several refinements, and the current version in use in several labs is proving to be effective for the study of ethanol effects on coordination in a wide variety of inbred strains. Four distinct steps in the process occurred.

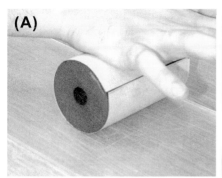

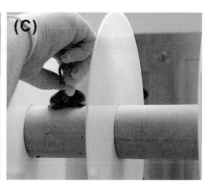

FIGURE 13.7
Techniques for applying 320 grit sandpaper to a 6 cm diameter rotarod. (A) The sandpaper is firmly attached using rubber cement. (B) The excess is trimmed from each end of a rod. (C) The rod is mounted for use with mice. Note that the seam in C is barely perceptible on the rod in the lane next to the mouse. A tight seam is essential for preventing grasping and wrapping on the rod. (*Photos courtesy of Andy Cameron and Dr. John C. Crabbe, Oregon Health & Science University.*)

First, a series of parametric studies was done by Rustay et al., (2003a,b) to find a way to force walking and prevent wrapping. Four refinements were devised (Figure 13.7).

1. The larger diameter rod commonly used for rats is actually the best choice for mice. Several diameters were evaluated, and 6 cm gave superior results. The 6 cm version can be machined from PVC plastic or aluminum and substituted for the rod provided by the manufacturer.
2. The rod should be covered with 320 grit sandpaper to provide traction for walking while preventing the mouse from grasping the rod.
3. The acceleration rate should be fairly rapid so that most mice fall within 60 sec without encountering fatigue. A rate of 20 rpm/min gives good results.
4. The fall height should be higher than in some mouse-sized models so that the mouse is more highly motivated to remain on the rod.

Second, when the rotarod was imported to the author's lab as part of a multisite study (Chapter 15), it was discovered that the technique to apply the sandpaper to the rod is critically important. Fall latencies were much longer in the Edmonton lab than in Portland or Albany (Wahlsten et al., 2003) because, as was learned after the study was completed, there was a narrow gap in the seam in Edmonton that allowed mice to get a good grip with one or two claws. A later visit by a technician from the author's lab to Portland solved this problem.

Third, a later study in the author's lab entailed four separate shipments of mice over two months. As shown in Figure 8.2, fall latencies became shorter in successive shipments in Edmonton. Thus, the sandpaper must be changed weekly. Otherwise, gradual wear from mouse claws and cleaning of the rod generate a smoother surface over time that makes it more difficult for the mouse to gain traction.

The fourth innovation was inspired by moving the author's lab and shipping rotarods from one place to another. We were not completely confident that the acceleration rate set on the device was indeed the true rate. Consequently, a simple means to calibrate acceleration rate was devised (Bohlen, Cameron, Metten, Crabbe, & Wahlsten, 2009). This proved to be valuable when, prior to publication of the report on the method, a spare rotarod was shipped from Portland to UNCG and smashed into pieces by the courier. We glued everything back together and then used the new method to calibrate the old rod.

There is one remaining difficulty with our current version: the mouse sometimes falls into the trough of bedding without breaking the photocell beams that stop a timer. Because the experimenter always watches the mice and makes notes on behavior on every trial, it is a simple matter to note the fall latency of each mouse using a stopwatch. Adding more photocells would solve the problem, but the current alternative is sufficient to yield good data.

The submerged platform water escape task

The Morris water maze has proven very effective for studies of spatial learning and memory in rats, but it has not been an unmitigated success when used with mice (Whishaw & Tomie, 1996). Some strains of mice are prone to passive floating if the animal cannot reach the submerged platform quickly, while others hug the walls and never find the platform at all (Gerlai, 2001; Gerlai, McNamara, Williams, & Phillips, 2002; Wolfer et al., 1998). Some mice learn to swim in an annulus a few centimeters away from the wall instead of utilizing spatial cues. Certain strains are prone to becoming waterlogged if more than three or four trials are given in one day. Some mice leap from the escape platform in an attempt to flee the entire situation (Wahlsten et al., 2005).

Given the stressful nature of the task, handling the animal very gently using a special scoop has been found to improve results. The investigator is urged not to adopt a method observed by Gerlai (1999) where the wet mouse was retrieved from the escape platform and then patted dry in a towel by the experimenter.

Another set of refinements altered the actual apparatus. The purpose of the task is to require use of spatial room cues rather than proximal cues in the maze to locate the hidden platform. Accordingly, several modifications were introduced to counteract some of the behaviors that defeated the investigator's aim of testing spatial memory (Wahlsten et al., 2005; see Figure 13.8).

1. Four clear plastic arms were fit into the circular tank, which prevented swimming in an annulus while allowing mice that hug a wall to reach the goal. The arms had to be sufficiently wide (at least 10 cm) because narrow arms prompted many mice to float passively if they could touch both walls at the same time.

2. In an initial version with four arms, mice often swam back and forth between two incorrect arms and found it difficult to turn a sharp corner while swimming rapidly. Accordingly, a larger central zone was created.

3. The submerged platform was located at the end of one arm so that it could be easily located if the correct arm was chosen using spatial room cues. This arrangement made it more likely that an animal would encounter the platform purely by chance early in training, which forestalled passive floating.

4. The platform was covered with a ventilated plastic lid that prevented jumping out of the apparatus, and the mouse remained there during the intertrial interval, reducing the aversive handling experience.

In a formal test, the refined apparatus was compared with the usual Morris configuration using identical room cues, the same escape platform, and the same video-tracking system in a well-balanced study. Not only were mice generally superior in learning the 4-arm version of the task but strain rank orders were altered, especially for BALB/cByJ. Only three strains showed a high concentration of search in the correct quadrant of the Morris task on probe trials, whereas all six strains with good eyes did well on the 4-arm task, including many 129S1/SvImJ and A/J mice that made no errors at all by the end of training. The 4-arm water escape task had higher face validity because many troublesome non-spatial behaviors were eliminated.

Prospects for refining other tests

Many other tests could benefit from small improvements. In the domain of motor behavior, the grip test is a prime candidate because so much depends on how rapidly and firmly the experimenter pulls the animal away from the strain gauge (Brooks & Dunnett, 2009). A mechanism that moves the strain gauge away from the stationary hand of the experimenter

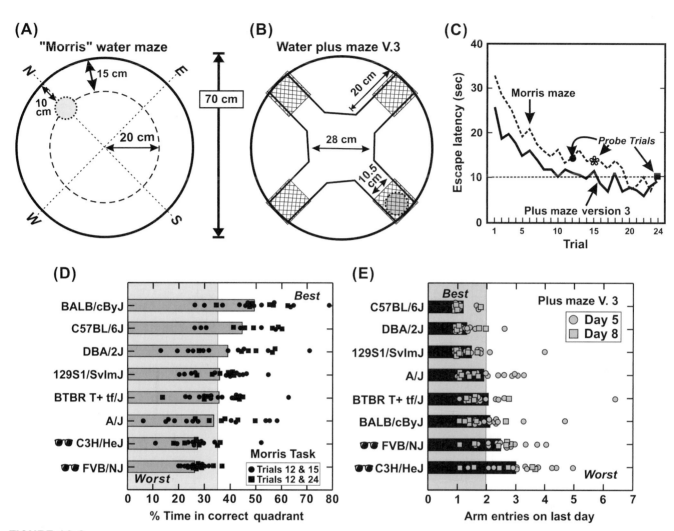

FIGURE 13.8

(A) The Morris water maze and (B) version 3 of a 4-arm maze that used the same tank, escape platform, and extramaze cues. (C) Escape latencies were generally faster in the 4-arm maze when averaged over all 8 strains. (E) More strains achieved good performance on the 4-arm maze than the Morris maze (D), and strain rank orders were different in the two tasks. Mice with retinal degeneration (dark glasses) did poorly on both tasks. Performance on two probe trials on the Morris maze was close to 25% for animals that failed to learn the task after either 15 or 24 training trials. For the 4-arm maze, excellent performance where only one arm was entered per trial was seen in many mice of four strains. *(Reproduced with permission from Wahlsten, Cooper, & Crabbe, 2005.)*

231

would probably give more consistent results. For the social recognition and preference test (Nadler et al., 2004), a protocol that requires less habituation time would aid throughput, and assessment of both recognition and preference in a single trial would be welcome. The 8-arm radial maze is beset by problems of serial or radial response patterns that are non-spatial (Crusio, 1999; Crusio et al., 1993). A means to disrupt those patterns while not alarming the mouse with noisy doors or gates would be helpful.

Several tests that have proven inadequate in some labs have been adapted successfully in others, albeit with considerable effort and ingenuity (O'Leary & Brown, 2008). The Barnes maze (Figure 3.2) requires a mouse to find a small hole in a large disk in order to escape from a bright light. In the author's lab, many mice remained immobile near the center of the disk or failed to enter the escape chamber at all, even after placing the snout into the opening. Other labs have found the Barnes maze serviceable and regard it as a less stressful alternative to the Morris water maze for assessing spatial memory (Harrison, Hosseini, & McDonald,

2009). Not all mouse strains adapt well to the Barnes maze (Patil et al., 2009), however, and a remaining challenge is to obtain valid data from a wider range of strains.

TEST STANDARDIZATION

A test should never be standardized on a set of parameters that is seriously flawed. Once a test has been carefully refined and evaluated in several labs, however, it is time to consider whether all labs should adopt that version as a standard (Blizard et al., 2007; van der Staay & Steckler, 2002; Wahlsten, 2001). If there is good reason for one lab to adopt a refined version for future studies, the same reason will apply to many labs. Because there is no governing body for mouse behavioral genetics and the field is diverse, nothing can be done to enforce a standard array of task parameters. The most likely fate of a seriously flawed version of a test is gradual extinction following critical comments by referees for journal articles and grant proposals. Informal peer-to-peer discussions at conferences can also be very effective in discouraging the use of certain methods, especially if bad versions are subjected to criticism.

Any further steps toward choosing a standard set of task parameters need to be taken very carefully because the specific parameters interact with genotype (Tucci et al., 2006). Which strain does best on a task can depend strongly on the details of apparatus configuration and timing (Wahlsten et al., 2005). It is conceivable that there is a different set of task parameters that yield optimal performance for each strain. A massive investigation would be needed to optimize more than one or two task parameters for several strains. Furthermore, if experiments on drug effects, for example, were then done so that each strain is tested with different versions of a task, this would almost certainly evoke objections that genetic differences in drug response are confounded by task differences. Within a study, good experimental design requires that the test given to every mouse be the same. With a different set of parameters, the patterns of genetic variation might be different. There is no easy escape from this conundrum.

One argument against test standardization is that it would deny us knowledge of the effects of a rich variety of alternatives for a test (Würbel, 2000; 2002). If standardization is done prematurely, before sufficient variants have been tried, much valuable information would indeed be missed. Now the central issue is how to deal with situations where abundant data on task variants do exist. As Figures 13.4 and 13.6 indicate for the Morris maze, variations in certain parameters can give rise to substantial differences in performance between labs. Certain combinations of parameters yield better results. There can be no cogent argument for continued use of features that are clearly inferior, such as cool water in the Morris maze (Figure 11.15).

Today the topic for debate is whether the field should seek to standardize *all* task parameters, including those that are not known to make a major difference in results. Standardizing a behavioral test requires adoption of specific values for every aspect of the physical apparatus and procedures. This practice requires some effort but not an unreasonable degree of diligence. Chapter 15 discusses several studies where the test is identical in multiple labs. As an example, Figure 13.5 highlights values of 12 parameters of the Morris maze that could serve well as a standard configuration. Some of the choices are dictated by the analysis of 100 studies in the literature or specific evidence from a controlled study, whereas others are chosen because they give decent results and are employed most often in different labs.

Two major advantages to adopting a standard set of values are apparent. First, a good standard will not include any values that are poor choices, and validity of results should therefore be high. Quality control is more or less assured. Second, because the test will be identical in many different labs, it will be far easier to identify aspects of the lab environment that may be giving rise to discrepancies in data. If each lab uses its own version of a test and also has an idiosyncratic lab environment, it will be almost impossible to find the causes of failures to replicate results. It seems likely that a standard would be most appreciated by those who are new to behavioral testing and do not care to go through the laborious process of choosing parameters on the basis of controlled experiments in their own labs. Newcomers are also less burdened by inertia from years of doing something a particular way.

There are probably experienced investigators who will persist in using their own kinds of apparatus and protocols, no matter what others decide. This fact does not decide the issue of standardization, because, unlike a political federation or union, there is no requirement that all investigators must do things the same way. The field can tolerate rebels and deviants. The non-conformists have a role to play, because they may stumble upon new and better ways of working. The objective of standardization is not to purge all variation. The hope is that the field will progress more rapidly if the extent of variation in how tests are constructed and conducted is less than the chaotic situation portrayed in Figure 13.5.

Among the major challenges to proper standardization are a paucity of scientists who specialize in the study of mouse behavior (Blizard et al., 2007) and a relative lack of funding for behavioral studies that are not focused on genetic or pharmacological factors. Psychologists have devised a sophisticated set of tools for refining and standardizing tests for use with humans, and many of those tools could be adapted for work with mice. It is not reasonable to expect molecular biologists who have created a novel mutation to then identify and ameliorate the shortcomings of a battery of behavioral tests before applying them for a phenotypic scan. The field of behavioral genetics could benefit greatly from increased attention by behavioral psychologists to methods for testing mouse behavior.

STANDARDIZING LAB ENVIRONMENT

The argument for test standardization cannot be applied readily to standardization of the lab environment. So many local factors influence the design and operation of animal facilities that uniformity is not a realistic option. For many factors, there is no compelling evidence that one way is better than another for the outcome of behavioral tests. Few details of husbandry and conditions in colony rooms are provided in most published reports, and the range of those that are described is often narrow (e.g., room temperature). In certain multi-lab studies described in Chapter 15, almost heroic efforts were made to standardize the lab environments, but many differences still intruded and they clearly influenced test results. For the mouse behavioral researcher, adopting a good standard for a test offers some real advantages, whereas efforts to standardize the lab environment do not seem worth the added burdens of cost and frustration.

Video Tracking

235

Photocells and microswitches are useful for monitoring many kinds of actions, but there are several situations where their use is limited. A mouse swimming in water cannot be recorded with photocells because most of the animal is submerged and waves generate visual noise. Photocell arrays work very well in simple devices such as the open field or light–dark box, but they present major difficulties when the device is complex. An 8-arm radial maze can be run with several photocells in each arm, but it cannot utilize the same photocell grid that works nicely with the square open field. Difficulties multiply when the environment becomes even more complex or has zones with opaque walls.

Video tracking can provide a convenient solution in many situations where a camera can view the entire scene from above (Noldus, Spink, & Tegelenbosch, 2001; Spink, Tegelenbosch, Buma, & Noldus, 2001). It is critically important that the mouse has no place to hide out of the view of the camera. There are now more than a dozen good video-tracking systems available for work with mice (Box 14.1), and software for performing many kinds of analyses is available (Wolfer et al., 2001). Open source software for tracking small animals has also been published (Aguiar, Mendonca, & Galhardo, 2007). Many reports have used video tracking to quantify behavior in a Morris water maze. Most systems can provide measures of the length of a path traveled by the mouse, latency to enter specific zones, time and distance in each zone, and number of entries into zones. Some systems claim to be able to distinguish between a mouse that is walking versus standing or grooming, and some can discriminate between very slow movement and genuine freezing or rigid immobility.

Several steps are involved. Most systems locate the mouse using background subtraction (Maddalena & Petrosino, 2008) and then calculate the center of the image.

Mouse Behavioral Testing. DOI: 10.1016/B978-0-12-375674-9.10014-X

14.1 COMMERCIAL VIDEO-TRACKING SYSTEMS

AccuScan Instruments	EZVideo
BIObserve GmbH	Viewer^2
Clever Sys	PhenoScan
Columbus Instruments	Videomex
Coulbourn Instruments	Big Brother
IITC Life Sciences	ANY-maze
Innovision Systems	MaxPRO
Leica Microsystems	Tracker
Neuralynx	CheetahVT
Noldus Information Tech	EthoVision XT
San Diego Instruments	ANY-maze
Stoelting Company	ANY-maze
TSE Systems	VideoMot2
ViewPoint Life Sciences	Videotrack

1. Before the mouse is introduced into the apparatus, a background image is stored. It is very important that lighting during this initial step be identical to that used when the mouse is present.
2. Then a mouse is added to the apparatus, every few milliseconds another image is saved, and the difference between the new image and the background is determined for every pixel in the array.
3. The entire difference array of N_X by N_Y pixels is scanned and, for pixels where the absolute difference exceeds some preset threshold, the X and Y values for that pixel are added to two counters (ΣX, ΣY) and the pixel counters (n_X, n_Y) are incremented by one unit. After the scan is complete, the coordinates of the weighted center of the image are found as $\Sigma X/n_X$ and $\Sigma Y/n_Y$.
4. A scale factor from an initial calibration of the image is then used to convert pixels to centimeters. Later, the vectors of scaled X and Y values for each image are used to determine path length, time in zones, and other measures.

BASIC FEATURES

Video tracking is a highly technical topic because of both the digital imaging device and the sophisticated software that is usually employed. Information about many aspects of a video camera and lenses can be found at the Web site maintained by The Imaging Source, whereas information on software is provided by each vendor. A few of the basic features of any system are discussed in the next sections.

Camera and lens

The essential features of a video tracking situation are shown in Figure 14.1. A solid-state digital camera with a charge-coupled device (CCD) detector and a lens is fastened rigidly above the apparatus so the entire apparatus is in the field of view. If the lens is D mm above the floor of the apparatus and the apparatus is W mm wide, the tangent of half the angle of view is $\tan \theta = (W/2)/D$. The required focal length of a lens (f) that can see the entire apparatus must be less than or equal to $f = (D \times CCD)/(W + CCD)$. For the situation in Figure 14.1, the lens should have a focal length of about 4 mm. If a longer focal length lens is to be used, the camera must be farther from the apparatus. The format of the CCD chip is an important consideration because the size of the chip and the focal length determine the angle of view. Common video formats in the United States include chips that are 1/4", 1/3", and 1/2", whose smaller widths are 2.4, 3.6, and 4.8 mm, respectively.

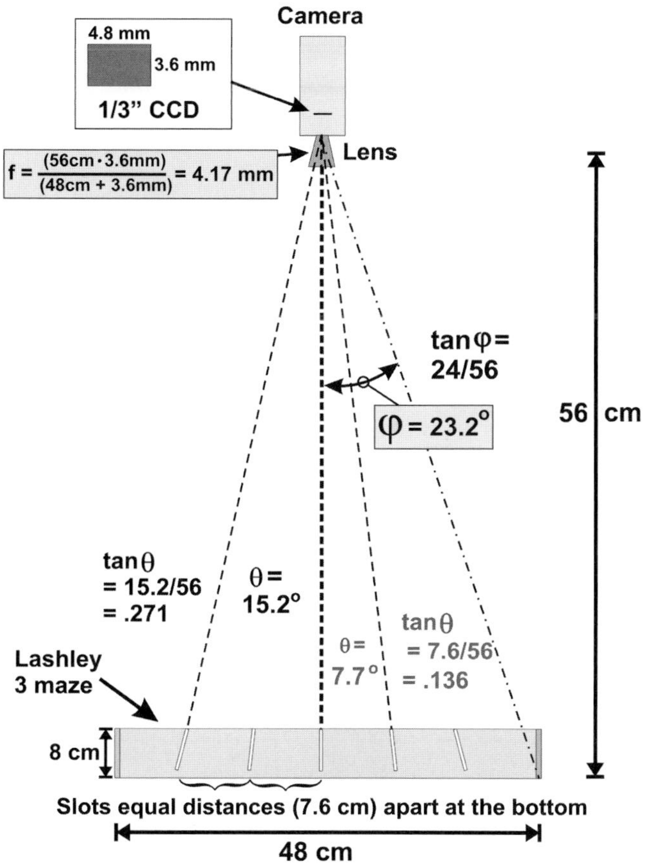

Camera

4.8 mm

3.6 mm

1/3" CCD

$f = \dfrac{(56cm \cdot 3.6mm)}{(48cm + 3.6mm)} = 4.17\ mm$

Lens

$\tan\varphi = 24/56$

$\varphi = 23.2°$

56 cm

$\tan\theta = 15.2/56 = .271$

$\theta = 15.2°$

$\tan\theta = 7.6/56 = .136$

$\theta = 7.7°$

Lashley 3 maze

8 cm

Slots equal distances (7.6 cm) apart at the bottom

48 cm

FIGURE 14.1

Arrangement of the video camera and a Lashley 3 maze. The focal length (f) of the lens needed to include the entire maze in the field of view is determined by the width of the CCD chip (3.6 mm for a 1/3″ camera in this example), the width of the maze, and the distance of the camera from the maze. The maze is divided into six alleys by five partitions inserted into slots, seen in an end view. Each alley has the same width at the level of the floor. Slots are cut at an angle so that each partition is aligned with the camera and does not obscure the view of the mouse.

Many kinds of lenses are available. Zoom lenses allow the user to set the field of view to frame any apparatus nicely, but there is a danger that the zoom setting may shift slightly with days and weeks of use, so that the scale factors may not be the same for all mice. It is much safer to use a fixed focal length lens. Some lenses have an aperture that is adjusted automatically by the camera to give the proper amount of light on the CCD, while others allow the user to set the opening manually. Because the lighting of a particular apparatus is usually constant across different animals, the manual option is effective and cheaper.

Spectral sensitivity and filters

Human and mouse eyes can see red light but not when the wavelength exceeds about 720 nm. The rods that are sensitive to low levels of light are especially effective with blue-green light and have little response to wavelengths longer than 650 nm (Salchow et al., 1999; Williams et al., 2005). The normal mouse retina has cones that extend spectral sensitivity beyond 600 nm but not beyond 700 nm. The typical CCD is sensitive to very long wavelength or infrared (IR) light extending to 1,000 nm (Figure 14.2A). This can be troublesome if there is an unwanted source of IR light in the field of view. Consumer video cameras often have a filter over the CCD surface that cuts out all IR light, which will improve image quality on broiling hot days in a desert. Rarely is there much ambient IR light in a mouse lab. Room lighting is usually done with fluorescent lights that emit strongly in the green part of the spectrum. IR light is abundant

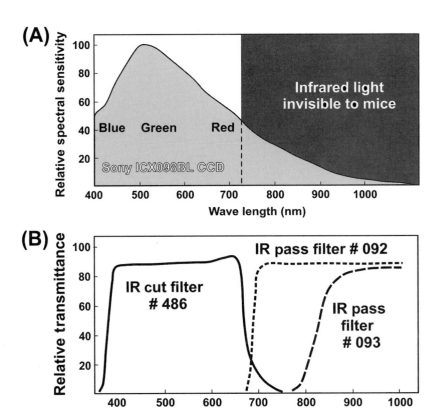

FIGURE 14.2

(A) Spectral sensitivity of a CCD chip made by Sony and used in DMK cameras. The CCD is sensitive to IR light that cannot be seen by mice. (B) Transmittance of filters that can pass visible but not IR light (#486) or block visible but not IR light (#092, 093). *(Adapted with permission from The Imaging Source; www.theimagingsource.com.)*

when a tungsten (incandescent) bulb is illuminated using low voltage. The investigator can demonstrate this effect by aiming a CCD camera without an IR cut filter at the tungsten bulb and then gradually increasing the voltage across the filament. The camera will sense the IR light coming from the bulb when the human eye cannot see any change.

It is highly recommended that the mouse researcher use a CCD device that has no built-in IR cut filter. If the application would benefit from suppressing all IR light, a separate cut filter can be attached temporarily to the lens (Figure 14.2B). The sensitivity of the CCD to IR light can be used to great advantage in many difficult applications of video tracking technology. In fact, if the part of the apparatus containing the mouse is deliberately illuminated with IR light, the camera can be made to sense only the IR and ignore all other room light. An IR pass filter such as #093 in Figure 14.2 allows no visible light to reach the CCD. Shadows from the experimenter moving about the room under fluorescent lights do not affect the image. The appearance of the mouse will be the same regardless of the phase of the light—dark cycle in the test room.

Focus of the lens

When the lens is focused on the mouse, many fine details can be resolved, such as the animal's whiskers. High resolution images with a CCD chip that has many pixels can be very impressive in a photographic exhibition, but they can generate behavioral noise in the image used for tracking the animal. Movement of the whiskers may not help to understand the behavior being studied, unless the goal is to assess freezing behavior. Video tracking in most kinds of apparatus is designed to detect gross motor movements of the entire animal. This may be better realized by having the lens slightly out of focus so that the mouse appears as a blob with soft edges. The researcher needs to experiment with different settings in each situation to find which ones give the best, most consistent video tracking results.

HARDWARE MODIFICATIONS FOR VIDEO TRACKING

Results can be enhanced if the apparatus is specially adapted for use with video. One of the greatest challenges is the contrast of the mouse's coat color with the background. Low contrast objects result in very small differences from background so they are difficult to track. In a study with several inbred strains or a hybrid cross segregating at several loci affecting coat color, there will be a wide range of object reflectances, and it is likely that one or two of them will be very close to background brightness. This problem is greatly compounded if the apparatus has a wide range of brightness levels, especially shadows. Certain colors of mouse may literally disappear in the camera view in some places.

Solid floors for uniform fields of view

The open field is a situation ideally suited to video tracking because there are no objects except the mouse, and diffuse lighting can be set so there are no shadows (Kafkafi et al., 2003). There will always be a floor, however, and this can generate trouble. Mice defecate, and a system with high resolution may detect boli and include their X, Y values in the computation of the mouse center. More sophisticated software has a routine to exclude these small "islands" of differences, but others do not. Even worse, some labs place bedding on the floor of the open field. Mice will dig and push the particles, and the video system will detect and track these changes from background. If a white plastic floor is used, albino mice will be difficult to track, while black mice will disappear against a black plastic background. In the author's lab, it has been found that *pink* butcher's wrapping paper placed under the apparatus provides good contrast with almost any mouse when the video system uses RGB color, and results are passable using gray scale, the possible exception being the light gray of C57L/J. The paper has a dull and a waxed, shiny side. The waxed side must face the camera, because urine spots on the porous paper of the dull side instantly turn dark and affect tracking.

Paint in the water tank

Water escape mazes typically have white tempera paint added so the mouse cannot see the escape platform that is usually submerged 5 mm or more below the surface. The result is that albino mice can be exceedingly difficult to track with most video systems. A recent study declared that albinos could not be tracked automatically (Lad et al., 2010). In several labs, the situation has been improved a little by adding blue tempera paint to the white (1 mL white plus 0.04 mL blue per L of water), just enough to provide contrast with a white mouse (Wahlsten et al., 2005). This can lead to problems finding very light gray or light tan mice, however.

Eliminating shadows and blind spots

With careful placement of diffused lights, mice can always be seen and shadows can be greatly reduced in apparatus that are radially symmetrical, such as the elevated plus maze or the 8-arm radial maze. Other devices can be troublesome. In the author's lab a rectangular Lashley 3 maze with six alleys was created using five partitions having holes to permit passage through the maze. Ordinarily the partitions would be aligned vertically, but that configuration meant that a mouse hugging the outermost partition would be almost completely obscured by the partition. There were so many tracking errors in the first version of the device that a new one was built with sloping partitions. It was constructed so that each alley was 7.6 cm wide at floor level. Then all but the center partition were sloped so that each partition was aligned with the view from the camera (Figure 14.1), giving the mouse no place to hide from the camera.

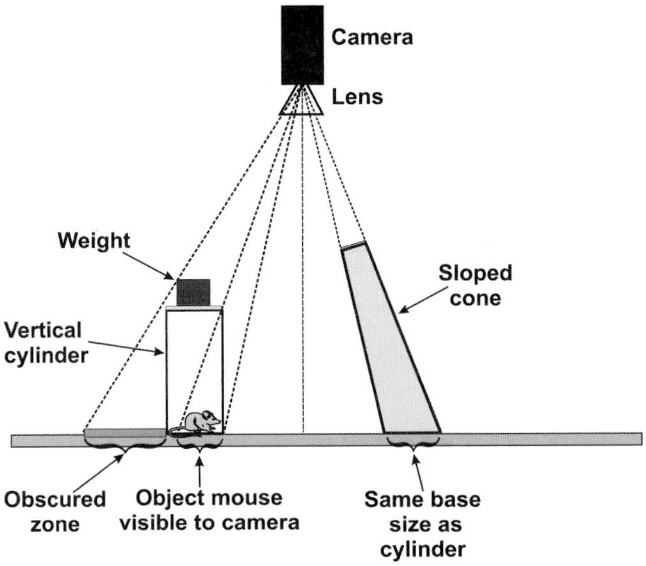

FIGURE 14.3

Social recognition apparatus viewed by a video camera. On the left is a clear plastic cylinder that houses an object mouse. The weight prevents the subject mouse from climbing onto the cylinder. With a vertical cylinder, the camera will not be able to view a subject mouse on the far side of the cylinder but will be able to view the object mouse in the cylinder, which will contaminate the video track of the subject mouse. On the right is a sloped cone made of clear plastic with an opaque cap. The camera will always have a clear view of the subject mouse but not the object mouse.

A similar problem arises in the social recognition or preference task where object mice are confined in a ventilated cylinder and time spent by the subject mouse near each object is determined (Macbeth, Edds, & Young, III, 2009; Moy et al., 2007; Page, Kuti, & Sur, 2009). The most common device for this test uses photocells to monitor entry into three chambers, two of which contain object mice (Nadler et al., 2004). This test can be adapted for use with video tracking, but the cylinder creates a hidden zone where the subject mouse cannot be seen, and a clear cylinder also allows the camera to see movements of the object mouse (Figure 14.3). These difficulties can be solved by making a sloped cone whose walls are aligned with the view from the camera. The cone needs to be tall enough so that a subject mouse cannot climb onto its top, and the top must be blocked by an opaque disk so that the subject mouse cannot be seen by the camera.

The experimenter's hand

Many behavioral tests begin when the experimenter places the mouse into the apparatus and releases the grip on its tail. This can create trouble if the video system is set to start a trial when an object suddenly appears in the field of view, because the largest object will be the experimenter's hand. Shadows cast by the experimenter standing near the apparatus can also affect video tracking for the first few seconds. Some systems allow the experimenter to start a trial by hitting a key on the keyboard or pressing a button on a switch with a cord extending to the computer, so that the trial is not started until the person has backed away from the apparatus. A different kind of problem can then be caused by eager mice that dash down an alley before the start signal is given. Indeed, in the Lashley 3 maze used in the author's lab with a push button start, many mice of the more active strains that learned quickly had already entered the second alley before the official trial began.

Thus, it is important to start the trial after the experimenter's hand is out of the field of view and in a way that does not allow the mouse to move in the field of view until the trial begins. A small, opaque start box that can be opened remotely by an electromagnet or cord can address these problems in some apparatus. The actuating mechanism should be very smooth and free from loud clicks or other noise that could alarm the animal. Such a device is not practical in a water maze, and the simple push button start should work well enough with a large water tank. A baffle can be positioned near the tank so that the experimenter's hand can be seen only when it is directly over the water.

DOUBTS ABOUT THE ACCURACY OF VIDEO TRACKING

Experience with three video tracking systems in the author's lab has revealed several short-comings in each and eventually led to a solution that should be applicable to a wide range of situations.

Tracking in Edmonton and Portland

As part of the Mouse Phenome Project, two labs ran experiments using the same 21 inbred strains in physically identical apparatus with a common test protocol. Three of the tests utilized video tracking: open field activity, elevated plus maze, and water escape maze. Both labs used the VideoScan system from AccuScan Inc. and took care to use the same interface cards, cameras, and lenses. The open field tests were done in an enclosed cubicle that provided the same fluorescent lighting in both labs, whereas the two mazes were exposed to room cues and had somewhat different lighting at the two sites. Results for distance traveled and speed of movement in the open field were almost identical (Wahlsten et al., 2006; see Figure 14.4A), which suggested that the video systems were giving the same readings in the two labs.

The situation was not as comforting in the 4-arm water maze. Escape latencies declined rapidly over trials, and they reached almost identical asymptotes in the two labs (Wahlsten et al., 2005; see Figure 14.4B). The measures of distance swum and swim speed, however, were almost 50% higher in Portland than Edmonton (Figure 14.4C). After a long e-mail correspondence, the only plausible source of the difference was the lighting of the water tank. The lighting was clearly different in the two labs owing to different kinds of ceiling and supplementary light sources, and room cues differed in many ways. There had been a long series of preliminary tests in each lab to find a configuration that yielded consistently good tracking with almost all mice, and the two labs arrived at somewhat different solutions. It appeared that the Portland arrangement allowed ripples caused by the swimming mouse to affect tracking, but we were never able to prove this.

241

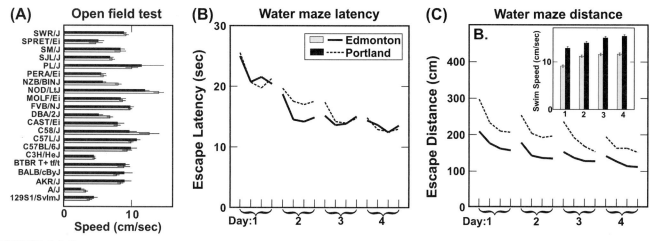

FIGURE 14.4
Results from video tracking of 21 inbred strains in laboratories in Edmonton, Alberta, and Portland, Oregon. (A) Results (mean ± standard error) in the open field were nearly identical in the two labs. (*Based on data from Wahlsten, Bachmanov, Finn, & Crabbe, 2006.*) (B) Escape latencies in a 4-arm water maze were also very similar in the two labs. (C) Estimates of swim distance and speed for the same mice as in panel B differed substantially between labs because of differences in lighting of the water maze. (*Based on water maze data reported in Wahlsten, Cooper, & Crabbe, 2005.*)

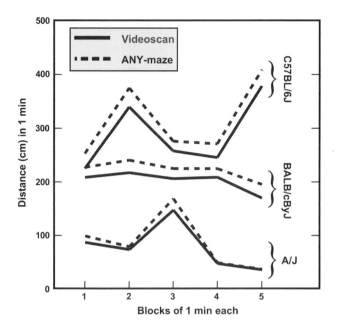

FIGURE 14.5

Average distance traveled per minute in a 5 min trial in an open field by mice of three inbred strains, estimated using two video-tracking systems receiving identical signals from the same camera. Each pair of curves represents the mean of two mice. *Data collected in the author's lab at the University of Alberta.*

Lighting of the apparatus is critically important for good video tracking (Spink et al., 2001), and any investigator who has worked extensively with these systems knows this very well. The usual procedure in any lab is to try various arrangements of lights and baffles until the paths visible on the computer screen look good enough to the experimenter. Whether the measured path length is the true one does not matter in studies of a treatment versus a control condition. Lighting will always be the same for both groups within a study, so that any artifact in the data affects the groups equally. Consequently, the field has not found a need to ensure accuracy of path lengths. Only when data are compared across labs does the discrepancy raise eyebrows.

VideoScan versus ANY-maze

The studies in Edmonton had all been done with the VideoScan system. We were confident in the data but were interested in finding a system that could cope with ripples in the water maze. ANY-maze software was used with the same hardware that worked with VideoScan. A test was done to compare the two systems. A video amplifier received input from one camera viewing one mouse in an evenly illuminated open field, and it sent identical signals to two computers running different video software. Results were similar but they were not identical (Figure 14.5). Path lengths were consistently longer for the ANY-maze program. Which system was correct? At the time, there was no way to determine this.

ACCURACY AND PRECISION FROM A MECHANICAL DEVICE

After several years of frustration while working with video tracking in water mazes and complex apparatus such as the Lashley 3 maze, we finally decided to study the video systems instead of the mice (Bailoo et al., 2010). The systems were sufficiently complex that they almost felt like living organisms to the experimenters who ran the tests. Indeed, for certain lighting and object conditions, measures were remarkably variable under the same conditions when ten consecutive trials were done. It was possible to discover conditions that yielded very precise and stable data as well as conditions where video tracking failed miserably.

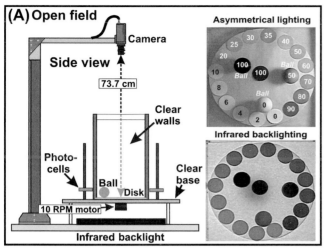

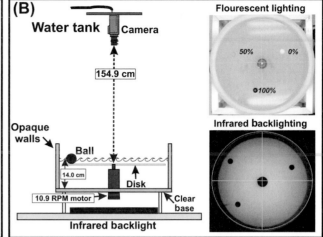

FIGURE 14.6

Apparatus used to determine the accuracy and precision of video tracking systems. (A) Coulbourn TrueScan open field was adapted by turning a plastic disk with a 10 rpm motor at the level of the floor. A paper disk ranging in percent gray from 0 to 100 or a ping pong ball (white, gray, black) was placed on the plastic disk and observed for 10 trials. Lighting with tungsten bulbs from one side of the apparatus generated strong shadows that impaired tracking, whereas IR backlighting provided very good contrast of all objects with the background. (B) Circular water maze with a submerged plastic disk driven by a motor. A white, gray, or black ping pong ball was mounted on the disk so that the water level was midway up the ball. Visibility of the ball was sometimes poor under fluorescent room lights but was very good with IR backlighting. *(Reprinted with permission from Bailoo, Bohlen, & Wahlsten, 2010.)*

The key to the methodology was the use of an inanimate object that moved in a known path at constant speed trial after trial. There is no way that a live mouse will ever repeat a path in any apparatus. If two video systems view the same mouse at the same time (e.g., Figure 14.5) and measures are different, there is no way to determine which is more precise. The solution is to move objects in perfect circles on motor-driven disks in an open field and a water tank (Figure 14.6). By measuring the diameter of the circle and motor speed with great accuracy, the true length of the path of the center of an object can be determined. Accuracy is then indicated by the deviation of the measurement from the true path length. The simple rotating disk can be duplicated in any lab and used to assess the qualities of any lighting system.

Trials were first run in an open field using three lighting conditions (Figure 14.6A): overhead fluorescent room lights that gave even illumination, asymmetrical lighting from tungsten lamps on one side of the apparatus that cast shadows, and IR backlighting. Two excellent video tracking systems were compared: ANY-maze and Ethovision XT. The first series of tests used non-glossy, flat paper disks to assess the importance of contrast with background. Then white, gray, or black ping pong balls were attached to the disk. For both video systems, there were shades of gray that simply could not be tracked precisely or consistently under fluorescent lighting (Figure 14.7A). Results with asymmetrical lighting (not shown) were much worse; both systems often lost track of certain kinds of gray disks in zones near a wall. With IR backlighting, on the other hand, results were generally very good and were independent of object reflectance.

Three shades of balls were moved in a circular path through the water (Figure 14.6B), and both video systems encountered severe difficulties with white or gray balls when lighting was asymmetrical (Figure 14.7B). Results with IR backlighting were outstanding, regardless of object brightness.

When live mice of different coat colors were tested in the most difficult situations, the water maze and the tunnel maze with five partitions under fluorescent room lights (Figure 14.7C),

243

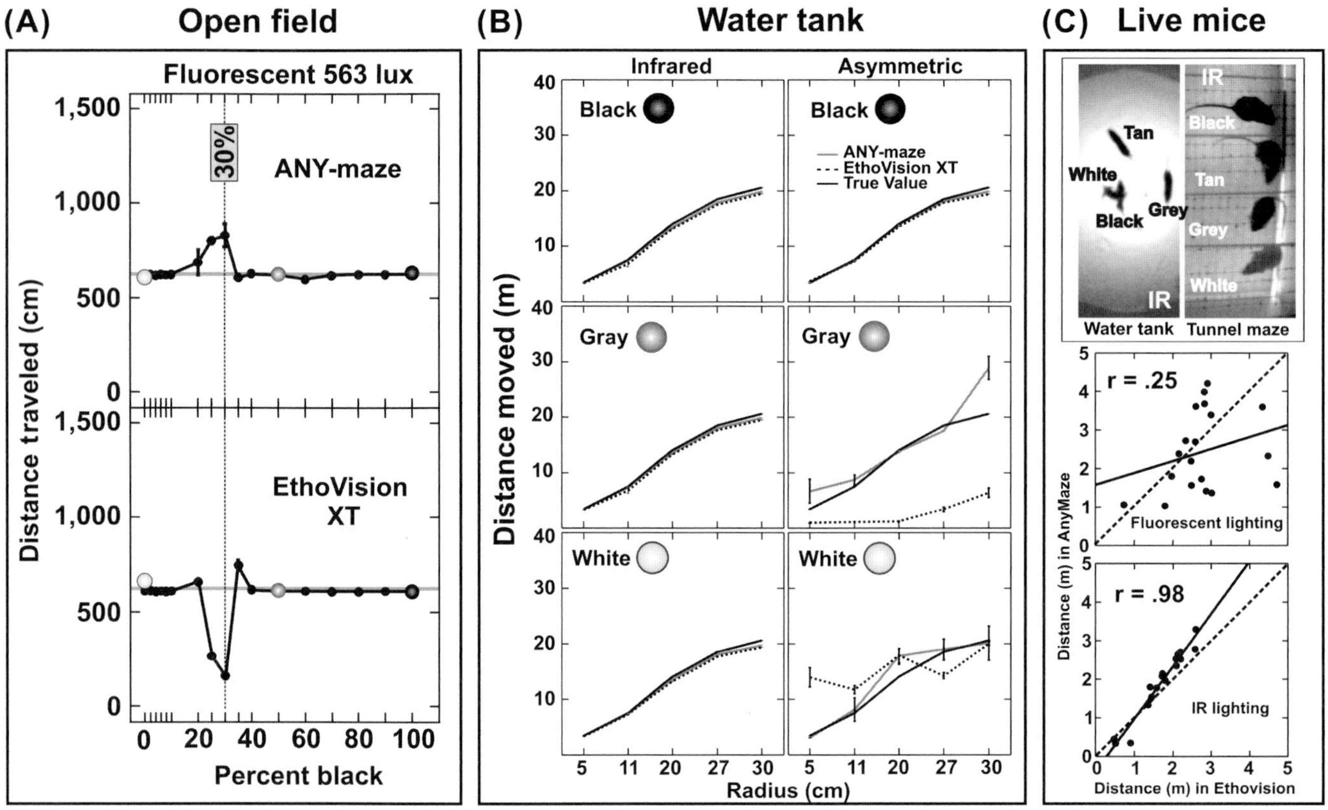

FIGURE 14.7

Results of trials using apparatus in Figure 14.6, where identical signals were fed to two computers running different tracking software. (A) In the open field, both video systems had difficulty tracking objects of 20 to 30% gray. The gray line is the true path length. (B) In the water tank, both video systems gave excellent results for a black ball but had many difficulties tracking gray and white objects when lighting was provided by tungsten bulbs from one side of the tank. Results with IR backlighting were excellent for all three balls and both video systems. (C) Trials were run in two apparatuses for 20 mice having black, white, tan, and gray coat colors. Under fluorescent ceiling lights, the two video systems gave markedly different results, whereas agreement was excellent with IR backlighting. *(Reprinted with permission from Bailoo, Bohlen, & Wahlsten, 2010.)*

there were remarkable discrepancies between the measures from ANY-maze and Ethovision XT. Results were very similar, however, with IR backlighting.

THE SOLUTION: IR BACKLIGHTING

Extensive experience with three video tracking systems employed with many kinds of apparatus has led to several conclusions. The solution to tracking problems will not be found in improved software, although improvements will no doubt give better results. Better apparatus design will not fix all the flaws, although better designs will eliminate some of the grosser errors. The only really effective solution in most situations is improved lighting. IR backlighting makes mice of all coat colors appear the same to the camera, and it greatly reduces problems arising from shadows and asymmetric incident lighting. It also eliminates the vexatious problem of ripples on the surface of a water maze.

For many of the common tasks used in neurobehavioral genetics where gross movement of the mouse through the apparatus is measured, IR backlighting yields excellent results with video tracking since mice are not influenced by low intensity IR light (Zurn, Hohmann, Dworkin, & Motai, 2005). Table 14.1 indicates tasks that are well suited to IR backlighting and video. Minor modifications of the usual designs will be needed in most cases. Many kinds of apparatus such as the mazes for testing anxiety and learning typically have an opaque floor that does not permit passage of IR light. The floors can be replaced with either clear lucite or

TABLE 14.1 Tests That Can be Adapted for Scoring with Video Tracking

Domain (Subdomain)	Abbreviation From Table 10.2
Anxiety models	DBT, EPM, ESQM, LD, NEOO, ZM
Depression models	LH, PFS, TS
Exploration	BT, HCA, OEX, OFA, YMSA
Learning (classical)	CPP, FCA, FCC
Learning (habituation)	OEX
Learning (maze)	HWM, L3M, LABM, TMF, TUM, WM, YMF
Learning (shock)	AA, GNG, IA, SDA, SHA, TMAA, YMA
Learning (spatial)	BM, MWM, RAD
Motor (fixed)	DT, GT, HW, ST
Sensory (vision)	VC
Social (fixed)	SRP

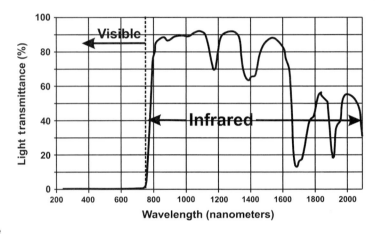

FIGURE 14.8
Transmittance of plastic (9C20 IRT) that blocks visible but not IR light. Many kinds of apparatus can be adapted for IR backlighting using this kind of plastic. *(Redrawn with permission from Acrylite GP technical bulletin, Evonik Cryo LLC; www.acrylite.net.)*

polycarbonate, which can be adapted by sanding the bottom surface so that it diffuses light and prevents the mouse seeing through the floor. Alternatively, the floor can be made from plastic that transmits IR but not visible light (Figure 14.8).

THE TRUE PATH LENGTH

A live mouse ambulates through a maze in what, from a distance, appears to be a smooth course. When it stops to sniff an object, there seems to be little motion except in its head. Viewed from closer range, however, there is lateral movement (wobble) of the center of the animal as it progresses and small movements of the center of its image when it pauses and only its head moves. The net result can be a center of the image that moves in a jagged line from frame to frame (Figure 14.9A). In most systems, there is also a level of noise in the video record that can affect the measured path length to a surprising extent. One study of mice in a large open field using EthoVision found that the sequence of raw, unfiltered XY coordinates for an anesthetized mouse indicated a total path length of 94 m in a 30 min trial (Hen, Sakov, Kafkafi, Golani, & Benjamini, 2004). The random fluctuations in X and Y coordinates were rarely more than 0.5% of the field width, but at a rate of 25 frames/second they added quickly to much larger values.

Software can smooth the jagged line to obtain a more general trend of movement through an environment. The smoothing algorithm will affect the measured path length

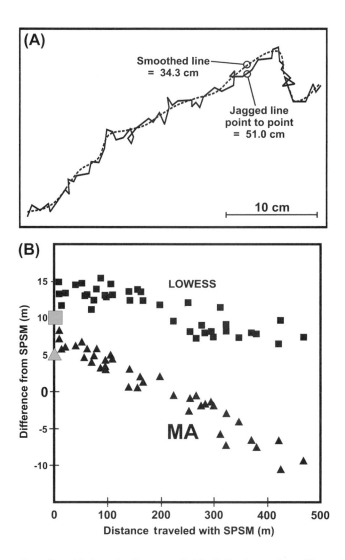

FIGURE 14.9

(A) Jagged line joining centers of an object moving in a maze that is digitized every few milliseconds. In this hypothetical example, the jagged line is 51.0 cm long. The smoothed line is substantially shorter. The amount of smoothing can have a strong effect on distance traveled by a mouse. (B) Comparison of three smoothing algorithms applied to paths of 40 mice from five inbred strains. The most sophisticated method (SPSM) is compared to a simple moving average (MA) and the polynomial regression method with outlier rejection (LOWESS). Large gray symbols represent values for an anesthetized DBA mouse. Only the SPSM represented that animal as not moving during the trial. (*Adapted with permission from Hen, Sakov, Kafkafi, Golani, & Benjamini, 2004.*)

(Figure 14.9A). If and how this is done is sometimes hidden deep within the software. The most appropriate degree of smoothing depends on what the investigator wants to know about the behavior in question. If there is interest in tremor in a neuromuscular mutant, the fine movements will matter. On the other hand, exploration of a radial maze is well described by the smooth, molar path. Much depends on the point of view. Hen and colleagues (2004) compared several smoothing algorithms with five inbred strains of mice. The simplest method was a moving average of contiguous points, whereas the Locally Weighted Scatter Plot Smoothing (LOWESS) method fit a polynomial equation to the points and rejected outliers. The SEE Path Smoother (SPSM) combines two methods, using one for active excursions and another for episodes of "lingering" when there are small movements. Especially for smaller and slower movements, the differences in path length between the methods were 10% or more (Figure 14.9B).

Behavior shares some features in common with inanimate entities whose measures depend on the scale of magnification. A classic example is the coastline of England. How long is that coastline? As Mandelbrot (1983) suggested: "coastline length turns out to be an elusive notion that slips between the fingers of one who wants to grasp it." An estimate can be made by measuring on a map or from a satellite image. If the magnification is then increased, many smaller irregularities will be seen, and tracing along these will greatly increase the total length. Zoom in further until large stones and wharves can be seen, and the length will zoom upwards, while uncertainty is churned by wave action of the sea against the shore. Behavior is similar in that the length of a path traveled by an animal is determined to some extent by the experimenter's point of view.

CHAPTER 15

The Laboratory Environment

CHAPTER OUTLINE

249

Environment in the broadest sense refers to things outside the organism that impinge on it. There is no doubt that many environmental factors can influence scores on a behavioral test. Studies of inbred strains invariably show that the variance within a strain is substantial compared with variance among strains, and much of that within-strain variance probably reflects environmental influences. Some of the environmental effects may arise prenatally from the uterine distribution of male and female fetuses (Ryan & Vandenbergh, 2002; vom Saal, 1981), a source of variation in behavior that can never be eradicated or reduced. Many other environmental influences can be controlled by the researcher, albeit with some difficulty.

TWO SOURCES OF ENVIRONMENTAL VARIANCE IN TEST SCORES

The score on a behavioral test can be influenced by factors that act during the test as well as features of the local lab environment that influence the animal before the test is given. The *test situation* includes the physical test apparatus and test procedures outlined in a protocol. These are things that could easily be equated across laboratories, should the investigators care to do so. The global *lab environment* includes everything that impinges on the animals before testing begins, some aspects of which also continue during the actual test (e.g., lab temperature and air). Like most formal dichotomies, the distinction between test situation and lab environment is challenged by a reality that seems to span categories: the experimenter. The experimenter who administers the test can influence the mouse only during the test, yet experimenters always differ among labs and can never be equated.

The lab environment has numerous dimensions, and is very different across laboratories. It is distinct from environmental influences that act differently on littermates, things such as uterine position and differential licking by the mother. The lab environment tends to impinge in the same way on all mice in the experiment.

Mouse Behavioral Testing. DOI: 10.1016/B978-0-12-375674-9.10015-1

ENVIRONMENTAL EFFECTS ON TEST SCORES: EARLY STUDIES

In a remarkable coincidence, the first studies on genetic strain differences in mouse behavior were undertaken independently by two investigators who chose to study the same phenotype in the same strains of mice (Ginsburg, 1967). Scott (1942) did a preliminary assessment of several strains and then focused his work on three: C57 black, C3H agouti, and the Bagg C albino, plus limited observations of strain A (standard strain names were not yet defined). Ginsburg and Allee (1942) also worked with C57 black, C3H agouti, and the Bagg C albino. Scott employed two test procedures: resident-intruder where a strange mouse was introduced into the home cage of the male to be tested, and neutral cage pairing, where two mice were placed into a novel cage for the first time (Maxson, 1992). Ginsburg used the neutral cage method. There were a number of other differences in testing methodology and husbandry, and they are summarized in Table 15.1. To the surprise of everyone, the studies reached opposite conclusions about which strain of mouse was the most socially aggressive and likely to win a mouse fight.

In the extensive study by Ginsburg and Allee (1942), mice were subjected to a controlled series of defeats or victories, and the experiences changed their conduct. An initially passive male that fled from an attacker and cringed in a corner could be transformed into a valiant warrior. Ginsburg (1967) later found that the way the males were handled prior to testing also had a large influence. In Scott's work, the mice were picked up by the tail with a forceps and carried to the test cage suspended from the forceps, whereas in Ginsburg's work the male was urged to enter a small box that was used for routine weighing, so that it was not directly handled by the experimenter prior to an encounter. Ginsburg found that handling by forceps substantially reduced the tendency of C57 mice to fight in the subsequent test. Thus, the first published studies on inbred mouse strain comparisons of behavior involved a very large heredity by environment interaction. After the work of Scott and Ginsburg, scientists working with fighting mice have been especially careful to control the social and testing environments when studying the role of genetic differences (Maxson, 1992).

The selective breeding of rats for high and low performance on a complex learning task by Tryon (1940) is legendary in behavior genetics. After 10 generations of selection, the maze Bright and Dull lines differed greatly, to the extent that there was very little overlap in the distributions of error scores of the lines. Many psychologists at the time interpreted the result as evidence that rat "intelligence" is hereditary. Tryon's student Searle (1949) evaluated this notion by testing the Tryon lines on other kinds of learning tasks. On the original Tryon alley maze the lines differed greatly (Figure 15.1), whereas the difference was considerably smaller on two elevated alley mazes having 14 or 16 units. There was virtually no line difference on a discrimination task, and the Dulls were markedly superior on an underwater T maze (Figure 15.1). Searle concluded: " … a 'general intelligence' factor, if it exists at all, may be regarded as of little or no importance. … differences in the maze-learning ability represent differences in *patterns* of behavior traits rather than in *degrees* of any single psychological capacity."

TABLE 15.1 Two Studies of Mouse Fighting Behavior

	Ginsburg at University of Chicago	Scott at Jackson Labs
Result: C3H vs Balb C	C3H 41% active winners, 7% defeated	C3H always attacked the stranger
Result: C57 vs Balb C	C57 61% active winners, 3% defeated	C57 never attacked the stranger
Handling to bring mouse to test cage	In small weighing box, not directly handled	Suspended by the tail with forceps
Criterion to end a trial	Stop if fight becomes severe	Stop as soon as fight starts

Sources: Ginsburg & Allee, 1942 and Scott, 1942.

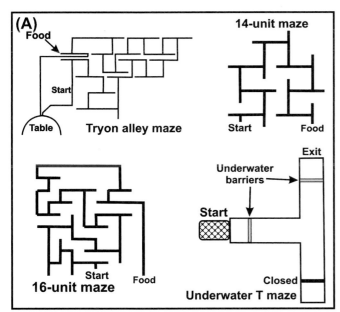

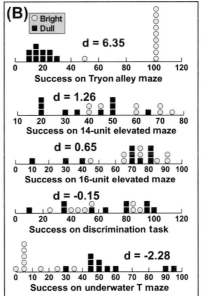

FIGURE 15.1

Data from Searle who used two lines of rats selectively bred by Tryon for performance on a multiple-unit alley maze and then trained them on different kinds of tasks shown in (A). For tasks less similar to the original Tryon maze, line differences were smaller or even reversed, as shown for scores by individual rats in (B). *(Redrawn with permission from Searle, 1949.)*

Two lines of rats were selectively bred at McGill University for high and low performance on the Hebb-Williams mazes (Rabinovitch & Rosvold, 1951), and they became known as the McGill Maze Bright and Maze Dull lines. Hughes and Zubek (1956) assessed rats from the 11th generation of selection in the usual lab environment and found a large line difference. Cooper and Zubek (1958) then reared independent samples of the two lines from the 13th generation of selection in either a restricted or an enriched lab environment. The effects were strongly dependent on the line (Figure 15.2). Impoverished rearing impaired the Bright line but not the Dulls, while enrichment benefited the Dulls but not the Brights. The net result was that there was no selected line difference in either the impoverished or enriched condition, only in the lab environment that was applied during the course of selection.

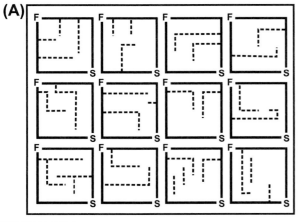

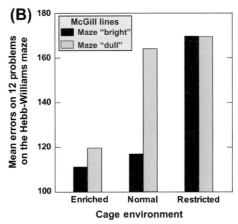

FIGURE 15.2

(A) The 12 problems on the Hebb-Williams maze *(adapted with permission from Rabinovitch & Rosvold, 1951)*. (B) Mean error scores of the McGill "bright" and "dull" selected lines of rats when housed in different kinds of lab environments. *Based on data in Hughes & Zubek (1956) and Cooper & Zubek (1958).*

The late 1960s and early 1970s witnessed a growing interest in quantitative genetic analysis of animal behavior. For example, Henderson (1970) compared six inbred strains and all 30 of their reciprocal F_1 hybrid crosses on a complex exploration task. Half of the animals were reared from birth in the usual small plastic mouse cage containing only bedding, while half were reared in larger enriched cages containing several objects, a maze, and a ramp. The genetic architecture of the exploration times depended greatly on the rearing environment: the F_1 hybrids were markedly superior to their inbred parent strains only when reared in the complex environment (Figure 15.3). In Henderson's parlance, impoverished rearing "obscured" the manifestation of genetic individual differences.

Maxson, Ginsburg, and Trattner (1979) sought to evaluate the possible role of genes on the Y chromosome of the male by backcrossing the Y chromosome of strain C57BL/10Bg onto the DBA/1Bg strain background so that the strain (D1.B10-Y) came to have DBA autosomes and X chromosomes, while the Y remained C57. He also backcrossed the DBA/1 Y onto the C57BL/10 strain background to generate the B10.D1-Y strain. The two strains differed in only a portion of the Y chromosome, the part that does not recombine with the X. After more than 10 generations of backcrossing, it became clear that the source of the Y chromosome did have a small but significant influence on mouse agonistic behavior (Maxson & Canastar, 2003), but the effect depended on the strain background. The D1.B10-Y males differed in agonistic behavior from DBA/1Bg, whereas B10.D1-Y males were very similar to C57BL/10Bg males.

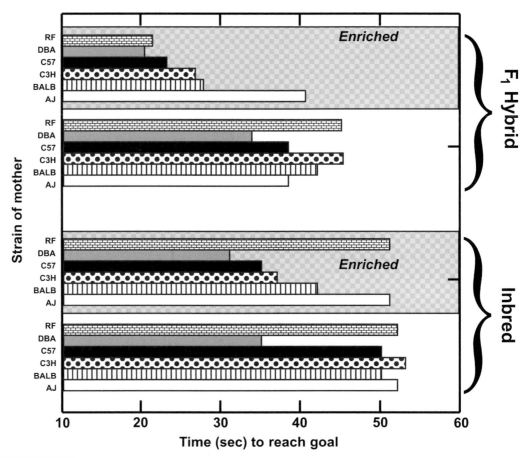

FIGURE 15.3

Mean time to reach the goal in a complex exploration task for food reward by six inbred strains of mice and their F_1 hybrid crosses. F_1 hybrids were only slightly better than their inbred parent strains when housed in standard lab cages, but they were substantially superior when reared in an enriched environment. *(Redrawn with permission from Henderson, 1970.)*

Meanwhile, the University of Connecticut decided to build a new mouse facility, a specific-pathogen free (SPF) lab with special filters on the air and water, and elaborate procedures for cleaning the people who entered the lab. When the sparkling new facility was ready, Maxson's strains were introduced to the new lab and before long were breeding well. Maxson tested their agonistic behavior, just to be sure the strain differences were the same in the new SPF lab as in the old conventional lab. The strain differences that arose exclusively from the Y chromosome had disappeared entirely. A frenzied search for the cause commenced. One peculiar feature of the SPF technique was the use of acidified water in the drinking bottles to kill bacteria, not just a little dash of acid but enough hydrochloric acid to reduce the pH to 2, something we would experience as extremely sour. The acidified water had a substantial impact on agonistic behavior. Switching to distilled drinking water restored the strains to levels of behavior seen before the big move (Maxson, 1992).

Two kinds of heredity—environment interaction were observed in those classic studies. In the work of Ginsburg, Scott, and Searle, animals differing in heredity were reared in equivalent environments but then were tested in more than one *test situation* that prevailed on the day of testing. In the studies of Cooper, Zubek, Henderson, and Maxson, the test situation was the same for all animals, whereas the pretest *laboratory environments* varied. Either kind of interaction could have a substantial impact on the manifestation of differences in heredity. Together, they demonstrated the necessity for devoting careful attention to the fine details of animal husbandry and test protocols in order to obtain consistent results from genetic experiments.

INTERACTIONS WITH LAB ENVIRONMENT: RECENT STUDIES

More recently, researchers in neurobehavioral genetics became interested in lab environment influences after a number of prominent failures to replicate results of genetic studies of phenotypes from newly created knockout mice. Several conferences were held to discuss approaches to phenotyping, beginning in 1995 with a meeting sponsored by the MacArthur Foundation on "Animal Models of Psychiatric Diseases." The Office of Behavioral and Social Science Research (OBSSR) at NIH made some seed money available for a "multi-center trial of a standardized battery of tests of mouse behavior," and in 1999 the National Institute of Mental Health issued a Request for Applications (RFA) aimed at exploring improved approaches to phenotyping. An OBSSR-funded project became the first major effort to evaluate the replicability of behavioral tests across laboratories (Crabbe et al., 1999). Three laboratories (Crabbe at Portland, Oregon; Wahlsten at Edmonton, Alberta; Dudek at Albany, New York) equated the test situation almost completely by building or borrowing identical copies of several kinds of test apparatus and shipping them to all labs, then negotiating common test protocols. Many features of the local lab environments were equated but some were not. For tests such as ethanol preference on the two-bottle test, results were very similar in the three labs (Figure 15.4B), but scores on the elevated plus maze and cocaine activation in the open field (Figure 15.4A) were substantially different, and significant lab by strain interactions were apparent.

That study provoked considerable discussion about the potential advantages of standardizing conditions across laboratories, and it was followed by several other multi-laboratory experiments (Kafkafi et al., 2005; Lewejohann et al., 2006; Mandillo et al., 2008; Rustay et al., 2003a; Salomé et al., 2002; Singer et al., 2005; Wahlsten et al., 2005; 2006; Wolfer et al., 2004). Table 15.2 summarizes features of recent studies with mice and rats that involved more than one laboratory administering very similar tests.

Only studies that rigorously equated the apparatus and test protocols provided conclusive tests of the *replicability* of genetic experiments. Others that allowed small differences

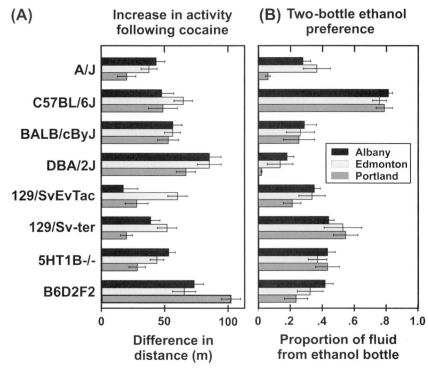

(A) Increase in activity following cocaine

(B) Two-bottle ethanol preference

Albany
Edmonton
Portland

0   50   100
Difference in distance (m)

0 .2 .4 .6 .8 1.0
Proportion of fluid from ethanol bottle

FIGURE 15.4

(A) Scores on a cocaine activation test of open field exploration and (B) two-bottle ethanol preference for eight genotypes tested simultaneously in three labs. Cocaine activation was much greater in the Edmonton lab than the Portland lab for most but not all genotypes, whereas mice tested in Edmonton and Albany were quite similar, with the marked exception of 129/SvEvTac. Ethanol preference, on the other hand, was remarkably similar at all three sites, such that rank ordering of the eight genotypes was the same. *(Redrawn with permission from Crabbe, Wahlsten, & Dudek, 1999.)*

between labs in apparatus and protocol tested the *robustness* of results in the presence of moderate procedural differences. Most of the studies examined open field activity, and it was possible to compare the same phenotype (distance traveled) using the same statistical method, analysis of variance, for six of the studies (Table 15.3). In all studies, the effect size for strain was far larger than for lab differences. Three of the studies showed no main effect of lab, whereas all but one reported a significant lab by strain interaction effect, with the interaction being quite substantial in two reports. The strain by lab interaction effect is the crucial indicator of generality and replicability. The interaction can be statistically significant, however, even when the results for the genetic variable are reasonably similar in the different labs. A significant interaction does not by itself denote a failure to replicate.

A recent study of open field and light–dark box behaviors of consomic strains (Singer et al., 2005) was done at Case Western Reserve University (CWRU) and the Whitehead Institute for Biomedical Research (WIBR). The authors presented their findings for open field activity as a two-way scatterplot (their Figure 5a), and they suggested that the correlation of strain means ($r = 0.7$) was high enough that the results could be considered very similar in the two labs. It is possible to do an ANOVA using strain means from their Figure 2a and Figure 4a (see Figure 5.8 in Chapter 5), and the variance within groups can be determined from the standard error bars, allowing an approximate ANOVA to be done. The ANOVA on 12 strains in two labs reveals a clearly significant main effect of genotype ($F = 13.9$, $df = 11/240$, $P < 0.000001$), a significant difference between labs ($F = 15.3$, $df = 1/240$, $P = 0.004$) as well as a significant and substantial strain by lab interaction effect ($F = 2.5$, $df = 11/240$ $P = 0.006$).

TABLE 15.2 Studies of Lab Environment Effects on Rodent Behavior

Study	# Labs	# Strains	Sex	# Tests	Factors Rigorously Equated			Tests Administered**
					Lab Environ.	Test Apparatus	Test Protocol	
(Crabbe, Wahlsten, & Dudek, 1999)	3	8	M, F	6	Yes*	Yes	Yes	OFA, EPM, WM, ALCPR,COCACT
(Rustay, Wahlsten, & Crabbe, 2003); (Wahlsten, Cooper, & Crabbe, 2005; Wahlsten, Bachmanov, Finn, & Crabbe, 2006)	2	21	M, F	5	No	Yes	Yes	OFA, EPM, ARR, 4WM, WILD
(Salomé et al., 2002)	3	2	M	2	No	No	No	EPM, LD
(Wolfer et al., 2004)	3	3	F	4	Yes*	Yes	Yes	OFA, ZM, OEX, WM
(Kafkafi, Benjamini, Sakov, Elmer, & Golani, 2005)	3	8	M	1	No	No	Yes	OFA
(Singer, Hill, Nadeau, & Lander, 2005)	2	12	M	2	No	?	?	OFA, LD
(Lewejohann et al., 2006)	2	1	F	3	Yes*	Yes	Yes	OFA, EPM
(Mandillo et al., 2008)	3-5	4	M	7	No	No	No	OFA, SHIRPA, GRIP, ARR, YMAZE, AS/PPI, HPTF

*Many aspects of husbandry and general lab conditions were equated, but there still remained many differences among the labs, such as technicians and animal caretakers.
**Abbreviations: ALCPR, alcohol preference; ARR, accelerating rotarod; COCACT, cocaine activation of open field activity; EPM, elevated plus maze; GRIP, grip strength; LD, light–dark exploration box; OBJEX, object exploration; OFA, open field activity; ZM, elevated zero maze; SHIRPA, battery of sensory and reflex tests; AS/PPI, startle reflex and prepulse inhibition; HPTF, WILD, wildness ratings; WM, water maze; YMSA, 3-arm Y maze alternation.

TABLE 15.3 Effect Sizes for Studies of Open Field Activity (Distance Traveled or Squares Entered)

Study	# Labs	# Strains	Sex	Effect Size (Partial ω^2)		
				Strain	Lab	Strain × Lab
Crabbe et al. (1999)	3	8	M, F	0.600	0.157	0.059
Rustay et al. (2003a); Wahlsten et al. (2005); Wahlsten et al. (2006)	2	21	M, F	0.618	0.006[NS]	0.029[NS]
Wolfer et al. (2004)	3	3	F	0.336	0[NS]	0.051
Kafkafi et al. (2005)	3	8	M	0.770	0.014[NS]	0.176
Singer et al. (2005)	2	12	M	0.360	0.030	0.060
Mandillo et al. (2008)	3-5	4	M	0.638	0.416	0.179

NS: not statistically significant, P > 0.05.

When comparing strains across two labs, the strain correlation is a useful indicator. The utility **Kstrains 2labs correl cohensf** allows the user to determine both the population strain correlation and effect sizes for strain, lab, and strain × lab interaction effects in an ANOVA. As an exercise, a series of means similar to those of Singer et al. (2005) were inserted into the appropriate cells, spacing the means five units apart across a range from 70 to 130. In the first case, means were identical in both labs, the strain correlation was r = 1.0, and there was no interaction effect. Means were then swapped between strains in Lab 2, so that both lab means remained at 102.5 and strain differences were substantial. Several cases were created such that the strain correlation ranged from 0 to 1.0, and the ratio of Cohen's effect size for the interaction effect f_{SxL} to the strain main effect f_S was determined. When there are two labs and K strains, the degrees of freedom for both the strain main effect and the interaction effect will be (K − 1), making their power functions the same in a fixed effects design. Figure 15.5 shows the relation between the strain correlation and the ratio (f_{SxL}/f_S). When the strain correlation is zero, the interaction effect has the same size as the strain main effect, whereas a strain correlation of r = 0.6 occurs when the interaction effect size is half the strain main effect size. Inspecting the scatterplots in Figure 15.5, it appears that the interaction when r = 0.6 is substantial and there are some

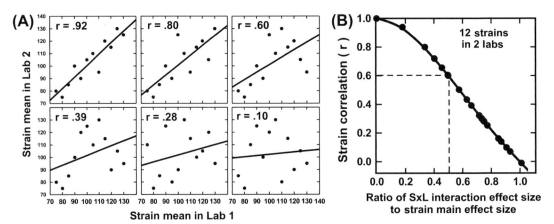

FIGURE 15.5

(A) Scatterplots of 12 hypothetical strain means in two labs, showing Pearson correlations between lab means. The strains in Lab 1 had means of 75, 80, 85, 90, 95, 100, 105, 110, 115, 120, 125, and 130 in each case. The strains in Lab 2 had the same 12 means but they were paired differently with the strains in Lab 1. (B) Effect sizes were determined for several cases for the strain main effect (f_S) and strain by lab interaction effect (f_{SxL}) that have identical degrees of freedom. The ratio (f_{SxL})/(f_S) expresses the relative size of the interaction. It has a consistent relation with the strain correlation between labs.

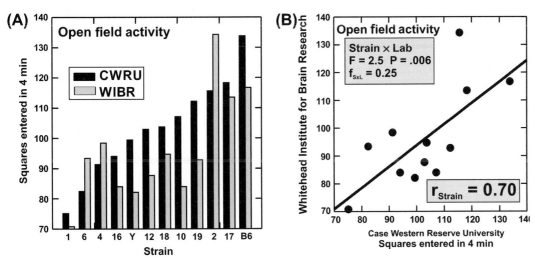

FIGURE 15.6

(A) Open field activity scores for 11 consomic strains and the progenitor C57BL/6 tested at two labs, plotted as a bar chart that shows the striking differences in rank orders of several strains at the two sites. (B) The same strain means portrayed as a scatterplot that shows the strain correlation (r = 0.7). Results of an ANOVA reveal a significant and substantial strain × lab interaction effect for the same data (see Table 15.3). *(Redrawn with permission from Singer, Hill, Nadeau, & Lander, 2005.).*

striking differences in rank orders of the strains in the two labs. Even when r = 0.8, certain strains can differ considerably between labs. Comparing the Singer et al. (2005) results with the others in Table 15.3, it is apparent that the study of 12 consomic strains in two labs found one of the largest strain × lab interaction effects relative to strain main effect that has appeared to date in the literature.

There is currently no widely accepted standard for what is a small, medium, or large interaction effect. More often than not, the investigator judges the importance of an interaction by inspecting a graph of the results. For the Singer et al. (2005) data, a bar chart (Figure 15.6A) suggests the strain × lab interaction was large because rank ordering of strains was substantially different in the two labs. The scatterplot (Figure 15.6B), on the other hand, emphasizes the generally positive correlation of strain means. Although journals generally do not publish two separate graphs showing identical data, it is possible to superimpose results of an ANOVA on the scatterplot that also shows the strain correlation. This option makes efficient use of the printed page while providing the reader with more information.

Comparisons of strain means for several behavioral phenotypes in studies done in different labs several years or decades apart reveal that strain correlations are often r = 0.8 or higher, which clearly qualifies as a high degree of replicability of results (Wahlsten et al., 2006). For the elevated plus maze, on the other hand, local lab environments often give rise to striking differences in strain rank orders, for example, the Wahlsten lab versus the Trullas and Skolnick data (1993), or when the lab environment changes within the same institution (Figure 15.7). Open-arm exploration in the elevated plus maze is sensitive to the level of anxiety experienced by a mouse, and many local environmental factors can influence anxiety-related behaviors.

How strongly the local lab environment influences test scores clearly depends on the particular test. No sweeping conclusions about the lab environment are warranted. It is not true that lab environment effects are generally small and therefore can be ignored. It is also not true that strain × lab interactions are so large that generality of behavioral experiments is dubious.

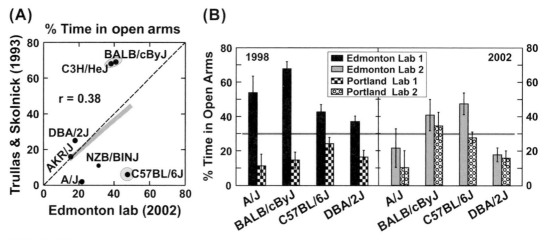

FIGURE 15.7

Elevated plus maze scores (percent time in the open arms) for several inbred strains. (A) Data from the Edmonton lab in 2002 versus data from Trullas & Skolnick (1993) show some remarkable differences in anxiety-related behavior for the same inbred strains. (B) Major changes in anxiety-related behavior when four inbred strains were tested at two lab locations four years apart at two universities. The results for Edmonton and Portland labs were much more similar in 2002 after both labs had been moved to another location; a different wing of the same building in Edmonton and a different building in Portland. *(Reproduced with permission from Wahlsten, Bachmanov, Finn, & Crabbe, 2006.)*

COMPLEXITY OF THE LAB ENVIRONMENT

If one wished to compare or possibly equate the lab environments at several sites, how many features would have to be considered? A discussion of the laboratory environment took place at the symposium on "Behavioral Phenotyping of Mouse Mutants" at the University of Cologne in 2000 (see www.medizin.uni-koeln.de/dekanat/fd/Transgenic_mice/schedule. html), and this resulted in the creation of the Mutant Mouse Behavior (MMB) network (Arndt & Surjo, 2001; Surjo & Arndt, 2001). One of the main activities of the now-defunct MMB was to create a list of features of the lab environment (http://www.medizin.uni-koeln. de/mmb-network/input-sheet), which became the basis for Table 1 in the discussion by Würbel (2002). That list lacked information about food and water. It was expanded recently by Philip et al., (2009) who provided extensive documentation on conditions at four different laboratories involved in testing 70 of the BXD RI strains. An expanded list is given in Table 15.4. To be useful, the information provided on each item needs to be sufficiently detailed so that another laboratory could duplicate those conditions. It would be wise for any lab to create such a list for future reference. If the field could agree on such a list, it would then be feasible to deposit relevant data in the Mouse Phenome Database in order to facilitate comparisons among test sites.

It is apparent from Table 15.4 that the lab environment has numerous features, each of which could potentially be the subject of a controlled experiment to evaluate its importance for behavioral test results of mice having different genotypes. Unraveling this complexity would be an immense undertaking in just one lab, and results from only one lab with just a few strains would not automatically apply to every other lab using other genotypes. It appears that it is simply not feasible to equate the lab environment across labs on more than a handful of factors, and it will not be possible to use controlled studies to determine which factors have the greatest impact on tests of behavior.

STANDARDIZING THE LAB ENVIRONMENT

Whether labs *should* seek to standardize many features of the environment across diverse sites is a controversial question. Spirited debate took place at the conference at the University of

TABLE 15.4 Features of the Lab Environment (Provide Enough Detail to Duplicate Conditions)

THE LABORATORY
    Director
    City
    Institution
    Building
    Room number(s) and layout
COLONY ROOM
    Dimensions
    Number of cages in room
    Number of mice in room
    Other species nearby
    Distance from testing room(s)
    Cleaning floor; frequency, chemical
    Cleaning walls; frequency, chemical
COLONY ROOM LIGHTING
    Lights on & off; transition time
    Brightness of light and dark phases;
    middle & corners of room
    Kind of lighting; color, temperature
COLONY ROOM NOISE
    Sources of ambient noise &
    vibration
    Intensity of ambient noise & vibr.
    Source of intermittent noise & vibr.
    Intensity of intermittent noise & vibr.
    Description of ultrasound levels
COLONY ROOM AIR
    Source of air
    Changes of room air per hour
    Temperature; mean, min, max
    Humidity; mean, min, max
    Odors in air
CAGE ENVIRONMENT
    Brand, description, material
    Size
    Transparency
    Kind of top or lid
    Air changes per hour in cage
    Humidity in cage

    Bedding material
    Nesting material
    Objects in cage; enrichment
    Number of adult mice in cage
    Frequency of cage changing
    Changing coordinated with testing?
    Cage changing procedures
    Washing of cages; chemical, rinse
DRINKING WATER
    Source
    Delivery mechanism
    Additives; pH
    Frequency of fresh water bottle
    Cleaning of bottles, spouts
FOOD
    Brand, formula or type
    Availability
    Delivery mechanism
    Fresh or reused from other cages
MARKING & HANDLING OF MICE
    Method of identification
    Frequency and method of weighing
    Method of handling
HEALTH MONITORING
    Colony type (conventional, SPF)
    Breeding (closed, shipped, etc.)
    Routing serology, parasitology
    Known bacteria, viruses, parasites
PERSONNEL
    Names of individuals
    Weekday visits to colony room
    Weekend visits to colony room
    Number of people per month
    Number of people changing cages
    Changing and testing by same
    people?
    Education of animal handlers
    Supervision of animal handlers

259

Cologne in 2000. Contributions to that discussion were subsequently published in special issues of *Behavioural Brain Research* and *Physiology and Behavior*. An article based on the keynote address to the conference pointed out a serious need for higher standards as well as standardization (Wahlsten, 2001). The case for standardization was argued effectively by van der Staay and Steckler (2002), while Würbel (2002) argued that standardization could have negative consequences.

In the 1999 study of identical tests in three labs (Crabbe et al., 1999), strenuous effort was made to equate the lab environments (Wahlsten et al., 2003). It was concluded from this experience that standardizing lab environments is not feasible and is not even a worthwhile objective. In the subsequent cross-laboratory comparisons (Rustay et al., 2003a; Wahlsten et al., 2006), researchers allowed the local lab environments to continue just as they were, while seeking to equate the test situations.

There is great diversity of conditions in labs around the world that work with mouse behavior. What the field needs to know is how strongly the results of genetic experiments are influenced by those environments. The data in Table 15.3 provide some indication of the sizes of interaction effects that can arise from ubiquitous differences in lab environments. Those studies probably underestimated interaction effects to some extent because they were done as part of a controlled experiment by self-selected collaborators. A more effective strategy would be to sample a wider range of labs in different countries and ship identical apparatus to them for testing identical mouse genotypes, while asking that nothing about their local conditions be changed to accommodate the study.

HETEROGENIZATION OF THE LAB ENVIRONMENT

It has been claimed that excessive standardization will actually increase the extent of differences in results among labs. This argument is counterintuitive and requires some fairly compelling empirical evidence if it is to be taken seriously. Richter and colleagues (2009) recently offered evidence based on the previously published study by Wolfer et al. (2004) that is summarized in Tables 15.2 and 15.3. By re-sampling from those data, they claimed to show that standardization on a single set of parameters would actually increase the false positive rate in tests of strain differences and accentuate differences between labs. As a remedy, they proposed that local conditions be *heterogenized* in the same way in all labs. For example, each lab might test mice of two different ages living in two kinds of cages (standard and enriched; see Figure 5.15). The paper provoked an editorial in *Nature Neuroscience* (Editorial, 2009), which questioned the replicability of behavioral data from mice at different sites.

A false positive result can occur only when the null hypothesis (no group difference in the populations) is true. The null condition will never be realized in empirical data because of ubiquitous sampling error. Suppose the null is true and in sample data the tests of group differences are not statistically significant at $\alpha = 0.05$. As an exercise, use the sample means in those data as the basis for generating random samples. Many more than 5% significant effects will be obtained because, in the population means based on sample data, the null is no longer true. False positive rates cannot be determined from empirical data. Thus, the issue must be addressed with a model of true population values. For this purpose, the utility **Simulation 6 factor experiment** presented in Chapter 5 can be utilized. Figure 5.15 shows a model that expresses a 5-way interaction. The mean of all 192 groups in that model is 32 sec fall latency on the accelerating rotarod. A null model entails the same groups where each group has a true mean $\mu = 32$, as shown in Figure 15.8. Using those identical means in the utility **Simulation 6 factor experiment** to obtain 100 random samples results in approximately 5% false positive

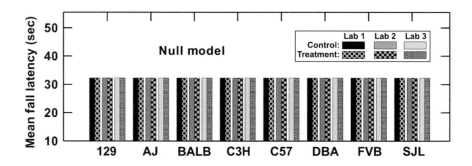

FIGURE 15.8

A null model where eight strains tested in three labs in either the control or treatment condition all have the same population mean of $\mu = 32$. In this model, unlike sample data, the null hypothesis is definitely true. ANOVAs on random samples drawn from this model provide a valid estimate of the false positive rate.

significant F tests when the strain, lab, and treatment effects are in the ANOVA (Table 15.5). Next, one of those random samples (run #10) was adopted as a model for population means, another 100 random samples were drawn, and 100 ANOVAs were done. This process was repeated for runs 20 and 30. "False positive" rates were considerably higher than 5% for all effects in the ANOVA (Table 15.5). The actual false positive rates for genetic and treatment effects will not be influenced by standardization or heterogenization, because a false positive can occur only when there are no such effects.

The second issue is whether heterogenization will reduce the tendency of different labs to obtain different results of genetic experiments and enhance replicability of findings in our field. This too can be addressed with the **Simulation 6 factor experiment** utility. In Chapter 5 a 5-factor model was presented in Figure 5.15 and results of 100 runs of a simulation based on that model were presented in Table 5.3. For this chapter, another model was created in which all three labs use mice of the same age all housed in standard cages. The results of 100 random samples for the standardized and heterogenized experiments were then analyzed with ANOVAs, first with all five factors in the analysis and then with just strain, lab, and treatment, which pooled variance arising from age and cage into the error term. Results were clear. For strain, lab, and treatment main effects and the strain × lab interaction that was very large, heterogenization had no effect, regardless of whether three or five factors were in the ANOVA; power was always 96% or higher. For the moderate strain × treatment and strain × lab × treatment interactions, however, heterogenizing the experiment reduced power to detect real interactions when a 3-factor ANOVA was done but not with a 5-factor analysis. This makes good sense: introducing greater variability into an experimental design and then including variance from two control factors in the error term should reduce power to detect real interactions, and in this simulation it does. Heterogenizing genetic studies does not increase replicability of results across labs; it simply makes it more difficult to detect real interactions with the lab environment by injecting noise into the study.

Added control factors in a design do offer some advantage in terms of the generality of results. For example, if a study employs both males and females, it has greater generality than a study that uses only one sex. Likewise, including separate groups of mice that are 8 and 12 weeks of age will increase generality. So too will the use of two kinds of caging in the colony room. This kind of enhanced generality is purchased at the cost of daunting logistics when the study must be properly balanced and randomized. If any control factor is included to increase generality, it must be properly balanced in the cage placement as well as behavioral test order. Sample size will also need to be increased in the entire study to retain the same level of power because so many degrees of freedom will be consumed by dividing a group of mice into subgroups. Suppose the original plan in a 8 × 3 × 2 design called for 8 mice in each of 48 groups. There would be 336 degrees of freedom for the within-group variance. If each group of 8 mice is instead divided into 4 heterogenized subgroups, the full design will have 192 groups with two mice each and there will be 192 degrees of freedom in the error term. Sample size can be increased only by one mouse at a time, so the researcher will then need to use n = 3 mice per group to achieve 384 degrees of freedom in the error term, and the entire study will grow from 384 to 576 mice, simply to add two control factors to the design.

When the full spectrum of potentially important features of the lab environment is considered (Table 15.4), it is apparent that systematic heterogenization of research designs in behavior genetics is not a practical alternative to standardization of the lab environment.

STRATEGIES FOR STANDARDIZING THE LAB ENVIRONMENT

While the lab environment can never be equated across laboratories, there are some advantages to measuring and reporting the lab environment in greater detail than is customary. If this were done, it would eventually become feasible to perform a formal data analysis on studies that observe the same kinds of behavioral phenotypes. A meta-analysis could first be

TABLE 15.5 Number of Significant F Tests ($\alpha = 0.05$) for 100 Runs of a Simulation (n = 4 Mice Per Group)

Population Means Based On	Effect in Analysis of Variances (ANOVA)							Average % False Positives
	Strain	Lab	Treatment	Strain × Lab	Strain × Treatment	Lab × Treatment	Strain × Lab × Treatment	
Null model (Figure 15.8)	4	7	3	7	6	1	4	4.6%
Re-sampling from random samples from the null model (Figure 15.8) that then serve as population means								
Run 10 of null	26	11	8	49	14	11	39	23%
Run 20 of null	59	11	7	34	72	18	49	36%
Run 30 of null	9	22	14	73	17	48	75	37%
Samples from model with 5-way interaction (see Figure 5.15); analyzed with 3 or 5 factors in ANOVA								
All 5 factors	100	100	100	97	100	72	21	
Only 3 factors	100	100	100	96	100	67	16	
Samples from model where 3 labs use same standard age and cage; analyzed with 3 or 5 factors in ANOVA								
All 5 factors	100	100	100	96	100	72	23	
Only 3 factors	100	100	100	97	100	72	25	

done to compare effect sizes across labs (Borenstein et al., 2009). If there is significant heterogeneity among sites, multiple regression analysis could then be used to search for features of the lab environment that correlate most strongly with results of behavioral tests, as is done for the Morris maze in Chapter 13.

There are three general kinds of lab environment effects that warrant different approaches in the future.

1. Certain practices are definitely superior. Once it becomes clear that a feature of the lab environment yields superior results when it is done one way instead of another, it can then be recommended that all labs adopt the better way. If the data supporting the claim are convincing, many labs will change their ways.
2. For many other features where data suggest that different values make only small differences in test scores, it might make good sense to adopt the better way. If some labs prefer to adhere to their current protocols, however, no great harm will be done to the field.
3. There will also be features of the lab environment where different parameters make no difference at all, and it would be pointless to press for standardization of these. At the present time, there are not sufficient data to place most features of the lab environment into one class or another. A systematic investigation of many of these factors would be helpful.

Utilities in Excel

The utilities are type *xls* files in the Microsoft Office 97–2003 format. Each sheet is protected with a password so that the user can enter new values into only certain cells. Other cells are locked and formulas are hidden. For most utilities, the cells that can be altered are yellow, except in the Design utilities where several colors are shown. Automatic recalculation is turned OFF, and the user needs to press F9 Recalculate to update any spreadsheet. Many of the utilities already have values entered and results calculated. The user can then enter new values and recalculate to obtain new results. Boxes 4.2 and 4.3 in the text provide rules for usage and additional tips about adapting the utilities to different situations.

File Name	Description of Features
Chapter 4 Designs	
Design 1way	Enter names for up to 24 groups; generates list with 10 replications
Design 2way	Enter names for up to 8 levels of factor A and 4 levels of factor B; 10 replications
Design 3way	Enter names for $A \times B \times C$ design with up to $8 \times 4 \times 3$ levels; 10 replications
Design 4way	Enter names for $A \times B \times C \times D$ design, $8 \times 4 \times 3 \times 2$ levels; 10 replications
Design 5way	Enter names for $A \times B \times C \times D \times E$ design, $8 \times 4 \times 3 \times 2 \times 2$ levels; 10 replications
Design 6way	Enter names for $A \times B \times C \times D \times E \times F$ design, $8 \times 4 \times 3 \times 2 \times 2 \times 2$ levels; generates separate lists for F_1 and F_2 conditions; user must append these to generate the complete list; 10 replications
Design Example 5way	Example with eight inbred strains tested in four labs under three drug conditions where half of mice were shipped from supplier and half were bred locally; also has equal numbers of males and females
Breeding list	Shows 16 breeding combinations of male and female genotypes; specific example is for reciprocal cross design shown in Table 4.3; other names can be substituted for use with other breeding designs; 10 replications
Chapter 5: Sample Size	
Normdist 10K	Enter true mean, standard deviation; generates 10,000 normally distributed scores; exactly normally distributed population, not a sample; rounded up to nearest whole number; for use, copy to another sheet

(Continued)

File Name	Description of Features
1sample 1group	Enter true mean, standard deviation; generates random sample of 100 normally distributed scores; finds sample mean and standard deviation for different sample sizes
1sample 2groups	Enter two true means, standard deviations; generates random sample of 100 normally distributed scores for each group; finds sample means and sample standard deviations for different sample sizes as well as difference between sample means and sample effect size d
1sample 8groups	Enter eight true means and standard deviations; generates eight random samples of 100 normally distributed scores; finds standard deviation of those eight sample means for sample sizes ranging from 2 to 100; second sheet (pop1_8) is for computations only, and results are transferred back to sheet 1
100samples 1group	Enter true mean and standard deviation; generates random sample of 50 normally distributed scores and finds sample means and standard deviation for different sample sizes; repeats the process 100 times and finds standard error of sample means for different sample sizes; samples sheet is for computations
100samples 2groups	Enter two true means and standard deviations; generates random sample of 50 normally distributed scores for each group and finds sample means and standard deviation for different sample sizes; finds difference between sample means; repeats the process 100 times and finds standard deviation of difference between sample means for different sample sizes (n); other sheets are for computations
Effect size from article	Finds effect size d for study with 2 groups from published t test and degrees of freedom or just the P value; find d from sample means and standard deviations; measure graph to find means and standard deviations; do a meta-analysis (sheet not protected) to estimate d from several studies; find omega-squared and Cohen's f for studies with more than 2 groups
Samplesize 2 2×2 Jgroups	For 2 groups, enter effect size delta; for 2 × 2 design, enter four means and standard deviation; other sheets find n to detect linear contrast when there are J = 3, 4, 5, 6, 7, 8, 10, 12 or 16 groups; n is given for different Type I and II errors
Samplesize omni Jgroups	Enter number of groups (J) in a one-way design and Cohen's f; plots power for sample sizes ranging from 2 to 100 for three values of Type I error; sheets J3 to J10 allow the user to enter true means and standard deviation in order to find Cohen's f for different numbers of groups in a one-way design, which can then be copied to the power sheet; sheet J12 is unprotected so the one-way design can have any number of groups
Cohens f J×2 design	Finds Cohen's effect size f for A and B main effects as well as the A×B interaction for a two-way factorial design with **2 levels** of Factor B and 2, 3, 4, 5, 6, 7, or 8 levels of Factor A; transfer the values of f and number of items to the spreadsheet *Samplesize J×K for Cohens f* to find power for different sample sizes

(Continued)

File Name	Description of Features
Cohens f J×3 design	Finds Cohen's effect size f for A and B main effects as well as the A×B interaction for a two-way factorial design with **3 levels** of Factor B and 3, 4, 5, 6, 7, or 8 levels of Factor A
Cohens f J×4 design	Does the same for a two-way factorial design with **4 levels** of Factor B and 4, 5, 6, 7, or 8 levels of Factor A
Cohens f J×5 design	Does the same for a two-way factorial design with **5 levels** of Factor B and 5, 6, 7, or 8 levels of Factor A
Cohens f J×6 design	Does the same for a two-way factorial design with **6 levels** of Factor B and 6, 7, or 8 levels of Factor A
Cohens f J×7 design	Does the same for a two-way factorial design with **7 levels** of Factor B and 7 or 8 levels of Factor A
Cohens f J×8 design	Does the same for a two-way factorial design with **8 levels** of Factor B and 8 levels of Factor A
Samplesize J×K for Cohens f	Plots power against sample size for two values of Type I error separately for A main effect, B main effect and A×B interaction for any J×K factorial design; user must insert number of levels of each factor and Cohen's effect size f; graphs plot power for n = 2 to 50, while table shows power for n up to 100
Simulation 6 factor experiment	User specifies standard deviation within groups and then groups means for a large factorial design with 6 factors; provides $8 \times 3 \times 2 \times 2 \times 2 \times 2$ design with two replications, for a total of 768 mice in the study; each RUN is a separate random sample from the user-specified population means; the values are then copied to another spreadsheet and imported into a statistical program for doing the ANOVAs and finding power

Chapter 7: Logistics

20strains Design	List of 20 inbred strains in a one-way design with n = 20 females per strain
20strains Shipping	Divides study into 5 shipments of 2 batches each; assigns cages numbers with 4 mice per cage for housing until mating; then assigns ID numbers and tail numbers in the order of unpacking; last sheet explains methods
20strains Mating	When mice are old enough to mate, assign to mating cages after randomizing order within a shipment and batch by sorting on random number
20strains Final	Has the three finished lists for each phase of the study; the design, the unpacking list, and the mating cage list, all properly balanced and randomized
3×2 Design	Two-way factorial design with 3 genotypes × 2 sexes with 12 replications
3×2 Final	Sort by replication, assign random #s, and sort by random # within squad to determine test order within a squad; notes sheet describes steps
4×2×3 Design	Design with 4 strains, 2 sexes, 3 treatments, 9 replications per group
4×2×3 Balanced	Enumeration of 12 possible sequences; application of Latin square to determine test order; sorting by random number to determine order within squad; final test order

(Continued)

File Name	Description of Features
4×2×3 Final	Unpacking sheet with ID codes, cage and tail numbers; final testing order
8×2×2×2 Design	8 strains × 2 housing conditions × 2 drug conditions × 2 sexes; 8 replications; as an exercise, this study design then can be divided into shipments, cages, squads, and then balanced and randomized to determine test order
Throughput schedules	Determining timing of tests when elevated plus maze (5 min) and open field (10 min) tests are given to each mouse in a squad in the same day; longer day with 4 per squad and shorter day with 8 per squad, overlapping tests
Alcohol Jax order	Schedule of 4 shipments of 20 strains and 2 sexes from Jackson Labs
Alcohol Timing	Overall schedule of testing; timing of events on Days 1, 2, 5, and 8
Alcohol Test order Ship1	Systematic order of mice in shipment 1 balanced for ethanol(E)/saline(S); balanced for E/S and randomized order of strains in each squad of 4 mice for 2 groups or batches

Chapter 9: Prelude to Data Analysis

Simulation delta	Enter true mean and standard deviation of the control group; generates random samples of 100 normally distributed scores when the treatment group differs by 0.5, 1.0, 1.5 or 2.0 standard deviations

Chapter 15: The Laboratory Environment

Kstrains 2labs correl cohensf	Enter the within-group standard deviation and then group means for 12 strain tested in two labs; fewer or more than 12 strains can be assessed by deleting or adding rows (use caution here); finds correlation (r) of strain means in the two labs and Cohen's f for the Strain and Lab main effects as well as the Strain×Lab interaction effect; use utility *Samplesize J×K for Cohens f* to find sample size needed for different levels of power

Abbey, H., & Howard, E. (1973). Statistical procedure in developmental studies on species with multiple offspring. *Developmental Psychobiology*, 6(4), 329–335.

Abizaid, A., Liu, Z. W., Andrews, Z. B., Shanabrough, M., Borok, E., Elsworth, J. D., et al. (2006). Ghrelin modulates the activity and synaptic input organization of midbrain dopamine neurons while promoting appetite. *American Society for Clinical Investigation*, 116(12), 3229–3239.

Adamec, R., Head, D., Blundell, J., Burton, P., & Berton, O. (2006). Lasting anxiogenic effects of feline predator stress in mice: Sex differences in vulnerability to stress and predicting severity of anxiogenic response from the stress experience. *Physiology & Behavior*, 88(1–2), 12–29.

Aguiar, P., Mendonca, L., & Galhardo, V. (2007). OpenControl: A free opensource software for video tracking and automated control of behavioral mazes. *Journal of Neuroscience Methods*, 166(1), 66–72.

al-Ramadi, B. K., Fernandez-Cabezudo, M. J., El-Hasasna, H., Al-Salam, S., Bashir, G., & Chouaib, S. (2009). Potent anti-tumor activity of systemically-administered IL2-expressing *Salmonella* correlates with decreased angiogenesis and enhanced tumor apoptosis. *Clinical Immunology*, 130(1), 89–97.

Alberts, B., Johnson, A., Lewis, J., Roberts, K., & Walter, P. (2008). *Molecular biology of the cell* (5th ed.). New York, NY: Garland Science.

Anagnostaras, S. G., Josselyn, S. A., Frankland, P. W., & Silva, A. J. (2000). Computer-assisted behavioral assessment of Pavlovian fear conditioning in mice. *Learning & Memory*, 7(1), 58–72.

Anisman, H., & Matheson, K. (2005). Stress, depression, and anhedonia: Caveats concerning animal models. *Neuroscience & Biobehavioral Reviews*, 29(4–5), 525–546.

Arakawa, H., Blanchard, D. C., Arakawa, K., Dunlap, C., & Blanchard, R. J. (2008). Scent marking behavior as an odorant communication in mice. *Neuroscience & Biobehavioral Reviews*, 32(7), 1236–1248.

Arndt, S. S., Laarakker, M. C., van Lith, H. A., van der Staay, F. J., Gieling, E., Salomons, A. R., et al. (2009). Individual housing of mice — impact on behaviour and stress responses. *Physiology & Behavior*, 97(3–4), 385–393.

Arndt, S. S., & Surjo, D. (2001). Methods for the behavioural phenotyping of mouse mutants. How to keep the overview. *Behavioural Brain Research*, 125(1–2), 39–42.

Bachmanov, A. A., Reed, D. R., Beauchamp, G. K., & Tordoff, M. G. (2002). Food intake, water intake, and drinking spout side preference of 28 mouse strains. *Behavioral Genetics*, 32(6), 435–443.

Bagg, H. J. (1916). Individual differences and family resemblances in animal behavior. *American Naturalist*, 50, 222–236.

Bailey, K. R., Rustay, N. R., & Crawley, J. N. (2006). Behavioral phenotyping of transgenic and knockout mice: Practical concerns and potential pitfalls. *Institute for Laboratory Animal Research Journal*, 47(2), 124–131.

Bailoo, J. D., Bohlen, M. O., & Wahlsten, D. (2010). The precision of video and photocell tracking systems and the elimination of tracking errors with infrared backlighting. *Journal of Neuroscience Methods*, 188(1), 45–52.

Barlind, A., Karlsson, N., Björk-Eriksson, T., Isgaard, J., & Blomgren, K. (2010). Decreased cytogenesis in the granule cell layer of the hippocampus and impaired place learning after irradiation of the young mouse brain evaluated using the IntelliCage platform. *Experimental Brain Research*, 201(4), 781–787.

Bauer, S., & Sokolowski, M. B. (1988). Automsomal and maternal effects on pupation behavior in *Drosophila melanogaster*. *Behavioral Genetics*, 18(1), 81–97.

Bausell, R. B., & Li, Y. F. (2002). *Power analysis for experimental research: A practical guide for the biological, medical, and social sciences*. New York, NY: Cambridge University Press.

Belknap, J. K., & Atkins, A. L. (2001). The replicability of QTLs for murine alcohol preference drinking behavior across eight independent studies. *Mammalian Genome*, 12(12), 893–899.

Belknap, J. K., Dubay, C., Crabbe, J. C., & Buck, K. J. (1997). Mapping quantitative trait loci for behavioral traits in the mouse. In K. Blum, E. P. Noble, R. S. Sparkes, T. H. J. Chen, & J. G. Cull (Eds.), *Handbook of Psychiatric Genetics* (1st ed.). (pp. 435–453) Boca Raton, FL: CRC Press, Taylor & Francis Group.

Belknap, J. K., Richards, S. P., O'Toole, L. A., Helms, M. L., & Phillips, T. J. (1997). Short-term selective breeding as a tool for QTL mapping: ethanol preference drinking in mice. *Behavioral Genetics*, 27(1), 55–66.

Belzung, C., & Griebel, G. (2001). Measuring normal and pathological anxiety-like behaviour in mice: A review. *Behavioural Brain Research, 125*(1–2), 141–149.

Benatar, M. (2007). Lost in translation: Treatment trials in the SOD1 mouse and in human ALS. *Neurobiology of Disease, 26*(1), 1–13.

Benefiel, A. C., Dong, W. K., & Greenough, W. T. (2005). Mandatory "enriched" housing of laboratory animals: The need for evidence-based evaluation. *Institute for Laboratory Animal Research Journal, 46*(2), 95–105.

Benefiel, A. C., & Greenough, W. T. (1998). Effects of experience and environment on the developing and mature brain: Implications for laboratory animal housing. *Institute for Laboratory Animal Research Journal, 39*(1), 5–11.

Benjamini, Y., Drai, D., Elmer, G., Kafkafi, N., & Golani, I. (2001). Controlling the false discovery rate in behavior genetics research. *Behavioural Brain Research, 125*(1–2), 279–284.

Benjamini, Y., & Hochberg, Y. (1995). Controlling the false discovery rate: A practical and powerful approach to multiple testing. *Journal of the Royal Statistical Society. Series B (Methodological), 57*(1), 289–300.

Benstaali, C., Mailloux, A., Bogdan, A., Auzeby, A., & Touitou, Y. (2001). Circadian rhythms of body temperature and motor activity in rodents: Their relationships with the light—dark cycle. *Life Sciences, 68* (24), 2645–2656.

Biddle, F. G., & Eales, B. A. (1999). Mouse genetic model for left-right hand usage: Context, direction, norms of reaction, and memory. *Genome, 42*(6), 1150–1166.

Biddle, F. G., Jones, D. A., & Eales, B. A. (2001). A two-locus model for experience-conditioned direction of paw usage in the mouse is suggested by dominant and recessive constitutive paw usage behaviours. *Genome, 44*(5), 872–882.

Blanchard, D. C., Griebel, G., & Blanchard, R. J. (2003). Conditioning and residual emotionality effects of predator stimuli: Some reflections on stress and emotion. *Progress in Neuro-Psychopharmacology & Biological Psychiatry, 27* (8), 1177–1185.

Blanchard, D. C., Griebel, G., Rodgers, R. J., & Blanchard, R. J. (1998). Benzodiazepine and serotonergic modulation of antipredator and conspecific defense. *Neuroscience & Biobehavioral Reviews, 22*(5), 597–612.

Blizard, D. A., Takahashi, A., Galsworthy, M. J., Martin, B., & Koide, T. (2007). Test standardization in behavioural neuroscience: A response to Stanford. *Journal of Psychopharmacology, 21*(2), 136–139.

Bohlen, M., Cameron, A., Metten, P., Crabbe, J. C., & Wahlsten, D. (2009). Calibration of rotational acceleration for the rotarod test of rodent motor coordination. *Journal of Neuroscience Methods, 178*(1), 10–14.

Bolles, R. C. (1971). Species-specific defense reactions. In F. R. Brush (Ed.), *Aversive conditioning and learning* (pp. 183–233). New York, NY: Academic Press Inc.

Borenstein, M., Cohen, J., Rothstein, H., Schoenfeld, D., Berlin, J., & Lakatos, E. (2001). *Power and precision.* Englewood, NJ: Biostat.

Borenstein, M., Hedges, L. V., Higgins, J. P., & Rothstein, H. R. (2009). *Introduction to Meta-analysis (Statistics in Practice).* New York, NY: Wiley.

Borenstein, M., Rothstein, H., Cohen, J., Schoenfeld, D., & Berlin, J. (2000). *SamplePower 2.0.* Chicago, IL: SPSS, Inc.

Bothe, G. W. M., Bolivar, V. J., Vedder, M. J., & Geistfeld, J. G. (2004). Genetic and behavioral differences among five inbred mouse strains commonly used in the production of transgenic and knockout mice. *Genes, Brain and Behavior, 3*(3), 149–157.

Bourin, M., Petit-Demouliere, B., Dhonnchadha, B. N., & Hascoet, M. (2007). Animal models of anxiety in mice. *Fundamental & Clinical Pharmacology, 21*(6), 567–574.

Bovet, D., Bovet-Nitti, F., & Oliverio, A. (1969). Genetic aspects of learning and memory in mice. *Science, 163*(863), 139–149.

Box, G. E. P., Hunter, J. S., & Hunter, W. G. (2005). *Statistics for Experimenters: Design, Innovation, and Discovery* (2nd ed.). New York, NY: Wiley.

Brain, P. (1975). What does individual housing mean to a mouse? *Life Sciences, 16*(2), 187–200.

Branchi, I., Santucci, D., Puopolo, M., & Alleva, E. (2004). Neonatal behaviors associated with ultrasonic vocalizations in mice (*mus musculus*): A slow-motion analysis. *Developmental Psychobiology, 44*(1), 37–44.

Broadhurst, P. (1961). Analysis of maternal effects in the inheritance of behavior. *Animal Behaviour, 9*(3-4), 129–141.

Brooks, S. P., & Dunnett, S. B. (2009). Tests to assess motor phenotype in mice: A user's guide. *Nature Reviews Neuroscience, 10*(7), 519–529.

Brooks, S. P., Pask, T., Jones, L., & Dunnett, S. B. (2004). Behavioural profiles of inbred mouse strains used as transgenic backgrounds. I: Motor tests. *Genes, Brain and Behavior, 3*(4), 206–215.

Brooks, S. P., Pask, T., Jones, L., & Dunnett, S. B. (2005). Behavioural profiles of inbred mouse strains used as transgenic backgrounds. II: Cognitive tests. *Genes, Brain and Behavior, 4*(5), 307–317.

Brown, R. E., & Wong, A. A. (2007). The influence of visual ability on learning and memory performance in 13 strains of mice. *Learning and Memory, 14*(3), 134–144.

270

Buccafusco, J. J. (2001). *Methods of behavior analysis in neuroscience*. Boca Raton, FL: CRC Press, Taylor & Francis Group.

Bulman-Fleming, B., & Wahlsten, D. (1988). Effects of hybrid maternal environment on brain growth and corpus callosum defects of inbred BALB/c mice: A study using ovarian grafting. *Experimental Neurology, 99*(3), 636–646.

Bult, C. J., Eppig, J. T., Kadin, J. A., Richardson, J. E., & Blake, J. A. (2008). The Mouse Genome Database (MGD): Mouse biology and model systems. *Nucleic Acids Research, 36*, D724–D728.

Calhoun, J. B. (1962). Population density and social pathology. *Scientific American, 306*, 139–148.

Capecchi, M. R. (2007). *Gene Targeting 1977–present*. Nobel lecture, Dec 7. 2007. See nobelprize.org.

Carlier, M., & Nosten, M. (1987). Interaction between genotype and pre or postnatal maternal environments: Examples from behaviors observed in inbred strains of mice. In T. Fujii, & P. M. Adams (Eds.), *Functional Teratogenesis* (pp. 27–36). Tokyo: Teikyo University Press.

Carlier, M., Nosten-Bertrand, M., & Michard-Vanhée, C. (1992). Separating genetic effects from maternal environmental effects. In D. Goldowitz, D. Wahlsten, & R. E. Wimer (Eds.), *Techniques for the Genetic Analysis of Brain and Behavior: Focus on the Mouse (Techniques in the Behavioral and Neural Sciences)* (pp. 111–126). New York, NY: Elsevier.

Carran, A. B., Yeudall, L. T., & Royce, J. R. (1964). Voltage level and skin resistance in avoidance conditioning in inbred strains of mice. *Journal of Comparative and Physiological Psychology, 58*, 427–430.

Caspi, A., Sugden, K., Moffitt, T. E., Taylor, A., Craig, I. W., Harrington, H., et al. (2003). Influence of life stress on depression: Moderation by a polymorphism in the 5-HTT gene. *Science, 301*(5631), 386–389.

Champagne, F. A. (2008). Epigenetic mechanisms and the transgenerational effects of maternal care. *Frontiers in Neuroendocrinology, 29*(3), 386–397.

Chesler, E. J., Wilson, S. G., Lariviere, W. R., Rodriguez-Zas, S. L., & Mogil, J. S. (2002). Identification and ranking of genetic and laboratory environment factors influencing a behavioral trait, thermal nociception, via computational analysis of a large data archive. *Neuroscience & Biobehavioral Reviews, 26*(8), 907–923.

Chorney, M. J., Chorney, K., Seese, N., Owen, M. J., McGuffin, P., Daniels, J., et al. (1998). A quantitative trait locus associated with cognitive ability in children. *Psychological Science, 9*(3), 159–166.

Cimadevilla, J. M., Fenton, A. A., & Bures, J. (2001). New spatial cognition tests for mice: Passive place avoidance on stable and active place avoidance on rotating arenas. *Brain Research Bulletin, 54*(5), 559–563.

Clamp, M., Fry, B., Kamal, M., Xie, X., Cuff, J., Lin, M. F., et al. (2007). Distinguishing protein-coding and noncoding genes in the human genome. *Proceedings of the National Academy of Sciences of the United States of America, 104*(49), 19428–19433.

Clark, M. G., Vasilevsky, S., & Myers, T. M. (2003). Air and shock two-way shuttlebox avoidance in C57BL/6J and 129X1/SvJ mice. *Physiology & Behavior, 78*(1), 117–123.

Clément, Y., Calatayud, F., & Belzung, C. (2002). Genetic basis of anxiety-like behaviour: A critical review. *Brain Research Bulletin, 57*(1), 57–71.

Cohen, J. (1988). *Statistical Power Analysis for the Behavioral Sciences* (2nd ed.). Mahwah, NJ: Lawrence Erlbaum.

Cohen, J. (1992). A power primer. *Psychological Bulletin, 112*(1), 155–159.

Coleman, D. L. (1979). Obesity genes: Beneficial effects in heterozygous mice. *Science, 203*(4381), 663–665.

Coleman, D. L., & Hummel, K. P. (1967). Studies with the mutation, diabetes, in the mouse. *Diabetologia, 3*(2), 238.

Collins, F. S., Rossant, J., & Wurst, W. (2007). A mouse for all reasons. *Cell, 128*(1), 9–13.

Collins, R. A. (1970). Aggression in mice selectively bred for brain weight. *Behavioral Genetics, 1*(2), 169–171.

Collins, R. L. (1975). When left-handed mice live in right-handed worlds. *Science, 187*(4172), 181–184.

Cook, M. N., Bolivar, V. J., McFadyen, M. P., & Flaherty, L. (2002). Behavioral differences among 129 substrains: Implications for knockout and transgenic mice. *Behavioral Neuroscience, 116*(4), 600–611.

Cooper, R. M., & Zubek, J. P. (1958). Effects of enriched and restricted early environments on the learning ability of bright and dull rats. *Canadian Journal of Psychology, 12*(3), 159–164.

Crabbe, J. C., Cotnam, C. J., Cameron, A. J., Schlumbohm, J. P., Rhodes, J. S., Metten, P., et al. (2003). Strain differences in three measures of ethanol intoxication in mice: The screen, dowel and grip strength tests. *Genes, Brain and Behavior, 2*(4), 201–213.

Crabbe, J. C., Metten, P., Cameron, A. J., & Wahlsten, D. (2005). An analysis of the genetics of alcohol intoxication in inbred mice. *Neuroscience & Biobehavioral Reviews, 28*(8), 785–802.

Crabbe, J. C., & Morris, R. G. M. (2004). Festina lente: Late-night thoughts on high-throughput screening of mouse behavior. *Nature Neuroscience, 7*(11), 1175–1179.

Crabbe, J. C., Wahlsten, D., & Dudek, B. C. (1999). Genetics of mouse behavior: Interactions with laboratory environment. *Science, 284*(5420), 1670–1672.

Crawley, J. N. (2000). *What's wrong with my mouse? Behavioral phenotyping of transgenic and knockout mice* (1st ed.). New York, NY: Wiley.

Crawley, J. N. (2008). Behavioral phenotyping strategies for mutant mice. *Neuron, 57*(6), 809–818.

Crawley, J. N., Belknap, J. K., Collins, A., Crabbe, J. C., Frankel, W., Henderson, N., et al. (1997). Behavioral phenotypes of inbred mouse strains: Implications and recommmendations for molecular studies. *Psychopharmacology, 132*(2), 107–124.

Crawley, J. N., Gerfen, C., McKay, R., Rogawski, M., Sibley, D., & Skolnick, P. (2005). *Current protocols in neuroscience.* Hoboken, NJ: Wiley.

Crawley, J. N., Gerfen, C. R., Rogawski, M. A., Sibley, D. R., Skolnick, P., & Wray, S. (2007). *Short protocols in neuroscience.* Hoboken, NJ: Wiley.

Crawley, J. N., & Paylor, R. (1997). A proposed test battery and constellations of specific behavioral paradigms to investigate the behavioral phenotypes of transgenic and knockout mice. *Hormones and Behavior, 31*(3), 197–211.

Crowcroft, P. (1973). *Mice all over.* Chicago, IL: Chicago Zoological Society.

Crusio, W. E. (1999). Methodological considerations for testing learning in mice. In W. E. Crusio, & R. Gerlai (Eds.), *Handbook of Molecular–Genetic Techniques for Brain and Behavior Research (Techniques in the Behavioral and Neural Sciences)* (pp. 638–651). Amsterdam: Elsevier.

Crusio, W. E. (2004). Flanking gene and genetic background problems in genetically manipulated mice. *Biological Psychiatry, 56*(6), 381–385.

Crusio, W. E., & Gerlai, R. (1999). *Handbook of Molecular–Genetic Techniques for Brain and Behavior Research (Techniques in the Behavioral and Neural Sciences).* Amsterdam: Elsevier

Crusio, W. E., Goldowitz, D., Holmes, A., & Wolfer, D. (2009). Standards for the publication of mouse mutant studies. *Genes, Brain and Behavior, 8*(1), 1–4.

Crusio, W. E., Schwegler, H., & Brust, I. (1993). Covariations between hippocampal mossy fibres and working and reference memory in spatial and non-spatial radial maze tasks in mice. *European Journal of Neuroscience, 5*(10), 1413–1420.

Cunningham, C. L., & Phillips, T. J. (2003). Genetic Basis of Ethanol Reward. In R. Maldonado (Ed.), *Molecular Biology of Drug Addiction* (pp. 263–294). Totowa, NJ: Humana Press, Inc.

D'Amato, M. R., & Schiff, D. (1964). Long-term discriminated avoidance performance in the rat. *Journal of Comparative and Physiological Psychology, 57,* 123–126.

Darvasi, A. (2005). Dissecting complex traits: The geneticists' "Around the world in 80 days." *Trends in Genetics, 21* (7), 373–376.

Dawson, P. A., Steane, S. E., & Markovich, D. (2005). Impaired memory and olfactory performance in NaSi-1 sulphate transporter deficient mice. *Behavioural Brain Research, 159*(1), 15–20.

de Visser, V. L., van den Bos, R., Kuurman, W. W., Kas, M. J., & Spruijt, B. M. (2006). Novel approach to the behavioural characterization of inbred mice: Automated home cage observations. *Genes, Brain and Behavior, 5* (6), 458–466.

Deacon, R. M. (2006). Housing, husbandry and handling of rodents for behavioral experiments. *Nature Protocols, 1,* 936–946.

DeFries, J. C., Gervais, M. C., & Thomas, E. A. (1978). Response to 30 generations of selection for open-field activity in laboratory mice. *Behavioral Genetics, 8*(3), 3–13.

Demers, G., Griffin, G., De, V. G., Haywood, J. R., Zurlo, J., & Bedard, M. (2006). Animal research. Harmonization of animal care and use guidance. *Science, 312*(5774), 700–701.

Donaldson, Z. R., Yang, S. H., Chan, A. W., & Young, L. J. (2009). Production of germline transgenic prairie voles (*Microtus ochrogaster*) using lentiviral vectors. *Biology of Reproduction, 81*(6), 1189–1195.

Dreary, I. J., Spinath, F. M., & Bates, T. C. (2006). Genetics of intelligence. *European Journal of Human Genetics, 14*(6), 690–700.

Dubreuil, D., Tixier, C., Dutrieux, G., & Edeline, J. M. (2003). Does the radial arm maze necessarily test spatial memory? *Neurobiology of Learning and Memory, 79*(1), 109–117.

Dunham, N. W., & Miya, T. S. (1957). A note on a simple apparatus for detecting neurological deficit in rats and mice. *Journal of the American Pharmaceutical Association. American Pharmaceutical Association (Baltimore), 46*(3), 208–209.

Editorial. (2009). Troublesome variability in mouse studies. *Nature Neuroscience, 12*(9), 1075.

Eisen, E. J. (2005). *The mouse in animal genetics and breeding research.* London, UK: Imperial College Press.

Eisenberg, J. F. (1968). Behaviour patterns. In J. A. King (Ed.), *Biology of Peromyscus (rodents).* Washington, DC: AIBS.

Epling, W. F. (1989). Rats. *Behavior Analyst, 12*(2), 251–253.

Erdfelder, E., Faul, F., & Buchner., A. (1996). GPOWER: A general power analysis program. *Behavior Research Methods, Instruments, & Computers, 28*(1), 1–11.

Falconer, D. S., & MacKay, T. F. (1996). *Introduction to Quantitative Genetics* (4th ed.). Harlow, UK: Longman.

Faul, F., Erdfelder, E., Lang, A. G., & Buchner, A. (2007). G*Power 3: A flexible statistical power analysis program for the social, behavioral, and biomedical sciences. *Behavior Research Methods, 39*(2), 175–191.

Fentress, J. C. (1992). Emergence of pattern in the development of mammalian movement sequences. *Journal of Neurobiology, 23*(10), 1529–1556.

Fernandez, R. C., Rezola, L. G., Moreno, O. V., & Sanchez, A. A. (2008). Effects of social stress on tumor development in dominant male mice with diverse behavioral profiles. *Psicothema, 20*(4), 818–824.

File, S. E. (2001). Factors controlling measures of anxiety and responses to novelty in the mouse. *Behavioural Brain Research, 125*(1–2), 151–157.

Fink, S., Excoffier, L., & Heckel, G. (2006). Mammalian monogamy is not controlled by a single gene. *Proceedings of the National Academy of Sciences of the United States of America, 103*(29), 10956–10960.

Finn, D. A., Rutledge-Gorman, M. T., & Crabbe, J. C. (2003). Genetic animal models of anxiety. *Neurogenetics, 4*(3), 109–135.

Flint, J. (2003). Analysis of quantitative trait loci that influence animal behavior. *Journal of Neurobiology, 54*(1), 46–77.

Flood, J. F., Smith, G. E., Bennett, E. L., Alberti, M. H., Orme, A. E., & Jarvik, M. E. (1986). Neurochemical and behavioral effects of catecholamine and protein synthesis inhibitors in mice. *Pharmacology, Biochemistry and Behavior, 24*(3), 631–645.

Foldi, C. J., Eyles, D. W., McGrath, J. J., & Burne, T. H. (2010). Advanced paternal age is associated with alterations in discrete behavioural domains and cortical neuroanatomy of C57BL/6J mice. *European Journal of Neuroscience, 31*(3), 556–564.

Fox, J., Davisson, M., Quimby, F., Barthold, S., Newcomer, C., & Smith, A. (2007). *The mouse in biomedical research.* Amsterdam: Elsevier.

Frazer, K. A., Eskin, E., Kang, H. M., Bogue, M. A., Hinds, D. A., Beilharz, E. J., et al. (2007). A sequence-based variation map of 8.27 million SNPs in inbred mouse strains. *Nature, 448*(7157), 1050–1053.

Froy, O., & Miskin, R. (2010). Effect of feeding regimens on circadian rhythms: Implications for aging and longevity. *Aging (Albany, NY), 2*(1), 7–27.

Fuller, J. L., & Thompson, W. R. (1960). *Behavior genetics.* New York, NY: Wiley.

Funahashi, H., Takenoya, F., Guan, J. L., Kageyama, H., Yada, T., & Shioda, S. (2003). Hypothalamic neuronal networks and feeding-related peptides involved in the regulation of feeding. *Anatomical Science International/Japanese Association of Anatomists, 78*(3), 123–138.

Galsworthy, M. J., Amrein, I., Kuptsov, P. A., Poletaeva, I. I., Zinn, P., Rau, A., et al. (2005). A comparison of wild-caught wood mice and bank voles in the Intellicage: Assessing exploration, daily activity patterns and place learning paradigms. *Behavioural Brain Research, 157*(2), 211–217.

Gaskill, B. N., Rohr, S. A., Pajor, E. A., Lucas, J. R., & Garner, J. P. (2009). Some like it hot: Mouse temperature preferences in laboratory housing. *Applied Animal Behaviour Science, 116*(2–4), 279–285.

Gawrylewski, A. (2007). The trouble with animal models. *The Scientist, 21*(7), 45–51.

Gelegen, C., Collier, D. A., Campbell, I. C., Oppelaar, H., & Kas, M. J. (2006). Behavioral, physiological, and molecular differences in response to dietary restriction in three inbred mouse strains. *American Journal of Physiology, Endocrinology and Metabolism, 291*(3), E574–E581.

Geller, A., Robustelli, F., Barondes, S. H., Cohen, H. D., & Jarvik, M. E. (1969). Impaired performance by post-trial injections of cyclohexamide in a passive avoidance task. *Psychopharmacologia, 14*(5), 371–376.

Gerlai, R. (1996). Gene-targeting studies of mammalian behavior: Is it the mutation or the background genotype? *Trends in Neurosciences, 19*(5), 177–181.

Gerlai, R. (1999). Ethological approaches in behavioral neurogenetic research. In W. E. Crusio, & R. Gerlai (Eds.), *Handbook of molecular-genetic techniques for brain and behavior research* (pp. 605–613). Amsterdam: Elsevier.

Gerlai, R. (2001). Behavioral tests of hippocampal function: Simple paradigms, complex problems. *Behavioural Brain Research, 125*(1–2), 269–277.

Gerlai, R. T., McNamara, A., Williams, S., & Phillips, H. S. (2002). Hippocampal dysfunction and behavioral deficit in the water maze in mice: An unresolved issue? *Brain Research Bulletin, 57*(1), 3–9.

Giampa, C., Middei, S., Patassini, S., Borreca, A., Marullo, F., Laurenti, D., et al. (2009). Phosphodiesterase type IV inhibition prevents sequestration of CREB binding protein, protects striatal parvalbumin interneurons and rescues motor deficits in the R6/2 mouse model of Huntington's disease. *European Journal of Neuroscience, 29*(5), 902–910.

Gilroy, D. J., Kauffman, K. W., Hall, R. A., Huang, X., & Chu, F. S. (2000). Assessing potential health risks from microcystin toxins in blue-green algae dietary supplements. *Environmental Health Perspectives, 108*(5), 435–439.

Ginsburg, B., & Allee, W. C. (1942). Some effects of conditioning on social dominance and subordination in inbred strains of mice. *Physiological Zoology, 15*(4), 485–506.

Ginsburg, B. E. (1967). Genetic parameters in behavioral research. In J. Hirsch (Ed.), *Behavior-genetic analysis* (pp. 135–153). New York, NY: McGraw-Hill.

Goddyn, H., Leo, S., Meert, T., & d'Hooge, R. (2006). Differences in behavioural test battery performance between mice with hippocampal and cerebellar lesions. *Behavioural Brain Research, 173*(1), 138–147.

Goios, A., Pereira, L., Bogue, M., Macaulay, V., & Amorim, A. (2007). mtDNA phylogeny and evolution of laboratory mouse strains. *Genome Research, 17*(3), 293–298.

Golani, I., & Fentress, J. C. (1985). Early ontogeny of face grooming in mice. *Developmental Psychobiology, 18*(6), 529–544.

Gottlieb, G. (2007). Probabilistic epigenesis. *Developmental Science, 10*(1), 1–11.

Gould, S. J. (1966). Allometry and size in ontogeny and phylogeny. *Biological Review, 41*(4), 587–640.

Goulding, E. H., Schenk, A. K., Juneja, P., MacKay, A. W., Wade, J. M., & Tecott, L. H. (2008). A robust automated system elucidates mouse home cage behavioral structure. *Proceedings of the National Academy of Sciences of the United States of America, 105*(52), 20575–20582.

Gray, S., & Hurst, J. L. (1995). The effects of cage cleaning on aggression within groups of male laboratory mice. *Animal Behaviour, 49*(3), 821–826.

Green, E. L. (1981). *Genetics and probability in animal breeding experiments*. New York, NY: Oxford University Press.

Greenman, D. L., Bryant, P., Kodell, R. L., & Sheldon, W. (1982). Influence of cage shelf level on retinal atrophy in mice. *Laboratory Animal Science, 32*(4), 353–356.

Greenough, W. T., Black, J. E., Klintsova, A., Bates, K. E., & Weiler, I. J. (1999). Experience and plasticity in brain structure: Possible implications of basic research findings for developmental disorders. In S. H. Broman, & J. M. Fletcher (Eds.), *The changing nervous system* (pp. 51–70). New York, NY: Oxford University Press.

Gregory, R. J. (2010). *Psychological testing. History, principles, and applications* (6th ed.). Englewood Cliffs, NJ: Prentice-Hall.

Grubb, S. C., Churchill, G. A., & Bogue, M. A. (2004). A collaborative database of inbred mouse strain characteristics. *Bioinformatics, 20*(16), 2857–2859.

Guillen, J. (2010). The use of performance standards by AAALAC International to evaluate ethical review in European institutions. *Lab Animal (NY), 39*(2), 49–53.

Harrington, G. M. (1985). Developmental perspectives, behavior-genetic analysis, and models of the individual. In J. L. McGaugh (Ed.), *Contemporary Psychology: Biological Processes and Theoretical Issues*. Amsterdam: North-Holland.

Harrison, F. E., Hosseini, A. H., & McDonald, M. P. (2009). Endogenous anxiety and stress responses in water maze and Barnes maze spatial memory tasks. *Behavioural Brain Research, 198*(1), 247–251.

Hascoet, M., Bourin, M., & Dhonnchadha, B. A. (2001). The mouse light—dark paradigm: A review. *Progress in Neuro-Psychopharmacology and Biological Psychiatry, 25*(1), 141–166.

Hays, W. L. (1988). *Statistics* (4th ed.). New York, NY: Holt, Rinehart, Winston.

Hedrich, H. J., & Bullock, G. (2004). *The laboratory mouse*. London, UK: Elsevier.

Hen, I., Sakov, A., Kafkafi, N., Golani, I., & Benjamini, Y. (2004). The dynamics of spatial behavior: How can robust smoothing techniques help? *Journal of Neuroscience Methods, 133*(1–2), 161–172.

Henderson, N. D. (1970). Genetic influences on the behavior of mice can be obscured by laboratory rearing. *Journal of Comparative and Physiological Psychology, 72*(3), 505–511.

Henderson, N. D., Turri, M. G., DeFries, J. C., & Flint, J. (2004). QTL analysis of multiple behavioral measures of anxiety in mice. *Behavior Genetics, 34*(3), 267–293.

Henry, K. R. (1984). Noise and the young mouse: Genotype modifies the sensitive period for effects on cochlear physiology and auidogenic seizures. *Behavioral Neuroscience, 98*(6), 1073–1082.

Henry, K. R., & Bowman, R. E. (1970). Behavior—genetic analysis of the ontogeny of acoustically primed audiogenic seizures in mice. *Journal of Comparative and Physiological Psychology, 70*(2), 235–241.

Heron, W. T. (1941). The inheritance of brightness and dullness in maze learning ability in the rat. *Journal of Genetic Psychology, 59*, 41–49.

Herzog, H. A. (1988). The moral status of mice. *American Psychologist, 43*, 473–474.

Hinkelman, K., & Kempthorne, O. (2008). *Design and analysis of experiments. Volume I: Introduction to experimental design*. New York, NY: Wiley.

Holliday, R. (2006). Epigenetics: A historical overview. *Epigenetics, 1*(2), 76–80.

Holmes, A. (2001). Targeted gene mutation approaches to the study of anxiety-like behavior in mice. *Neuroscience & Biobehavioral Reviews, 25*(3), 261–273.

Holmes, A., & Rodgers, R. J. (1999). Influence of spatial and temporal manipulations on the anxiolytic efficacy of chlordiazepoxide in mice previously exposed to the elevated plus-maze. *Neuroscience & Biobehavioral Reviews, 23*(7), 971–980.

Holmes, A., Yang, R. J., Lesch, K. P., Crawley, J. N., & Murphy, D. L. (2003). Mice lacking the serotonin transporter exhibit 5-HT(1A) receptor-mediated abnormalities in tests for anxiety-like behavior. *Neuropsychopharmacology, 28*(12), 2077–2088.

Hood, K. E., & Cairns, R. B. (1989). A developmental—genetic analysis of aggressive behavior in mice: 4. Genotype—environment interaction. *Aggressive Behavior, 15*(5), 361—380.

Hubel, D. H., & Wiesel, T. N. (1959). Receptive fields of single neurones in the cat's striate cortex. *Journal of Physiology, 148*(3), 574—591.

Hughes, K. R., & Zubek, J. P. (1956). Effect of glutamic acid on the learning ability of bright and dull rats: I. Administration during infancy. *Canadian Journal of Psychology, 10*(3), 132—138.

Hyde, J. S., Lindberg, S. M., Linn, M. C., Ellis, A. B., & Williams, C. C. (2008). Gender similarities characterize math performance. *Science, 321*(5888), 494—495.

Iivonen, H., Nurminen, L., Harri, M., Tanila, H., & Puoliväli, J. (2003). Hypothermia in mice tested in Morris water maze. *Behavioural Brain Research, 141*(2), 207—213.

Jablonka, E., & Raz, G. (2009). Transgenerational epigenetic inheritance: Prevalence, mechanisms, and implications for the study of heredity and evolution. *Quarterly Review of Biology, 84*(2), 131—176.

Jackson, I. J., & Abbott, C. M. (2000). *Mouse genetics and transgenics.* Oxford, UK: Oxford University Press.

Jazin, E., & Cahill, L. (2010). Sex differences in molecular neuroscience: From fruit flies to humans. *Nature Reviews Neuroscience, 11,* 9—17.

Johnson, A. W., Bannerman, D. M., Rawlins, N. P., Sprengel, R., & Good, M. A. (2005). Impaired outcome-specific devaluation of instrumental responding in mice with a targeted deletion of the AMPA receptor glutamate receptor 1 subunit. *Journal of Neuroscience, 25*(9), 2359—2365.

Jones, B. C., & Mormede, P. (2007). *Neurobehavioral genetics: Methods and applications* (2nd ed.). Boca Raton, FL: CRC Press, Taylor & Francis Group.

Jones, B. J., & Roberts, D. J. (1968). Quantitative measurement of motor incoordination in naive mice using an accelerating rotarod. *Journal of Pharmacy and Pharmacology, 20*(4), 302—304.

Kafkafi, N., Benjamini, Y., Sakov, A., Elmer, G. I., & Golani, I. (2005). Genotype-environment interactions in mouse behavior: A way out of the problem. *Proceedings of the National Academy of Sciences of the United States of America, 102*(12), 4619—4624.

Kafkafi, N., Lipkind, D., Benjamini, Y., Mayo, C. L., Elmer, G. I., & Golani, I. (2003). SEE locomotor behavior test discriminates C57BL/6J and DBA/2J mouse inbred strains across laboratories and protocol conditions. *Behavioral Neuroscience, 117*(3), 464—477.

Kalueff, A. V., Aldridge, J. W., LaPorte, J. L., Murphy, D. L., & Tuohimaa, P. (2007). Analyzing grooming microstructure in neurobehavioral experiments. *Nature Protocols, 2*(10), 2538—2544.

Kalueff, A. V., Wheaton, M., & Murphy, D. L. (2007). What's wrong with my mouse model? Advances and strategies in animal modeling of anxiety and depression. *Behavioural Brain Research, 179*(1), 1—18.

Kaplan, R. M., & Saccuzzo, D. P. (2009). *Psychological testing. Principles, applications and issues* (7th ed.). Belmont, CA: Wadsworth.

Kas, M. J., de Mooij-van Malsen, J. G., de, K. M., van Gassen, K. L., van Lith, H. A., Olivier, B., et al. (2009). High-resolution genetic mapping of mammalian motor activity levels in mice. *Genes, Brain and Behavior, 8*(1), 13—22.

Kas, M. J., Kaye, W. H., Foulds, M. W., & Bulik, C. M. (2009). Interspecies genetics of eating disorder traits. *American Journal of Medical Genetics (Neuropsychiatric Genetics), 150B*(3), 318—327.

Kazdoba, T. M., Del Vecchio, R. A., & Hyde, L. A. (2007). Automated evaluation of sensitivity to foot shock in mice: Inbred strain differences and pharmacological validation. *Behavioural Pharmacology, 18*(2), 89—102.

Kempermann, G., Gast, D., & Gage, F. H. (2002). Neuroplasticity in old age: Sustained fivefold induction of hippocampal neurogenesis by long-term environmental enrichment. *Annals of Neurology, 52*(2), 135—143.

King, J. A., & Mavromatis, A. (1956). The effect of a conflict situation on learning ability in two strains of mice. *Journal of Comparative and Physiological Psychology, 49*(5), 465—468.

Klapdor, K., & van der Staay, F. J. (1996). The Morris water-escape task in mice: Strain differences and effects of intra-maze contrast and brightness. *Physiology & Behavior, 60*(5), 1247—1254.

Koike, H., Ibi, D., Mizoguchi, H., Nagai, T., Nitta, A., Takuma, K., et al. (2009). Behavioral abnormality and pharmacologic response in social isolation-reared mice. *Behavioural Brain Research, 202*(1), 114—121.

Kopp, C. (2001). Locomotor activity rhythm in inbred strains of mice: Implications for behavioural studies. *Behavioural Brain Research, 125*(1—2), 93—96.

Koubova, J., & Guarente, L. (2003). How does calorie restriction work? *Genes & Development, 17,* 313—321.

Kraemer, H. C., & Thiemann, S. (1987). *How many subjects? Statistical power analysis in research.* Newbury Park, CA: Sage Publications.

Lad, H. V., Liu, L., Paya-Cano, J. L., Parsons, M. J., Kember, R., Fernandes, C., et al. (2010). Behavioural battery testing: Evaluation and behavioural outcomes in 8 inbred mouse strains. *Physiology & Behavior, 99*(3), 301—316.

Lagerspetz, K. M., & Lagerspetz, K. Y. (1971). Changes in the aggressiveness of mice resulting from selective breeding, learning and social isolation. *Scandinavian Journal of Psychology, 12*(4), 241—248.

Lander, E. S., Linton, L. M., Birren, B., Nusbaum, C., Zody, M. C., Baldwin, J., et al. (2001). Initial sequencing and analysis of the human genome. *Nature, 409*(6822), 860–921.

Lee, S. M., & Bressler, R. (1981). Prevention of diabetic nephropathy by diet control in the db/db mouse. *Diabetes, 30*(2), 106–111.

Lehner, P. N. (1996). *Handbook of ethological methods.* New York, NY: Cambridge University Press.

Lehrman, D. S. (1970). Semantic and conceptual issues in the nature-nurture problem. In L. R. Aronson, D. S. Lehrman, E. Tobach, & J. S. Rosenblatt (Eds.), *Development and evolution of behavior* (pp. 17–52). San Francisco, CA: Freeman.

Lewejohann, L., Reinhard, C., Schrewe, A., Brandewiede, J., Haemisch, A., Gortz, N., et al. (2006). Environmental bias? Effects of housing conditions, laboratory environment and experimenter on behavioral tests. *Genes, Brain and Behavior, 5*(1), 64–72.

Liao, J. F., Hung, W. Y., & Chen, C. F. (2003). Anxiolytic-like effects of baicalein and baicalin in the Vogel conflict test in mice. *European Journal of Pharmacology, 464*(2–3), 141–146.

Liebetanz, D., Baier, P. C., Paulus, W., Meuer, K., Bahr, M., & Weishaupt, J. H. (2007). A highly sensitive automated complex running wheel test to detect latent motor deficits in the mouse MPTP model of Parkinson's disease. *Experimental Neurology, 205*(1), 207–213.

Lipkind, D., Sakov, A., Kafkafi, N., Elmer, G. I., Benjamini, Y., & Golani, I. (2004). New replicable anxiety-related measures of wall vs center behavior of mice in the open field. *Journal of Applied Physiology, 97*(1), 347–359.

Lipp, H.-P., Amrein, I., Slomankia, L., & Wolfer, D. P. (2007). Natural genetic variation of hippocampal structures and behavior — an update. In B. C. Jones, & P. Mormède (Eds.), *Neurobehavioral genetics* (2nd ed.). (pp. 389–410) Boca Raton, FL: CRC Press, Taylor & Francis Group.

Little, C. C., Snell, G. D., Bittner, J. J., Cloudman, A. M., Fekete, E., Heston, W. E, et al. (1941). *Biology of the Laboratory Mouse.* New York, NY: Dover.

Liu, X., & Gershenfeld, H. K. (2003). An exploratory factor analysis of the Tail Suspension Test in 12 inbred strains of mice and an F2 intercross. *Brain Research Bulletin, 60*(3), 223–231.

Longo, V. D., & Fabrizio, P. (2002). Regulation of longevity and stress resistance: A molecular strategy conserved from yeast to humans? *Cellular and Molecular Life Sciences, 59*(6), 903–908.

Lorenz, K. (1965). *Evolution and modification of behavior.* Chicago, IL: University of Chigaco Press.

Lorenz, K. (1981). *The Foundations of Ethology.* New York, NY: Touchstone.

Lynch, C. J. (1969). The so-called Swiss mouse. *Laboratory Animal Care, 19*(2), 214–220.

Macbeth, A. H., Edds, J. S., & Young, W. S., III (2009). Housing conditions and stimulus females: A robust social discrimination task for studying male rodent social recognition. *Nature Protocols, 4*, 1574–1581.

Maddalena, L., & Petrosino, A. (2008). A self-organizing approach to background subtraction for visual surveillance applications. *IEEE Transactions on Image Processing, 17*(7), 1168–1177.

Mair, W., & Dillin, A. (2008). Aging and survival: The genetics of life span extension by dietary restriction. *Annual Review of Biochemistry, 77*, 727–754.

Mamiya, T., Noda, Y., Noda, A., Hiramatsu, M., Karasawa, K., Kameyama, T., et al. (2000). Effects of sigma receptor agonists on the impairment of spontaneous alternation behavior and decrease of cyclic GMP level induced by nitric oxide synthase inhibitors in mice. *Neuropharmacology, 39*(12), 2391–2398.

Mandelbrot, B. B. (1983). *The fractal geometry of nature.* New York, NY: Freeman.

Mandillo, S., Tucci, V., Holter, S. M., Meziane, H., Banchaabouchi, M. A., Kallnik, M., et al. (2008). Reliability, robustness, and reproducibility in mouse behavioral phenotyping: A cross-laboratory study. *Physiological Genomics, 34*(3), 243–255.

Mangiarini, L., Sathasivam, K., Seller, M., Cozens, B., Harper, A., Hetherington, C., et al. (1996). Exon 1 of the HD gene with an expanded CAG repeat is sufficient to cause a progressive neurological phenotype in transgenic mice. *Cell, 87*(3), 493–506.

Martin, A. L., & Brown, R. E. (2010). The lonely mouse: Verification of a separation-induced model of depression in female mice. *Behavioural Brain Research, 207*(7), 196–207.

Martin, B., Ji, S., Maudsley, S., & Mattson, M. P. (2010). "Control" laboratory rodents are metabolically morbid: Why it matters. *Proceedings of the National Academy of Sciences of the United States of America, 107*, 6127–6133.

Martin, P., & Bateson, P. (2007). *Measuring behavior. An introductory guide* (3rd ed.). Cambridge, UK: Cambridge University Press.

Mathiasen, L., & Mirza, N. R. (2005). A comparison of chlordiazepoxide, bretazenil, L838,417 and zolpidem in a validated mouse Vogel conflict test. *Psychopharmacology (Berl), 182*(4), 475–484.

Matzel, L. D., Han, Y. R., Grossman, H., Karnik, M. S., Patel, D., Scott, N., et al. (2003). Individual differences in the expression of a "general" learning ability in mice. *Journal of Neuroscience, 23*(16), 6423–6433.

Maxson, S. C. (1992). Methodological issues in genetic analysis of an agonistic behavior (offense) in male mice. In D. Goldowitz, D. Wahlsten, & R. E. Wimer (Eds.), *Techniques for the genetic analysis of brain and behavior: Focus on the mouse* (pp. 349–373). Amsterdam: Elsevier.

Maxson, S. C., & Canastar, A. (2003). Conceptual and methodological issues in the genetics of mouse agonistic behavior. *Hormones and Behavior, 44*(3), 258–262.

Maxson, S. C., Ginsburg, B. E., & Trattner, A. (1979). Interaction of Y-chromosomal and autosomal gene(s) in the development of intermale aggression in mice. *Behavior Genetics, 9*(3), 219–226.

McClearn, G. E. (1982). Selected uses of the mouse in behavioral research. In H. L. Foster, J. D. Small, & J. G. Fox (Eds.), *Experimental biology and oncology* (pp. 37–49). New York, NY: Academic Press.

McClearn, G. E., & Rodgers, D. A. (1959). Differences in alcohol preference among inbred strains of mice. *Quarterly Journal for the Study of Alcoholism, 20*, 691–695.

McDougall, W. (1927). An experiment for the testing of the hypothesis of Lamarck. *Journal of Psychology, 17*, 267–304.

McIlwain, K. L., Merriweather, M. Y., Yuva-Paylor, L. A., & Paylor, R. (2001). The use of behavioral test batteries: Effects of training history. *Physiology & Behavior, 73*(5), 705–717.

McKay, T. F. C., Stone, E. A., & Ayroles, J. F. (2009). The genetics of quantitative traits: Challenges and prospects. *Nature Reviews Genetics, 10*(8), 565–577.

Mechan, A. O., Wyss, A., Rieger, H., & Mohajeri, M. H. (2009). A comparison of learning and memory characteristics of young and middle-aged wild-type mice in the IntelliCage. *Journal of Neuroscience Methods, 180*(1), 43–51.

Merali, Z., Levac, C., & Anisman, H. (2003). Validation of a simple, ethologically relevant paradigm for assessing anxiety in mice. *Biological Psychiatry, 54*(5), 552–565.

Metten, P., Best, K. L., Cameron, A. J., Saultz, A. B., Zuraw, J. M., Yu, C. H., et al. (2004). Observer-rated ataxia: Rating scales for assessment of genetic differences in ethanol-induced intoxication in mice. *Journal of Applied Physiology, 97*(1), 360–368.

Michel, G. F. (2010). Behavioral science, engineering, and poetry revisited. *Journal of Comparative Psychology*. In-Press.

Minana-Solis, M. C., Angeles-Castellanos, M., Feillet, C., Pevet, P., Challet, E., & Escobar, C. (2009). Differential effects of a restricted feeding schedule on clock-gene expression in the hypothalamus of the rat. *Chronobiology International, 26*(5), 808–820.

Mineur, Y. S., Belzung, C., & Crusio, W. E. (2006). Effects of unpredictable chronic mild stress on anxiety and depression-like behavior in mice. *Behavioural Brain Research, 175*(1), 43–50.

Mitsui, S., Osako, Y., Yokoi, F., Dang, M. T., Yuri, K., Li, Y., et al. (2009). A mental retardation gene, motopsin/neurotrypsin/prss12, modulates hippocampal function and social interaction. *European Journal of Neuroscience, 30*(12), 2368–2378.

Mogil, J. S. (2009). Animal models of pain: Progress and challenges. *Nature Reviews Neuroscience, 10*(4), 283–294.

Mogil, J. S., Wilson, S. G., Bon, K., Lee, S. E., Chung, K., Raber, P., et al. (1999a). Heritability of nociception I: Responses of 11 inbred mouse strains on 12 measures of nociception. *Pain, 80*(1), 67–82.

Mogil, J. S., Wilson, S. G., Bon, K., Lee, S. E., Chung, K., Raber, P., et al. (1999b). Heritability of nociception II. 'Types' of nociception revealed by genetic correlation analysis. *Pain, 80*(1), 83–93.

Mohammed, A. H., Zhu, S. W., Darmopil, S., Hjerling-Leffler, J., Ernfors, P., Winblad, B., et al. (2002). Environmental enrichment and the brain. *Progress in Brain Research, 138*, 109–133.

Monahan, E. J., & Maxson, S. C. (1998). Y chromsome, urinary chemosignals, and an agonistic behavior (offense) of mice. *Physiology & Behavior, 64*(2), 123–132.

Morgan, D. K., & Whitelaw, E. (2008). The case for transgenerational epigenetic inheritance in humans. *Mammalian Genome, 19*(6), 394–397.

Morse, H. C., III (2007). Building a better mouse: One hundred years of genetics and biology. In J. G. Fox, S. Barthold, M. T. Davisson, C. Newcomer, F. Quimby, & A. Smith (Eds.), *The mouse in biomedical research* (pp. 1–12). Amsterdam: Elsevier.

Moy, S. S., Nadler, J. J., Young, N. B., Perez, A., Holloway, L. P., Barbaro, R. P., et al. (2007). Mouse behavioral tasks relevant to autism: Phenotypes of 10 inbred strains. *Behavioural Brain Research, 176*(1), 4–20.

Murphy, K. R., & Myors, B. (2004). *Statistical power analysis: A simple and general model for traditional and modern hypothesis tests* (2nd ed.). Mahwah, NJ: Lawrence Erlbaum Associates.

Nadeau, J. H., & Frankel, W. N. (2000). The road from phenotypic variation to gene discovery: Mutagenesis versus QTLs. *Nature Genetics, 25*(4), 381–384.

Nadler, J. J., Moy, S. S., Dold, G., Trang, D., Simmons, N., Perez, A., et al. (2004). Automated apparatus for quantitation of social approach behaviors in mice. *Genes, Brain and Behavior, 3*(5), 303–314.

Nakagawa, S., & Cuthill, I. C. (2007). Effect size, confidence interval and statistical significance: A practical guide for biologists. *Biological Reviews of the Cambridge Philosophical Society, 82*(4), 591–605.

277

Nguyen, P. V. (2006). Comparative plasticity of brain synapses in inbred mouse strains. *The Journal of Experimental Biology, 209*(Pt. 12), 2293–2303.

Noldus, L. P., Spink, A. J., & Tegelenbosch, R. A. (2001). EthoVision: A versatile video tracking system for automation of behavioral experiments. *Behavior Research Methods, Instruments, & Computers, 33*, 398–414.

Nowakowski, S. G., Swoap, S. J., & Sandstrom, N. J. (2009). A single bout of torpor in mice protects memory processes. *Physiology & Behavior, 97*(1), 115–120.

O'Leary, T. P., & Brown, R. E. (2008). The effects of apparatus design and test procedure on learning and memory performance of C57BL/6J mice on the Barnes maze. In A. J. Spink, M. R. Ballintijn, N. D. Bogers, F. Grieco, L. W. Loijens, L. P. Noldus, G. Smit, & P. H. Zimmerman (Eds.), *6th International Conference on Methods and Techniques in Behavioral Research.* The Netherlands: Noldus Information Technology.

Odom, D. T., Dowell, R. D., Jacobsen, E. S., Gordon, W., Danford, T. W., MacIsaac, K. D., et al. (2007). Tissue-specific transcriptional regulation has diverged significantly between human and mouse. *Nature Genetics, 39*(6), 730–732.

Overton, J. M., & Williams, T. D. (2004). Behavioral and physiologic responses to caloric restriction in mice. *Physiology & Behavior, 81*(5), 749–754.

Padeh, B., Wahlsten, D., & De Fries, J. C. (1974). Operant discrimination learning and operant bar-pressing rates in inbred and heterogeneous laboratory mice. *Behavior Genetics, 4*(4), 383–393.

Page, D. T., Kuti, O. J., & Sur, M. (2009). Computerized assessment of social approach behavior in mouse. *Frontiers in Behavioral Neuroscience, 3*, 48.

Palermo-Neto, J., Fonseca, E. S., Quinteiro-Filho, W. M., Correia, C. S., & Sakai, M. (2008). Effects of individual housing on behavior and resistance to Ehrlich tumor growth in mice. *Physiology & Behavior, 95*(3), 435–440.

Panda, S., Antoch, M. P., Miller, B. H., Su, A. I., Schook, A. B., Straume, M., et al. (2002). Coordinated transcription of key pathways in the mouse by the circadian clock. *Cell, 109*(3), 307–320.

Papaioannou, V. E., & Behringer, R. R. (2005). *Mouse phenotypes. A handbook of mutation analysis.* Cold Spring Harbor, NY: Cold Spring Harbor Laboratory Press.

Patil, S. S., Sunyer, B., Hoger, H., & Lubec, G. (2009). Evaluation of spatial memory of C57BL/6J and CD1 mice in the Barnes maze, the Multiple T-maze and in the Morris water maze. *Behavioural Brain Research, 198*(1), 58–68.

Paylor, R. (2008). Simultaneous behavioral characterizations: Embracing complexity. *Proceedings of the National Academy of Sciences of the United States of America, 105*(52), 20563–20564.

Paylor, R., Spencer, C. M., Yuva-Paylor, L. A., & Pieke-Dahl, S. (2006). The use of behavioral test batteries, II: Effect of test interval. *Physiology & Behavior, 87*(1), 95–102.

Peeler, D. F. (1995). Shuttlebox performance in BALB/cByJ, C57BL/6ByJ, and CXB recombinant inbred mice: Environmental and genetic determinants and constraints. *Psychobiology, 23*, 161–170.

Pesold, C., & Treit, D. (1994). The septum and amygdala differentially mediate the anxiolytic effects of benzodiazepines. *Brain Research, 638*(1–2), 295–301.

Peters, L. L., Robledo, R. F., Bult, C. J., Churchill, G. A., Paigen, B. J., & Svenson, K. L. (2007). The mouse as a model for human biology: A resource guide for complex trait analysis. *Nature Reviews Genetics, 8*(1), 58–69.

Petkov, P. M., Ding, Y. M., Cassell, M. A., Zhang, W. D., Wagner, G., Sargent, E. E., et al. (2004). An efficient SNP system for mouse genome scanning and elucidating strain relationships. *Genome Research, 14*(9), 1806–1811.

Petree, A. D., Haddad, N. F., & Berger, L. H. (1992). A simple and sensitive method for monitoring running-wheel movement. *Behavior Research Methods, Instruments, & Computers, 24*, 412–413.

Pettitt, S. J., Liang, Q., Rairdan, X. Y., Moran, J. L., Prosser, H. M., Beier, D. R., et al. (2009). Agouti C57BL/6N embryonic stem cells for mouse genetic resources. *Nature Methods, 6*(7), 493–495.

Philip, V. M., Duvvuru, S., Gomero, B., Ansah, T. A., Blaha, C. D., Cook, M. N., et al. (2010). High-throughput behavioral phenotyping in the expanded panel of BXD recombinant inbred strains. *Genes, Brain and Behavior, 9*(2), 129–159.

Porsolt, R. D., Brossard, G., Hautbois, C., & Roux, S. (2007). Forced swimming test in the mouse. In J. N. Crawley, C. R. Gerfen, M. Rogawski, D. Sibley, P. Skolnick, & S. Wray (Eds.), *Short protocols in neuroscience. Systems and behavioral methods* (pp. 3-55–3-56). Hoboken, NJ: Wiley.

Posthuma, D., & de Geus, E. J. C. (2006). Progress in the molecular-genetic study of intelligence. *Current Directions in Psychological Science, 15*(4), 151–155.

Powell, C. M. (2006). Gene targeting of presynaptic proteins in synaptic plasticity and memory: Across the great divide. *Neurobiology of Learning and Memory, 85*(1), 2–15.

Powell, S. B., Zhou, X., & Geyer, M. A. (2009). Prepulse inhibition and genetic mouse models of schizophrenia. *Behavioural Brain Research, 204*(2), 282–294.

Prusky, G. T., West, P. W. R., & Douglas, R. M. (2000). Reduced visual acuity impairs place but not cued learning in the Morris water task. *Behavioural Brain Research, 116*(2), 135–140.

Rabinovitch, M. S., & Rosvold, H. E. (1951). A closed-field intelligence test for rats. *Canadian Journal of Psychology, 5*(3), 122–128.

Rampon, C., Jiang, C. H., Dong, H., Tang, Y.-P., Lockhart, D. J., Schultz, P. G., et al. (2000a). Effects of environmental enrichment on gene expression in the brain. *Proceedings of the National Academy of Sciences of the United States of America, 97*(23), 12880–12884.

Rampon, C., Tang, Y. P., Goodhouse, J., Shimizu, E., Kyin, M., & Tsien, J. Z. (2000b). Enrichment induces structural changes and recovery from nonspatial memory deficits in CA1 NMDAR1-knockout mice. *Nature Neuroscience, 3*(3), 238–244.

Ramsden, E., & Adams, J. (2009). Escaping the laboratory: The rodent experiments of John B. Calhoun and their cultural influence. *Journal of Social History, 42*(3), 761–792.

Richter, S. H., Garner, J. P., & Würbel, H. (2009). Environmental standardization: Cure or cause of poor reproducibility in animal experiments? *Nature Methods, 6*(4), 257–261.

Rikke, B. A., Battaglia, M. E., Allison, D. B., & Johnson, T. E. (2006). Murine weight loss exhibits significant genetic variation during dietary restriction. *Physiological Genomics, 27*(2), 122–130.

Ripoll, N., David, D. J., Dailly, E., Hascoet, M., & Bourin, M. (2003). Antidepressant-like effects in various mice strains in the tail suspension test. *Behavioural Brain Research, 143*(2), 193–200.

Risch, N., Herrell, R., Lehner, T., Liang, K. Y., Eaves, L., Hoh, J., et al. (2009). Interaction between the serotonin transporter gene (5-HTTLPR), stressful life events, and risk of depression: A meta-analysis. *The Journal of the American Medical Association, 301*(23), 2462–2471.

Roder, J. K., Roder, J. C., & Gerlai, R. (1996). Memory and the effect of cold shock in the water maze in S100 beta transgenic mice. *Physiology & Behavior, 60*(2), 611–615.

Rodgers, R. J. (1997). Animal models of "anxiety": Where next? *Behavioural Pharmacology, 8*(6-7), 477–496.

Rogers, D. C., Fisher, E. M. C., Brown, S. D. M., Peters, J., Hunter, A. J., & Martin, J. E. (1997). Behavioral and functional analysis of mouse phenotype: SHIRPA, a proposed protocol for comprehensive phenotype assessment. *Mammalian Genome, 8*(10), 711–713.

Roubertoux, P. L., Sluyter, F., Carlier, M., Marcet, B., Maarouf-Veray, F., Cherif, C., et al. (2003). Mitochondrial DNA modifies cognition in interaction with the nuclear genome and age in mice. *Nature Genetics, 35*(1), 65–69.

Rudenko, O., Tkach, V., Berezin, V., & Bock, E. (2009). Detection of early behavioral markers of Huntington's disease in R6/2 mice employing an automated social home cage. *Behavioural Brain Research, 203*(2), 188–199.

Russell, W. M. S., & Burch, R. L. (1959). *The principles of humane experimental technique*. London, UK: Methuen.

Rustay, N. R., Wahlsten, D., & Crabbe, J. C. (2003a). Assessment of genetic susceptibility to ethanol intoxication in mice. *Proceedings of the National Academy of Sciences of the United States of America, 100*(5), 2917–2922.

Rustay, N. R., Wahlsten, D., & Crabbe, J. C. (2003b). Influence of task parameters on rotarod performance and sensitivity to ethanol in mice. *Behavioural Brain Research, 141*(2), 237–249.

Ryan, B. C., & Vandenbergh, J. G. (2002). Intrauterine position effects. *Neuroscience & Biobehavioral Reviews, 26*(6), 665–678.

Salchow, D. J., Gouras, P., Doi, K., Goff, S. P., Schwinger, E., & Tsang, S. H. (1999). A point mutation (W70A) in the rod PDE-gamma gene desensitizing and delaying murine rod photoreceptors. *Investigative Ophthalmology & Visual Science, 40*(13), 3262–3267.

Salomé, N., Viltart, O., Darnaudéry, M., Salchner, P., Singewald, N., Landgraf, R., et al. (2002). Reliability of high and low anxiety-related behaviour: Influence of laboratory environment and multifactorial analysis. *Behavioural Brain Research, 136*(1), 227–237.

Sarkar, S. (1999). From the Reaktionsnorm to the adaptive norm: The norm of reaction, 1909–1960. *Biology and Philosophy, 14*, 235–252.

Schalomon, P. M., & Wahlsten, D. (2002). Wheel running behavior is impaired by both surgical section and genetic absence of the mouse corpus callosum. *Brain Research Bulletin, 57*(1), 27–33.

Schellinck, Cyr, & Brown, R. E. (2010). How many ways can mouse behavioral experiments go wrong? Confounding variables in mouse models of neurodegenerative diseases and how to control them. *Advances in the Study of Behavior, 41*, 225–271.

Schiff, M., Duyme, M., Dumaret, A., & Tomkiewicz, S. (1982). How much could we boost scholastic achievement and IQ scores? A direct answer from a French adoption study. *Cognition, 12*(2), 165–196.

Schmidt-Nielsen, K. (1984). *Scaling: Why is Animal Size so Important?* Cambridge, UK: Cambridge University Press.

Scott, J. P. (1942). Genetic differences in the social behavior of inbred strains of mice. *Journal of Heredity, 33*(1), 11–15.

Searle, L. V. (1949). The organization of herditary maze-brightness and maze-dullness. *Genetic Psychology Monographs, 39*, 279–325.

Selman, C., Tullet, J. M., Wieser, D., Irvine, E., Lingard, S. J., Choudhury, A. I., et al. (2009). Ribosomal protein S6 kinase 1 signaling regulates mammalian life span. *Science, 326*(5949), 140–144.

Service, R. F. (2009). A dark tale behind two retractions. *Science, 326*(5960), 1610–1611.

Severo, N. C., & Zelen, M. (1960). Normal approximation to the chi-square and noncentral F probability functions. *Biometrika, 47*(3–4), 411–416.

Shyu, B. C., Andersson, S. A., & Thorén, P. (1984). Spontaneous running in wheels. A microprocessor assisted method for measuring physiological parameters during exercise in rodents. *Acta Physiologica Scandinavica, 121* (2), 103–109.

Silver, L. M. (1995). *Mouse Genetics: Concepts and Applications.* Oxford, UK: Oxford University Press.

Simpson, E. M., Linder, C. C., Sargent, E. E., Davisson, M. T., Mobraaten, L. E., & Sharp, J. J. (1997). Genetic variation among 129 substrains and its importance for targeted mutagenesis in mice. *Nature Genetics, 16*(1), 19–27.

Singer, J. B., Hill, A. E., Burrage, L. C., Olszens, K. R., Song, J., Justice, M., et al. (2004). Genetic dissection of complex traits with chromosome substitution strains of mice. *Science, 304*(5669), 445–448.

Singer, J. B., Hill, A. E., Nadeau, J. H., & Lander, E. S. (2005). Mapping quantitative trait loci for anxiety in chromosome substitution strains of mice. *Genetics, 169*(2), 855–862.

Small, W. S. (1901). Experimental study of the mental processes of the rat. II. *American Journal of Psychology, 12*(2), 206–239.

Smith, B. K., Andrews, P. K., & West, D. B. (2000). Macronutrient diet selection in thirteen mouse strains. *American Journal of Physiology — Regulatory, Integrative and Comparitive Physiology, 278*(4), R797–R805.

Smith, S. M., Brown, H. O., Toman, J. E. P., & Goodman, L. S. (1947). The lack of cerebral effects of *d*-tubocuarine. *Anesthesiology, 8*(1), 1–14.

Solberg, L. C., Valdar, W., Gauguier, D., Nunez, G., Taylor, A., Burnett, S., et al. (2006). A protocol for high-throughput phenotyping, suitable for quantitative trait analysis in mice. *Mammalian Genome, 17*(2), 129–146.

Solinas, M., Thiriet, N., El, R. R., Lardeux, V., & Jaber, M. (2009). Environmental enrichment during early stages of life reduces the behavioral, neurochemical, and molecular effects of cocaine. *Neuropsychopharmacology, 34*(5), 1102–1111.

Solomon, R. L., & Corbit, J. D. (1974). An opponent-process theory of motivation. I: Temporal dynamics of affect. *Psychological Review, 81*(2), 119–145.

Solomon, R. L., & Wynne, L. C. (1953). Traumatic avoidance learning. Acquisition in normal dogs. *Psychological Monographs, 67*(4). (Whole No 354).

Southwick, C. H., & Clark, L. H. (1968). Interstrain differences in aggressive behavior and exploratory activity of inbred mice. *Communications in Behavioral Biology, 1A,* 49–59.

Spink, A. J., Tegelenbosch, R. A. J., Buma, M. O. S., & Noldus, L. P. J. J. (2001). The EthoVision video tracking system: A tool for behavioral phenotyping of transgenic mice. *Physiology & Behavior, 73*(5), 731–744.

Stanford, L., & Brown, R. E. (2003). MHC-congenic mice (C57BL/6J and B6-H-2K) show differences in speed but not accuracy in learning the Hebb-Williams Maze. *Behavioural Brain Research, 144*(1–2), 187–197.

Stanford, S. C. (2007). The Open Field Test: Reinventing the wheel. *Journal of Psychopharmacology, 21*(2), 134–135.

Stasko, M. R., & Costa, A. C. (2004). Experimental parameters affecting the Morris water maze performance of a mouse model of Down syndrome. *Behavioural Brain Research, 154*(1), 1–17.

Stavnezer, A. J., Hyde, L. A., Bimonte, H. A., Armstrong, C. M., & Denenberg, V. H. (2002). Differential learning strategies in spatial and nonspatial versions of the Morris water maze in the C57BL/6J inbred mouse strain. *Behavioural Brain Research, 133*(2), 261–270.

Stiedl, O., Radulovic, J., Lohmann, R., Birkenfeld, K., Palve, M., Kammermeier, J., et al. (1999). Strain and substrain differences in context- and tone-dependent fear conditioning of inbred mice. *Behavioural Brain Research, 104* (1–2), 1–12.

Suckow, M. A., Danneman, P., & Brayton, C. (2001). *The laboratory mouse.* Boca Raton, FL: CRC Press, Taylor & Francis Group.

Surjo, D., & Arndt, S. S. (2001). The Mutant Mouse Behaviour network, a medium to present and discuss methods for the behavioural phenotyping. *Physiology & Behavior, 73,* 691–694.

Surwit, R. S., Kuhn, C. M., Cochrane, C., McCubbin, J. A., & Feinglos, M. N. (1988). Diet-induced type II diabetes in C57BL/6J mice. *Diabetes, 37*(9), 1163–1167.

Taft, R. A., Davisson, M., & Wiles, M. V. (2006). Know thy mouse. *Trends in Genetics, 22*(12), 649–653.

Takahashi, J. S. (1995). Molecular neurobiology and genetics of circadian rhythms in mammals. *Annual Review of Neuroscience, 18,* 531–553.

Takahashi, J. S., Pinto, L. H., & Vitaterna, M. H. (1994). Forward and reverse genetic approaches to behavior in the mouse. *Science, 264*(5166), 1724–1733.

Talbot, C. J., Radcliffe, R. A., Fullerton, J., Hitzemann, R., Wehner, J. M., & Flint, J. (2003). Fine scale mapping of a genetic locus for conditioned fear. *Mammalian Genome, 14*(4), 223–230.

Tankersley, C. G., Irizarry, R., Flanders, S., & Rabold, R. (2002). Circadian rhythm variation in activity, body temperature, and heart rate between C3H/Hej and C57BL/6J inbred strains. *Journal of Applied Physiology, 92*(2), 870–877.

Tarantino, L. M., & Bucan, M. (2000). Dissection of behavior and psychiatric disorders using the mouse as a model. *Human Molecular Genetics, 9*(6), 953–965.

Tecott, L. H., & Nestler, E. J. (2004). Neurobehavioral assessment in the information age. *Nature Neuroscience, 7*(5), 462–466.

ten Cate, C. (2009). Niko Tinbergen and the red patch on the herring gull's beak. *Animal Behaviour, 77*(4), 785–794.

Thompson, W. R. (1953). The inheritance of behavior: Behavioral differences in fifteen mouse strains. *Canadian Journal of Psychology, 7*(4), 145–155.

Threadgill, D. W., Yee, D., Matin, A., Nadeau, J. H., & Magnuson, T. (1997). Genealogy of the 129 inbred strains: 129/SvJ is a contaminated inbred strain. *Mammalian Genome, 8*(6), 441–442.

Tomiyoshi, M. Y., Sakai, M., Baleeiro, R. B., Stankevicius, D., Massoco, C. O., Palermo-Neto, J., et al. (2009). Cohabitation with a B16F10 melanoma-bearer cage mate influences behavior and dendritic cell phenotype in mice. *Brain, Behavior, and Immunity, 23*(4), 558–567.

Toye, A. A., & Cox, R. (2001). Behavioral genetics: Anxiety under interrogation. *Current Biology, 11*(12), R473–R476.

Trainor, B. C., Sweeney, C., & Cardiff, R. (2009). Isolating the effects of social interactions on cancer biology. *Cancer Prevention Research (Phila, Pa), 2*(10), 843–846.

Treit, D. (1994). Animal models of anxiety and anxiolytic drug action. In J. A. Inden Boer, & J. M. Ad Sitzen (Eds.), *Handbook of Depression and Anxiety: A biological approach* (pp. 201–224). New York, NY: Marcel Dekker.

Trullas, R., & Skolnick, P. (1993). Differences in fear motivated behaviors among inbred mouse strains. *Psychopharmacology (Berl), 111*(3), 323–331.

Tryon, R. C. (1940). Genetic differences in maze learning ability in rats. *Thirty Ninth Yearbook of the National Society for Studies of Education, Part I,* 111–119.

Tucci, V., Lad, H. V., Parker, A., Polley, S., Brown, S. D., & Nolan, P. M. (2006). Gene-environment interactions differentially affect mouse strain behavioral parameters. *Mammalian Genome, 17*(11), 1113–1120.

Turner, J. G., Parrish, J. L., Hughes, L. F., Toth, L. A., & Caspary, D. M. (2005). Hearing in laboratory animals: Strain differences and nonauditory effects of noise. *Comparative Medicine, 55*(1), 12–23.

Turri, M. G., Datta, S. R., DeFries, J., Henderson, N. D., & Flint, J. (2001). QTL analysis identifies multiple behavioral dimensions in ethological tests of anxiety in laboratory mice. *Current Biology, 11*(10), 725–734.

Turri, M. G., Henderson, N. D., DeFries, J. C., & Flint, J. (2001). Quantitative trait locus mapping in laboratory mice derived from a replicated selection experiment for open-field activity. *Genetics, 158*(3), 1217–1226.

Tyynismaa, H., & Suomalainen, A. (2009). Mouse models of mitochondrial DNA defects and their relevance for human disease. *European Molecular Biology Organization Report, 10*(2), 137–143.

Ukai, M., Watanabe, Y., & Kameyama, T. (2000). Effects of endomorphins-1 and -2, endogenous mu-opioid receptor agonists, on spontaneous alternation performance in mice. *European Journal of Pharmacology, 395*(3), 211–215.

Umezu, T., Nagano, K., Ito, H., Kosakai, K., Sakaniwa, M., & Morita, M. (2006). Anticonflict effects of lavender oil and identification of its active constituents. *Pharmacology, Biochemistry and Behavior, 85*(4), 713–721.

van Alphen, B., Winkelman, B. H., & Frens, M. A. (2010). Three-dimensional optokinetic eye movements in the C57BL/6J mouse. *Investigative Ophthalmology & Visual Science, 51*(1), 623–630.

van de Weerd, H. A., Baumans, V., Koolhaas, J. M., & van Zutphen, L. F. (1994). Strain specific behavioural response to environmental enrichment in the mouse. *Journal of Experimental Animal Science, 36*(4–5), 117–127.

van de Weerd, H. A., Van Loo, P. L., van Zutphen, L. F., Koolhaas, J. M., & Baumans, V. (1997). Nesting material as environmental enrichment has no adverse effects on behavior and physiology of laboratory mice. *Physiology & Behavior, 62*(5), 1019–1028.

van der Staay, F. J., Arndt, S. S., & Nordquist, R. E. (2009). Evaluation of animal models of neurobehavioral disorders. *Behavioral and Brain Functions, 5,* 11.

van der Staay, F. J., & Steckler, T. (2002). The fallacy of behavioral phenotyping without standardisation. *Genes, Brain and Behavior, 1*(1), 9–13.

van Praag, H., Christie, B. R., Sejnowski, T. J., & Gage, F. H. (1999). Running enhances neurogenesis, learning, and long-term potentiation in mice. *Proceedings of the National Academy of Sciences of the United States of America, 96*(23), 13427–13431.

van Praag, H., Shubert, T., Zhao, C., & Gage, F. H. (2005). Exercise enhances learning and hippocampal neurogenesis in aged mice. *Journal of Neuroscience, 25*(38), 8680–8685.

Vitaterna, M. H., Pinto, L. H., & Takahashi, J. S. (2006). Large-scale mutagenesis and phenotypic screens for the nervous system and behavior in mice. *Trends in Neurosciences, 29*(4), 233–240.

Vogel, R. W., Ewers, M., Ross, C., Gould, T. J., & Woodruff-Pak, D. S. (2002). Age-related impairment in the 250-millisecond delay eyeblink classical conditioning procedure in C57BL/6 mice. *Learning and Memory, 9*(5), 321–336.

Voikar, V., Polus, A., Vasar, E., & Rauvala, H. (2005). Long-term individual housing in C57BL/6J and DBA/2 mice: assessment of behavioral consequences. *Genes, Brain and Behavior, 4*(4), 240–252.

vom Saal, F. S. (1981). Variation in phenotype due to random intrauterine positioning of male and female fetuses in rodents. *Journal of Reproduction and Fertility, 62*(2), 633–650.

Wahlsten, D. (1972a). Genetic experiments with animal learning: A critical review. *Behavioral Biology, 7*(2), 143–182.

Wahlsten, D. (1972b). Phenotypic and genetic relations between initial response to electric shock and rate of avoidance learning in mice. *Behavior Genetics, 2*(2), 211–240.

Wahlsten, D. (1974). A developmental time scale for postnatal changes in brain and behavior of B6D2F$_2$ mice. *Brain Research, 72*(2), 251–264.

Wahlsten, D. (1979). A critique of the concepts of heritability and heredity in behavioral genetics. In J. R. Royce, & L. Mos (Eds.), *Theoretical Advances in Behavioral Genetics* (pp. 425–481). Alphen aan den Rijn, The Netherlands: Sijthoff and Noordhoff.

Wahlsten, D. (1990). Insensitivity of the analysis of variance to heredity-environment interaction. *Behavioral and Brain Sciences, 13*, 109–120.

Wahlsten, D. (1991). Sample size to detect a planned contrast and a one degree-of-freedom interaction effect. *Psychological Bulletin, 110*(3), 587–595.

Wahlsten, D. (2001). Standardizing tests of mouse behavior: Reasons, recommendations, and reality. *Physiology & Behavior, 73*(5), 695–704.

Wahlsten, D. (2003). Genetics and the development of brain and behavior. In J. Valsiner, & K. J. Connolly (Eds.), *Handbook of developmental psychology* (pp. 18–47). London, UK: Sage.

Wahlsten, D. (2007). Sample size requirements for experiments on laboratory animals. In B. C. Jones, & P. Mormede (Eds.), *Neurobehavioral genetics: Methods and applications* (2nd ed.). (pp. 149–168) Boca Raton, FL: CRC Press, Taylor & Francis Group.

Wahlsten, D., Bachmanov, A. A., Finn, D. A., & Crabbe, J. C. (2006). Stability of inbred mouse strain differences in behavior and brain size between laboratories and across decades. *Proceedings of the National Academy of Sciences of the United States of America, 103*(44), 16364–16369.

Wahlsten, D., Blom, K., Stefanescu, R., Conover, K., & Cake, H. (1987). Lasting effects on mouse brain growth of 24 hr postpartum deprivation. *International Journal of Developmental Neuroscience, 5*(1), 71–75.

Wahlsten, D., & Bulman-Fleming, B. (1987). The magnitudes of litter size and sex effects on brain growth of BALB/c mice. *Growth, 51*(2), 240–248.

Wahlsten, D., Cole, M., Sharp, D., & Fantino, E. (1968). Facilitation or bar-press avoidance by handling during the intertrial interval. *Journal of Comparative and Physiological Psychology, 65*(1), 170–175.

Wahlsten, D., Cooper, S. F., & Crabbe, J. C. (2005). Different rankings of inbred mouse strains on the Morris maze and a refined 4-arm water escape task. *Behavioural Brain Research, 165*(1), 36–51.

Wahlsten, D., & Crabbe, J. C. (2007). Behavioral testing. In J. G. Fox, S. Barthold, M. T. Davisson, C. Newcomer, F. Quimby, & A. Smith (Eds.), *The mouse in biomedical research, Vol. 3, Normative biology, husbandry, and models* (pp. 513–534). Amsterdam: Elsevier.

Wahlsten, D., & Crabbe, J. C. (2010). Replicability and reliability of behavioral tests. In W. E. Crusio, F. Sluyer, & R. Gerlai (Eds.), *Handbook of Behavioral Genetics of the Mouse*. Cambridge, UK: Cambridge University Press (in press).

Wahlsten, D., Metten, P., & Crabbe, J. C. (2003a). A rating scale for wildness and ease of handling laboratory mice: results for 21 inbred strains tested in two laboratories. *Genes, Brain and Behavior, 2*(2), 71–79.

Wahlsten, D., Metten, P., & Crabbe, J. C. (2003b). Survey of 21 inbred mouse strains in two laboratories reveals that BTBR T/+ tf/tf has severely reduced hippocampal commissure and absent corpus callosum. *Brain Research, 971* (1), 47–54.

Wahlsten, D., Metten, P., Phillips, T. J., Boehm, S. L., II, Burkhart-Kasch, S., Dorow, J., et al. (2003). Different data from different Labs: lessons from studies of gene-environment interaction. *Journal of Neurobiology, 54*(1), 283–311.

Wahlsten, D., Rustay, N. R., Metten, P., & Crabbe, J. C. (2003). In search of a better mouse test. *Trends in Neurosciences, 26*(3), 132–136.

Wainwright, P. E., Huang, Y. S., Bulman-Fleming, B., Levesque, S., & McCutcheon, D. (1994). The effects of dietary fatty acid composition combined with environmental enrichment on brain and behavior in mice. *Behavioural Brain Research, 60*(2), 125–136.

Wainwright, P. E., Leatherdale, S. T., & Dubin, J. A. (2007). Advantages of mixed effects models over traditional ANOVA models in developmental studies: A worked example in a mouse model of fetal alcohol syndrome. *Developmental Psychobiology, 49*(7), 664–674.

Wainwright, P. E., Levesque, S., Krempulec, L., Bulman-Fleming, B., & McCutcheon, D. (1993). Effects of environmental enrichment on cortical depth and Morris-maze performance in B6D2F2 mice exposed prenatally to ethanol. *Neurotoxicology and Teratology, 15*(1), 11–20.

Walf, A. A., & Frye, C. A. (2007). The use of the elevated plus maze as an assay of anxiety-related behavior in rodents. *Nature Protocols, 2*(2), 322–328.

Wall, P., & Messier, C. (2000). U-69, 593 microinjection in the infralimbic cortex reduces anxiety and enhances spontaneous alternation memory in mice. *Brain Research, 856*(1–2), 259–280.

Washburn, M. F. (1926). Hunger and speed of running as factors in maze learning in mice. *Journal of Comparative Psychology, 6*, 181–187.

Waterston, R. H., Lindblad-Toh, K., Birney, E., Rogers, J., Abril, J. F., Agarwal, P., et al. (2002). Initial sequencing and comparative analysis of the mouse genome. *Nature, 420*(6915), 520–562.

Wehner, J. M., Bowers, B. J., & Paylor, R. (1996). The use of null mutant mice to study complex learning and memory processes. *Behavior Genetics, 26*(3), 301–312.

Wendt, K. D., Lei, B., Schachtman, T. R., Tullis, G. E., Ibe, M. E., & Katz, M. L. (2005). Behavioral assessment in mouse models of neuronal ceroid lipofuscinosis using a light-cued T-maze. *Behavioural Brain Research, 161*(2), 175–182.

Whishaw, I. Q., & Kolb, B. (2005). *The behavior of the laboratory rat.* Oxford, UK: Oxford University Press.

Whishaw, I. Q., & Tomie, J.-A. (1996). Of mice and mazes: Similarities between mice and rats on dry land but not water mazes. *Physiology & Behavior, 60*(5), 1191–1197.

Williams, A. H., Valdez, G., Moresi, V., Qi, X., McAnally, J., Elliott, J. L., et al. (2009). MicroRNA-206 delays ALS progression and promotes regeneration of neuromuscular synapses in mice. *Science, 326*(5959), 1549–1554.

Williams, G. A., Daigle, K. A., & Jacobs, G. H. (2005). Rod and cone function in coneless mice. *Visual Neuroscience, 22*(6), 807–816.

Willis-Owen, S. A., & Flint, J. (2007). Identifying the genetic determinants of emotionality in humans; insights from rodents. *Neuroscience & Biobehavioral Reviews, 31*(1), 115–124.

Willott, J. F., Tanner, L., O'Steen, J., Johnson, K. R., Bogue, M. A., & Gagnon, L. (2003). Acoustic startle and prepulse inhibition in 40 inbred strains of mice. *Behavioral Neuroscience, 117*(4), 716–727.

Wilson, M. D., Barbosa-Morais, N. L., Schmidt, D., Conboy, C. M., Vanes, L., Tybulewicz, V. L., et al. (2008). Species-specific transcription in mice carrying human chromosome 21. *Science, 322*(5900), 434–438.

Wilson, S. G., & Mogil, J. S. (2001). Measuring pain in the (knockout) mouse: Big challenges in a small mammal. *Behavioural Brain Research, 125*(1–2), 65–73.

Winer, B. J., Brown, D. R., & Michels, K. M. (1991). *Statistical principles in experimental design* (3rd ed.). New York, NY: McGraw-Hill.

Wohr, M., Dahlhoff, M., Wolf, E., Holsboer, F., Schwarting, R. K., & Wotjak, C. T. (2008). Effects of genetic background, gender, and early environmental factors on isolation-induced ultrasonic calling in mouse pups: An embryo-transfer study. *Behavior Genetics, 38*(6), 579–595.

Wolfer, D. P., Crusio, W. E., & Lipp, H. P. (2002). Knockout mice: Simple solutions to the problems of genetic background and flanking genes. *Trends in Neurosciences, 25*(7), 336–340.

Wolfer, D. P., Litvin, O., Morf, S., Nitsch, R. M., Lipp, H. P., & Würbel, H. (2004). Laboratory animal welfare: Cage enrichment and mouse behaviour. *Nature, 432*(7019), 821–822.

Wolfer, D. P., Madani, R., Valenti, P., & Lipp, H.-P. (2001). Extended analysis of path data from mutant mice using the public domain software Wintrack. *Physiology & Behavior, 73*(5), 745–753.

Wolfer, D. P., Stagljar-Bozicevic, M., Errington, M., & Lipp, H.-P. (1998). Spatial memory and learning in transgenic mice: Fact or artifact? *News in Physiological Sciences, 13*, 118–122.

Würbel, H. (2000). Behaviour and the standardization fallacy. *Nature Genetics, 26*, 263.

Würbel, H. (2002). Behavioral phenotyping enhanced — beyond (environmental) standardization. *Genes, Brain and Behavior, 1*(1), 3–8.

Wynne-Edwards, V. C. (1965). Self-regulating systems in populations of animals. *Science, 147*, 1543–1548.

Yanai, J., & McClearn, G. E. (1972). Assortative mating in mice and the incest taboo. *Nature, 238*, 281–282.

Yerkes, R. M. (1907). *The dancing mouse.* New York, NY: Macmillan.

Youngson, N. A., & Whitelaw, E. (2008). Transgenerational epigenetic effects. *Annual Review of Genomics and Human Genetics, 9*, 233–257.

Yu, X., Gimsa, U., Wester-Rosenlof, L., Kanitz, E., Otten, W., Kunz, M., et al. (2009). Dissecting the effects of mtDNA variations on complex traits using mouse conplastic strains. *Genome Research, 19*(1), 159–165.

Zhang, Z., Gildersleeve, J., Yang, Y. Y., Xu, R., Loo, J. A., Uryu, S., et al. (2004). A new strategy for the synthesis of glycoproteins. *Science, 303*(5656), 371–373.

Zurn, J. B., Hohmann, D., Dworkin, S. I., & Motai, Y. (2005). *A Real-Time Rodent Tracking System for Both Light and Dark Cycle Behavior Analysis wacv-motion.* pp. 87–92, *Seventh IEEE Workshops on Application of Computer Vision (WACV/MOTION'05)—Volume 1, 2005.* Los Alamitos, CA: IEEE Computer Society Press.